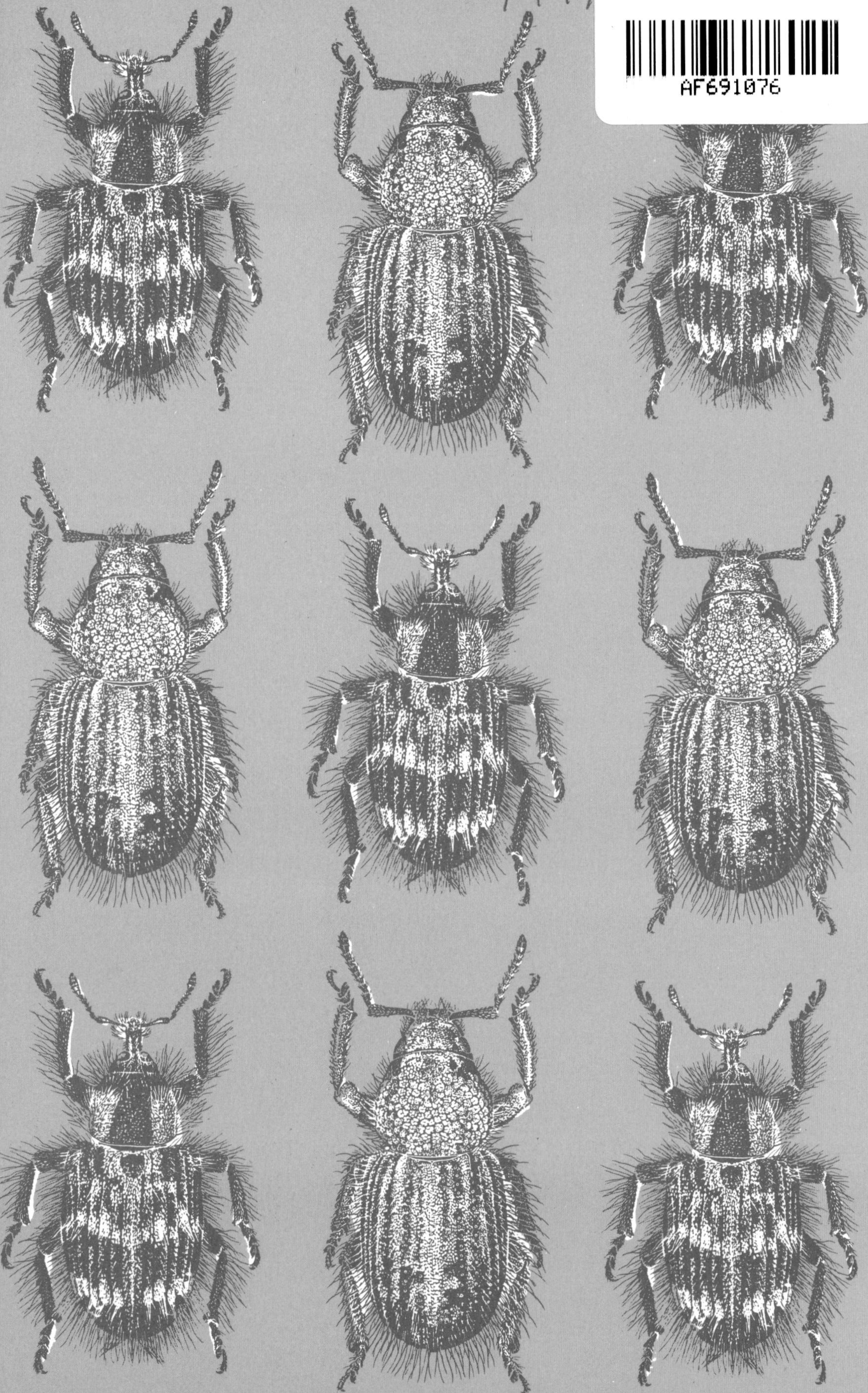

7497
AF691076

AUSTRALIAN WEEVILS

VOLUME V

COLOUR PLATES 1–304

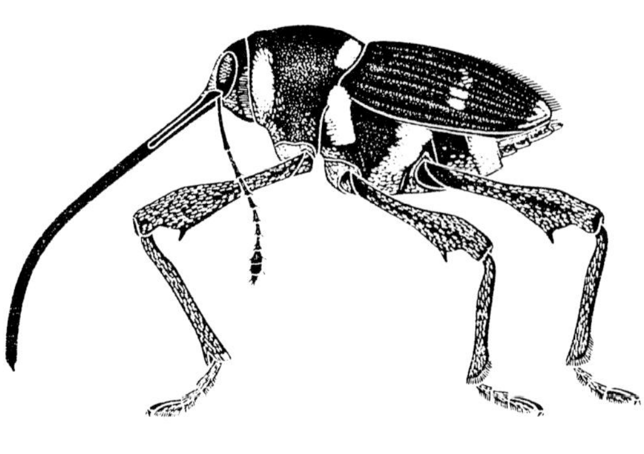

Australian Weevils

Volume	I	Anthribidae to Attelabidae
Volume	II	Brentidae, Eurhynchidae, Apionidae, Immature Stages
Volume	III	Rhynchophoridae, Erirhinidae, Amycterinae, Bibliography
Volume	IV	Adelognatha
Volume	V	Colour Plates 1–304
Volume	VI	Colour Plates 305–632
Volume	**VII**	Phanerognatha I
Volume	**VIII**	Phanerognatha II

AUSTRALIAN WEEVILS

(Coleoptera : Curculionoidea)

ELWOOD C. ZIMMERMAN

BSc (Berkeley), DIC, PhD, DSc (London)

Curator of Weevils, Emeritus
Australian National Insect Collection
CSIRO Division of Entomology
Canberra, Australia

Honorary Associate, British Museum
(Natural History)

VOLUME V

COLOUR PLATES 1–304

CSIRO
AUSTRALIA

Illustrations © CSIRO Australia 1991
Text © Elwood C. Zimmerman 1991
Except for private study or research, review or criticism, no part of this work may be reproduced or distributed by any means or stored in any retrieval system or data base without prior written permission from the publisher, CSIRO Australia.

Volume V published on 1 October, 1991

National Library of Australia Cataloguing-in-Publication Entry
Zimmerman, Elwood C. (Elwood Curtin), 1912 –

Australian weevils
Volume V

Includes indexes
ISBN 0 643 05144 9 (set).
ISBN 0 643 05145 7 (v. 1).
ISBN 0 643 05146 5 (v. 2).
ISBN 0 643 05147 3 (v. 3).
ISBN 0 643 05148 1 (v. 4).
ISBN 0 643 05149 X (v. 5).
ISBN 0 643 05150 3 (v. 6).
ISBN 0 643 05144 0 (v. 7).
ISBN 0 643 05145 9 (v. 8).

1. Coleoptera (Beetles) – Australia. 2. Curculionoidea (Weevils) – Australia.
I. CSIRO. II. Title.

595.7680994

Cover Illustration
Curculio fraudator Zimmerman
Drawn by Se Kim
CSIRO Division of Entomology, Canberra

End Papers
Agriochaeta crinita Pascoe
Essolithna villosa Lea
Drawn by Sybil Monteith, Brisbane

This Series is available from:
CSIRO Publications
PO Box 89 (314 Albert Street)
East Melbourne, Victoria 3002
Australia

Telephone: (03) 418 7217
Fax: (03) 419 0459

Entomological Society of America
9301 Annapolis Road
Lanham, Maryland 20706
USA

Telephone: (301) 731 4535
Fax: (301) 731 4538

This volume does not constitute formal publication for the purpose of the *International Code of Zoological Nomenclature,* Third Edition, 1985. All new names, new combinations and new synonymies are published in *Australian Weevils,* Volumes I – IV, VII, VIII.

Set in 10/11 Stone Serif using a Macintosh IIci running Quark Xpress 3.0 and a VT 600W Laser Printer
Design and production by CSIRO Editorial Services, Melbourne
Printed in Hong Kong by Bookbuilders Ltd.

Preface to Volumes V and VI

Nature, in her infinite wisdom, has found it right to have created these marvelous organisms for her mysterious purposes; it is therefore right that naturalists should consider it their obligation to make them better known to mankind. These volumes contribute to that end.

Although many illustrations have been prepared for this monograph, many more, especially of morphological features, should have been made more clearly to reveal details of the animals under study. It was my goal to illustrate every species, but, without time or help, that has not been possible; *non omnia possumus omnes* (we cannot all do everything).

The figures in colour presented here convey better than words the gross external features of the species. I hope that they will enable readers to identify without difficulty most of the insects illustrated. I have been told repeatedly that coloured illustrations are 'too expensive' to publish, but I disagree. Except, perhaps, for one or two illustrations the cost of equivalent detailed drawings made by skilled artists (which are few) is much greater, and drawings are artists' interpretations and subject to error. I leave it to those who use this monograph to be the judges. Are these marvels of creation, these wondrous products of the ages unworthy of having their portraits appear in natural colour on the printed page? Consider the savings made by authors in being able to say "facies, colour and pattern as on Figure X" instead of spending time in futile efforts preparing inadequate wordy descriptions. Consider the value to non-specialists, agriculturists, foresters and naturalists who may be able to identify specimens from the illustrations alone and then to obtain information on hosts, habits and distribution from the written text. No doubt these illustrations shall stimulate amateurs to observe weevils more closely and to contribute notes on their observations to the common pool of knowledge, and they may encourage students to follow my path and become trained specialists. Most of these coloured figures are of type specimens or specimens compared with types, and the illustrations will be preserved in various libraries throughout the world long after many of the specimens may have been damaged or destroyed. These are reasons for private funding the reproductions when the vision of administrators is obfuscated.

Most of us are neither linguists nor grammarians. I have had difficulty to master partly the English language, and I feel often that what I have written is inadequate to convey to others precisely what I wish them to understand. Much of the taxonomic literature that I read convinces me that clarity of exposition is especially difficult and is achieved by few authors. When I must consult works in languages foreign to me the difficulties of comprehension are compounded. How often it is that I find published taxonomic work incomprehensible, and I say "Oh, had the author only illustrated his text!". When good illustrations are given even those who do not understand the written word will have easily understood, pertinent details revealed. The impressions made by illustrations on a person's mind are far more lasting than are written descriptions that may convey different things to different readers. More than 40 years ago I viewed the Mona Lisa in the Louvre, but in my mind I can still see clearly that portrait. A written description of it would long since have disappeared from mind. I can immediately recognize illustrations in publications I used when I was a school boy, and it was illustrations that stimulated and led me during my formative years.

To those who make excuses for not illustrating their publications adequately, I would ask: If one cannot play the instrument, should one insist on being a member of the orchestra?

To authors I would plead: Write less, illustrate more — illustrate, *illustrate,* ILLUSTRATE!

To editors I would plead: Make regulations that, to be published, all descriptions of new species must be accompanied by meaningful illustrations. If an organism is considered new to science, does it not deserve to be illustrated?

Taxonomic literature 'overflows' with pages of 'dry-as-dust', inadequate, unillustrated, difficult or impossible-to-understand, all too often inaccurate descriptive matter that leave me comprehending only part of what was intended or feeling 'blind', and I have frequent reason to recall John Saxe's poem:

The Blind Men and the Elephant

It was six men of Indostan
To learning much inclined,
Who went to see the elephant
(Though all of them were blind),
That each by observation
Might satisfy his mind.

The first approached the elephant
And, happening to fall
Against his broad and sturdy side,
At once began to bawl,
"God bless me! but the elephant
Is very like a wall!"

The second feeling of the tusk
Cried: "Ho! what have we here
So very round and smooth and sharp?
To me 'tis mighty clear
This wonder of an elephant
Is very like a spear!"

The third approached the animal,
And, happening to take
The squirming trunk within his hands,
Thus boldly up and spake:
"I see," quoth he, "the elephant
Is very like a snake!"

The fourth reached out his eager hand,
And felt about the knee;
"What most this wondrous beast is like
Is mighty plain," quoth he;
"'Tis clear enough the elephant
Is very like a tree."

The fifth, who chanced to touch the ear,
Said: "E'en the blindest man
Can tell what this resembles most.
Deny the fact who can,
This marvel of an elephant
Is very like a fan!"

The sixth no sooner had begun
About the beast to grope,
Than, seizing on the swinging tail
That fell within his scope,
"I see," quoth he, "the elephant
Is very like a rope!

And so these men of Indostan
Disputed loud and long
Each in his own opinion
Exceeding stiff and strong,
Though each was partly in the right,
And all were in the wrong!

So, oft in theologic wars
The disputants, I ween,
Rail on in utter ignorance
Of what each other mean,
And prate about an elephant
Not one of them has seen!

John Saxe
Reproduced by kind permission of
Houghton Mifflin Co., Boston

I have personally removed the specimens illustrated from their old pins and mounts, cleaned them (many pinned specimens were damaged by vertigris), made repairs where possible and remounted them. Lea removed the heads of many of the specimens he described, including holotypes, especially of Baridinae and Cryptorhynchinae, and glued them separately to the card mounts. In that unfortunate process, many of the specimens were damaged, and some of the separated heads have since been lost. Returning separated heads to their rightful positions has been a time-consuming, often difficult and exhausting process that required utmost patience. Many specimens were glued to cards in the unsatisfactory 'European manner' with their ventral surfaces concealed, and when they were pressed down upon the cards during the original mounting process many appendages were broken or other damage was done. Many setose, hirsute and squamose specimens were glued to cards on their sides or backs, and when they were soaked off the cards the specimens were partly denuded and their appearances altered. These faults are reflected on the illustrations.

Caution: If a species cannot be found illustrated here, do not conclude that it is undescribed. Only *some* of the Australian weevils are illustrated on these colour plates. Consult the main texts of the volumes for further details and also see the indices in both Volumes V and VI. Illustrations of a number of species are out of systematic order, and numerous illustrations that should have been included in Volume V appear in volume VI.

Note: Unless otherwise stated, all measurements of specimens illustrated on the colour plates are total lengths from apex of rostrum or head, which ever is farther forward, to the apex of the elytra as the specimens appear in side view on the illustrations.

The correct number of plates in this volume is 304; there is an extra Plate numbered 29A.

And striving to be man, the worm
Mounts through all the spires of form.
Emerson

Funds to produce this Volume
were given by

Elwood and Hannah Zimmerman
and
Jim and Janette Bain

The stimulating delight of the systematist is that he never knows what remarkable unknown organism or structure or form or function or habit may come to his attention tomorrow, or the next day or the next; the tragedy of the systematist is that he knows that his work can never be completed.

Coleoptera : Curculionoidea

Colour Plates 1 – 304

The specimens illustrated hereafter were selected and prepared by the author and photographed by

John P. Green

Staff Photographer, CSIRO Division of Entomology, Canberra

Plate 1 "Good" Weevils

Control of the water weed *Salvinia molesta* in Australia by CSIRO-introduced Brazilian *Cyrtobagous* weevils.

Top figure shows extensive *Salvinia* infestation in Lake Moondarra, the water supply reservoir for Mount Isa, Queensland.

Bottom figure shows the same area after *Cyrtobagous* weevils were introduced by CSIRO and had eliminated the vegetation by natural means with a minimum of expense and without the use of chemicals. An estimated 30 000 tonnes of the weed were destroyed by the weevil in less than one year.

Salvinia molesta is an aggressive aquatic fern and is one of the world's worst weeds. It is a native of Brazil where it is kept under control by native herbivores. It has been introduced to many areas of the world which lack its natural control agents and where environmental damage caused by its unrestricted increase has been enormous. It chokes lakes, reservoirs, slow-moving rivers, irrigation systems, rice paddies, fishponds, etc. with such continuous, metre-thick mats of dense vegetation that they are rendered mostly useless for normal purposes. Moreover, its infestation leads to increases in mosquito populations. The damage it does is of particular importance to countries of the Third World where the contribution made by CSIRO entomologists in controlling the pest is especially significant.

See Plate 4:1, for an illustration of the small weevil used for the biological control of *Salvinia*. The weevils may develop populations of more than 1000 individuals per square surface metre of plants at the peak of their attack, and the destruction of the vegetation then becomes rapid.

Plate 2 "Good" Weevils

Control of the water weed *Salvinia molesta* in an overseas locality by CSIRO-introduced Brazilian *Cyrtobagous* weevils.

Top figure shows the river at Imbuando Village, East Sepik, Papua New Guinea, completely blocked by a great mat of *Salvinia* on 22 October 1983. The residents, being deprived of the use of the resources of the river, and unable to use it for navigation, abandoned the village.

Bottom figure shows the same area photographed on 22 January 1984 after *Cyrtobagous* weevils were introduced by CSIRO entomologists and which, spectacularly, had destroyed the weed infestation, cleared the waterway and enabled the villagers to reoccupy their land.

The success of the CSIRO *Salvinia* control project in Australia and overseas is one of the most remarkable events in the history of biological control. The financial benefits attained by this work have been great and reduce to comparative insignificance the funds expended to support the required research. The benefits are of particular importance to populations of underdeveloped countries of tropical Africa, East Asia and the Indo-Pacific.

Two species of *Cyrtobagous* weevils, originally thought to be only one kind, were introduced to Australia in the early stages of the biological control program. Study by CSIRO entomologists discovered that one of the two species, which differs only in microscopic ways from the other, was not an effective control agent, whereas the other, previously unrecognized kind, achieved remarkable control. CSIRO entomologist P.M. Room, who directed the *Salvinia* program, noted in an article in *Ecology and Evolution* 5:77, 1990, that "Biological control was achieved only after the true identity of *Salvinia* had been recognized, its native range found, and a previously unknown herbivore discovered. This illustrates vividly the crucial importance of sound taxonomy for understanding and solving ecological problems."

For further reading, see the Room article mentioned above and also J.A. Doeleman, "Biological Control of *Salvinia molesta* in Sri Lanka; An assessment of Costs and Benefits" in Technical Report 12 issued by the Australian Centre for International Agricultural Research, Canberra.

Plate 3 Weevil Pests of Grain

Top figure shows the granary weevil, *Sitophilus granarius* (Linnaeus), on wheat.

Bottom figure shows the maize weevil, *Sitophilus zeamais* Motschulsky, on maize.

The *Sitophilus* grain weevils cause extraordinary losses to stored grains throughout the world, and controlling their ravages is difficult and costly. The weevils were well known to the Ancients, and they no doubt have been a plague to man since he first stored grain. Vast sums are expended continuously to protect grain from their attacks and in research to develop better methods to control them and to devise more efficient storage methods. The adaptable weevils, however, continue to threaten or attack stored grain.

Plate 4 Good and Bad Weevils

"Good" Weevils

1 *Cyrtobagous salviniae* Calder and Sands, a *Salvinia* weevil. This very small, 1.5 to 2 mm long, CSIRO-introduced, Brazilian weevil is responsible for freeing thousands of hectares of rivers, lakes and waterways in Australia, the Indo-Pacific and Africa from dense infestations of the pest water fern, *Salvinia molesta*. See Plates 1 and 2.

2 *Rhinocyllus conicus* (Froelich), a thistle seed-head weevil. CSIRO has introduced this European weevil to aid in the control of nodding thistle, *Carduus nutans* Linnaeus, which infests vast areas of Australian grazing land at great cost to agricultural production .

"Bad" Weevils

3 *Sitona discoideus* (Gyllenhal), an introduced European pest of lucerne which does serious damage in Australia. The weevil is shown eating a lucerne leaf, and beside it is a braconid wasp (*Microtonus aethiopoides* Loan) which parasitizes the adult weevils and which was introduced to Australia by CSIRO to aid in its biological control. The weevil larvae destroy nitrogen-fixing lucerne roots, and the adults defoliate the plants.

4 *Cylas formicarius* (Fabricius), the sweetpotato weevil. At left, a female bores a hole in a sweetpotato into which to deposit her egg; at right a male stands alert. This Oriental weevil is now dispersed to most sweetpotato-growing regions. It is a pest of major importance, especially in underdeveloped countries. The larvae tunnel the tubers and render them inedible.

5 *Cosmopolites sordidus* (Germar), the banana root weevil. This Indonesian pest has followed the banana to most areas where bananas are grown.

6 *Rhabdoscelus obscurus* (Boisduval), the New Guinea sugarcane weevil, is one of the major pests of sugarcane in Australia, Fiji, and elsewhere in the Pacific and has cost sugar industries great financial loss.

7 *Listroderes difficilis* Germain, the vegetable weevil. This is one of several weevil pests accidentally imported from South America. It damages a large number of vegetable, flower and garden plants. It reproduces parthenogenetically (that is, without the need of males, and all specimens in Australia are females).

8 *Xyleborus perforans* (Wollaston), the island pinhole borer. This is one of many introduced bark beetles of economic importance. It is an abundant, widely distributed pest species. It perforates many species of trees; its burrows seriously degrade valuable timber and render useless otherwise cabinet- and veneer-grade woods.

1	2
3	4
5	6
7	8

Plate 5 Anthribidae

1, 2 *Ancylotropis waterhousei* Jekel
Brisbane, Queensland
Male; 13 mm

3, 4 *Phaulimia albirostris* (Frieser)
Cairns, Queensland
Holotype female; 6 mm

5, 6 *Apatenia insignis* Jordan
Eubenangee, north Queensland
Female; 8.5 mm

7 Face of holotype of *Phaulimia albirostris* (Frieser)

8 Face of *Apatenia insignis* Jordan

1	2
3	4
5	6
7	8

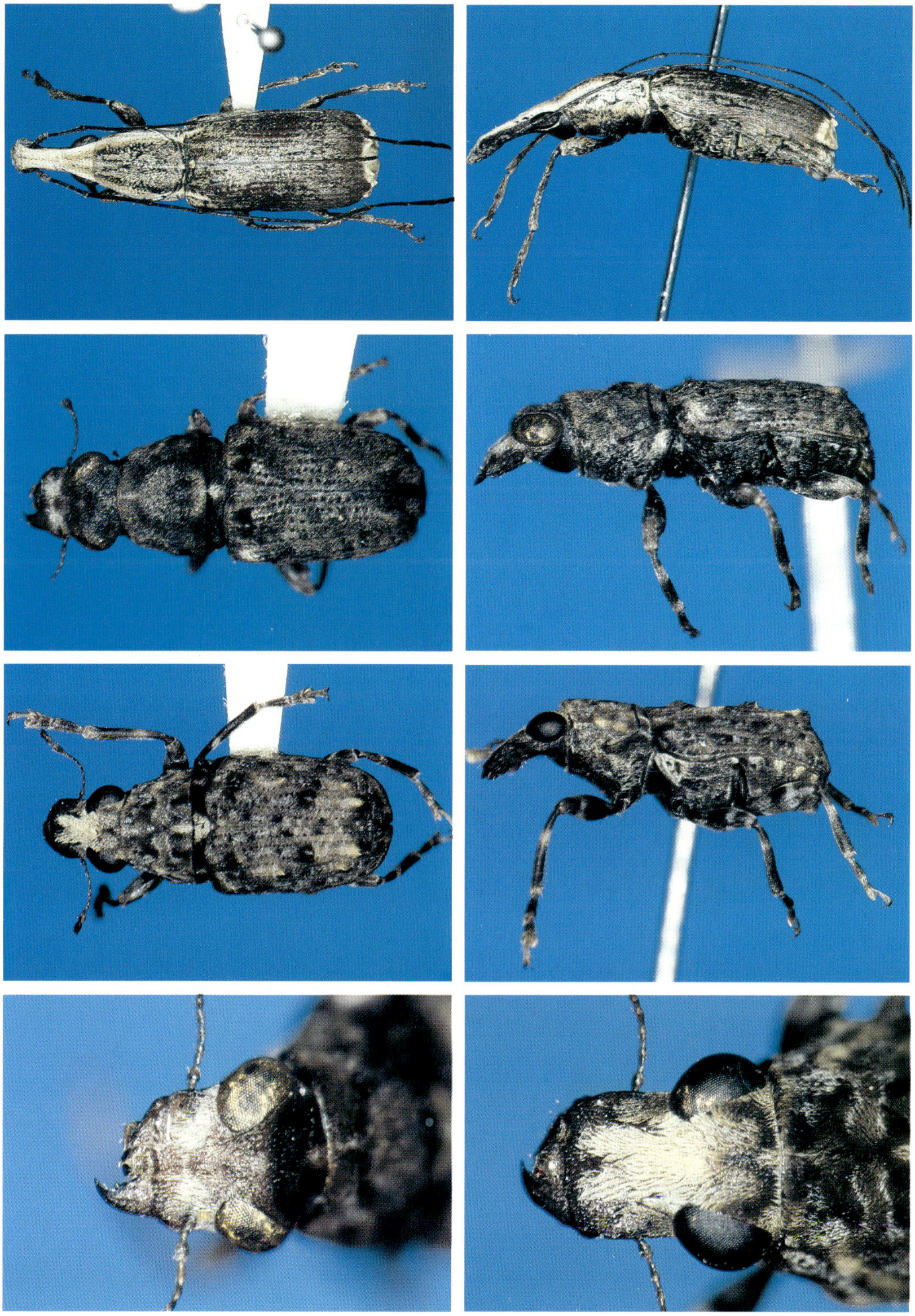

Plate 6 Anthribidae

1, 2 *Araecerus asperulus* Blackburn
Port Lincoln, South Australia
Paratype female; 3.3 mm

3, 4 *Araecerus bicristatus* Blackburn
Mackay, Queensland
Female; 3.7 mm

5, 6 *Araecerus eurorus* (Jordan)
Cooper Creek, Northern Territory
Male; 5.6 mm

7, 8 *Araecerus lindensis* Blackburn
Huon River, Tasmania
Male; 2.6 mm

1	2
3	4
5	6
7	8

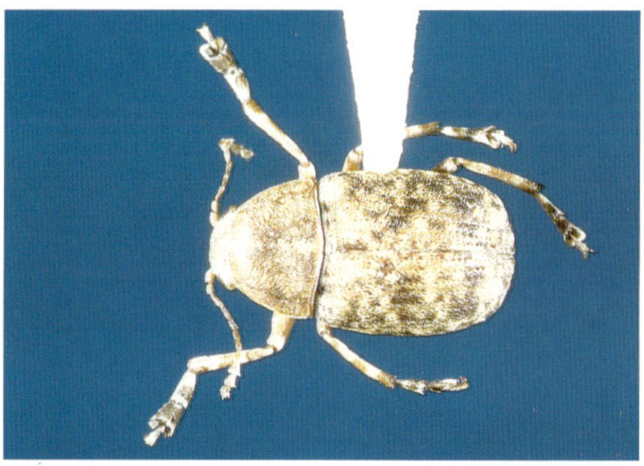

Plate 7 Anthribidae

1, 2 *Araecerus lutatus* (Fairmaire)
Edge Hill, Queensland
Male; 3.4 mm

3, 4 *Araecerus palmaris* (Pascoe)
Canberra, Australian Capital Territory
Female; 5.8 mm
Reared from gall on *Exocarpus*

5 *Araecerus* new species
Mabuiag Island, Torres Strait
Male; 8 mm

6 The same, female; 7.5 mm

7, 8 *Basitropis ingrata* Pascoe
Mount Carbine, north Queensland
Male; 7.5 mm

1	2
3	4
5	6
7	8

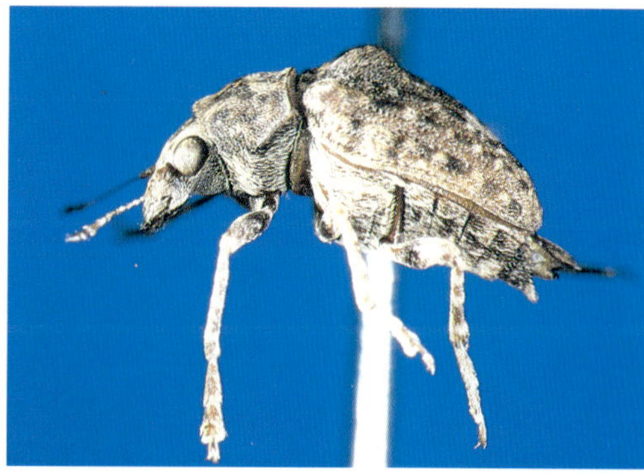

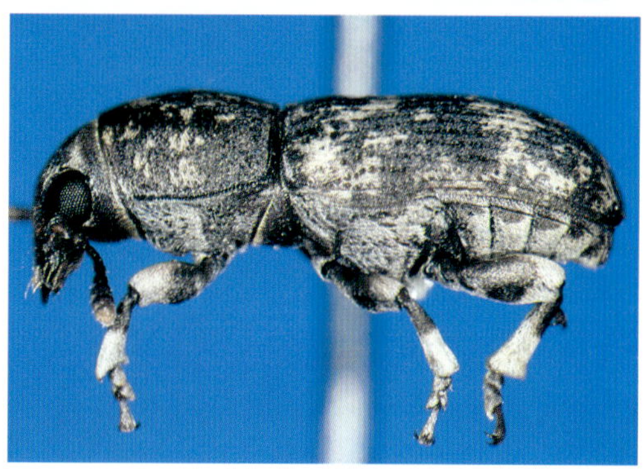

Plate 8 Anthribidae

1, 2 *Basitropis relicta* Blackburn
Mount Spec National Park, Queensland
Female; 6.7 mm

3 *Basitropis relicta* Blackburn, pale form
Cairns, Queensland
Sex ?; 8.5 mm

4 *Basitropis peregrinus* Pascoe
Batten Point, Northern Territory
Male; 8.5 mm

5, 6 *Basitropis solitaria* Pascoe
Mourilyan, north Queensland
Male; 5.1 mm

7, 8 *Basitropis* new species 1
Near The Craters, north of Ravenshoe, Queensland
Female; 8.1 mm

1	2
3	4
5	6
7	8

Plate 9 Anthribidae

1, 2 *Commista bispina* (Erichson)
Hobart, Tasmania
Male; 7 mm

3 *Commista bispina* (Erichson)

4 *Phloeobius gigas* (Fabricius)

5, 6 *Commista grisea* (Jordan)
Cromarty, north Queensland
Male; 6.5 mm

7, 8 *Commista latifrons* Jordan
Cairns, Queensland
Male; 10 mm

1	2
3	4
5	6
7	8

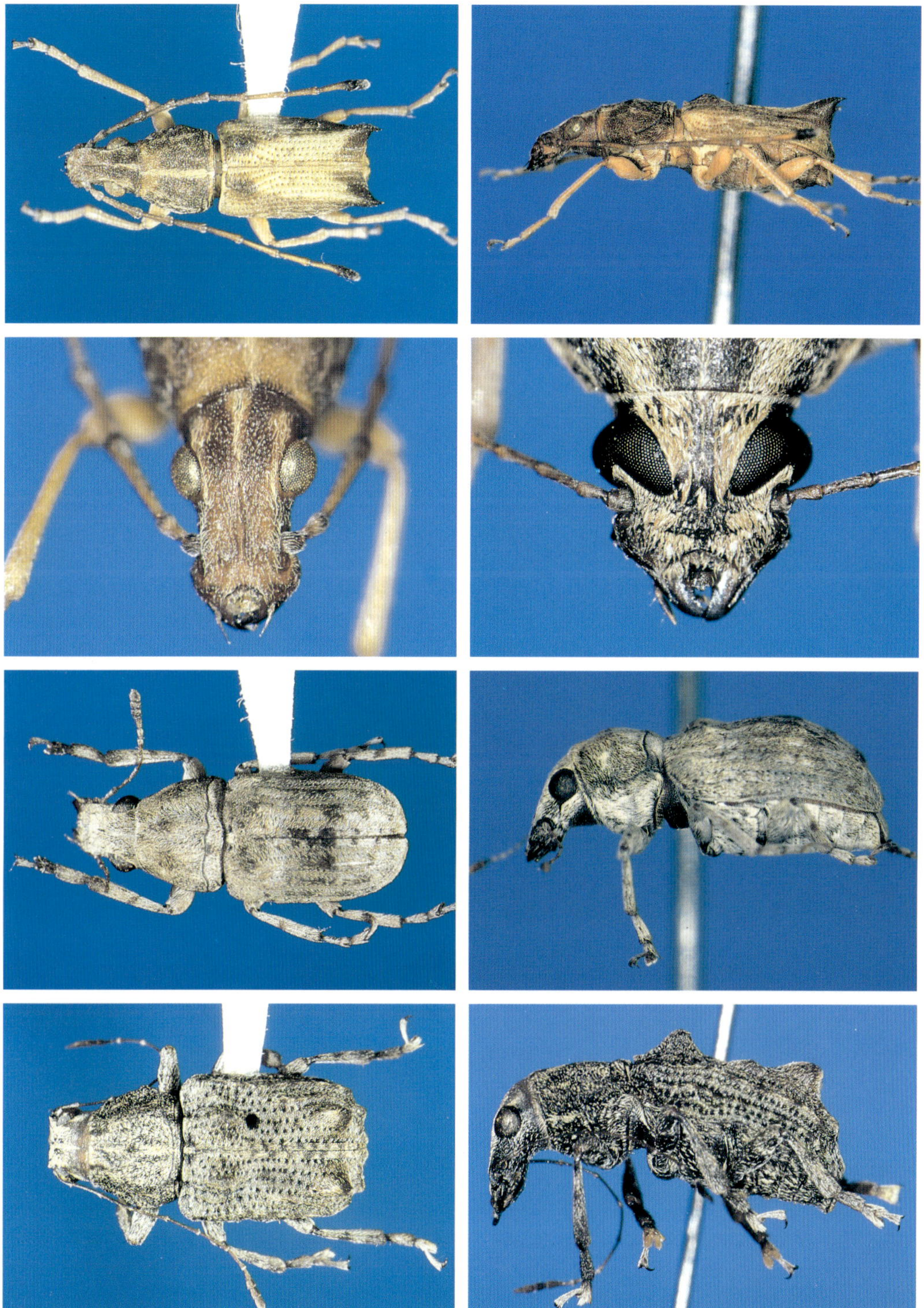

Plate 10 Anthribidae

1, 2 *Commista* new species 1
Polblue Swamp, Barrington Tops State Forest, New South Wales
Male; 6.5 mm

3, 4 *Dendropemon albinasus* (Jordan)
Dawson District, Queensland
Female; 8.7 mm

5 Abdomen of *Dendropemon albopictus* (Pascoe) showing remarkable thrichobothria of male.
Cardstone, Queensland

6 Face of the same specimen

7, 8 *Dendropemon virgatus* (Jordan)
Townsville, Queensland
Male; 13.2 mm

1	2
3	4
5	6
7	8

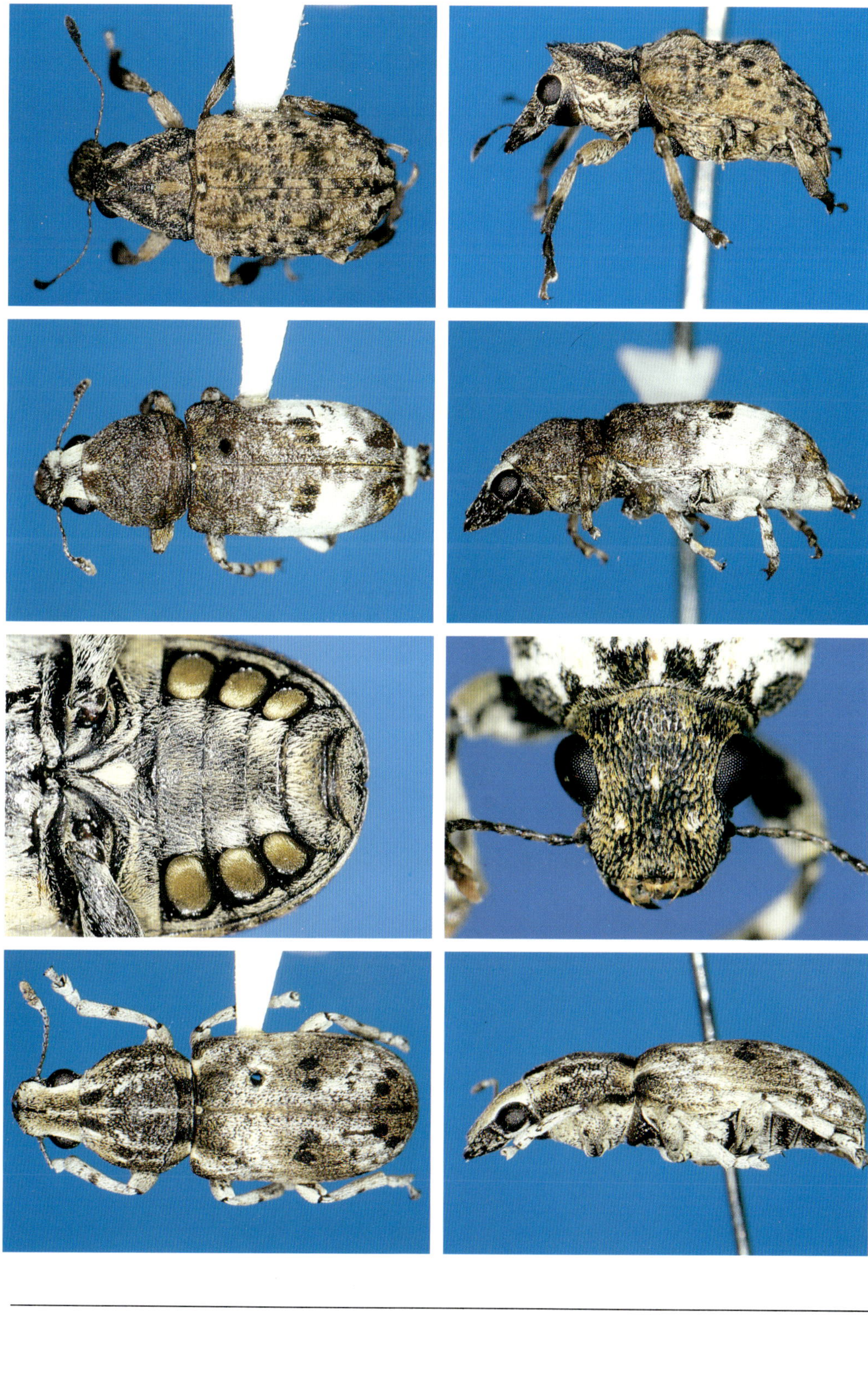

Plate 11 Anthribidae

1, 2 *Eczesaris atomaria* Pascoe
Banks Island, Queensland
Male; 11 mm

3, 4 *Epargemus marmoratus* Blackburn
New South Wales
Male 4.7 mm

5, 6 *Erichsonocis albuginosus* (Erichson)
Hobart, Tasmania
2.9 mm

7, 8 *Ethneca bakewelli* Pascoe
Beaconsfield, Victoria
6.3 mm

1	2
3	4
5	6
7	8

Plate 12 Anthribidae

1, 2 *Eucorynus crassicarnis* (Fabricius)
Cardstone, north Queensland
Male; 11 mm

3 *Eupanteos bifasciata* (Lea)
Kialoa, New South Wales
3.4 mm; compared with holotype

4 *Eupanteos doddi* Jordan
Kuranda, Queensland
From a paratype of *Allochromicis coccineus* Lea; 6.27 mm

5 *Eupanteos montana* (Oke)
Warburton, Victoria
5.1 mm

6 *Eupanteos ornata* Jordan
Richmond River, New South Wales
From a male paratype of *Allochromicis picticornis* Lea; 7.6 mm

7, 8 *Euparius lateripictus* (Jordan)
Cairns District, Queensland
Male; 13.6 mm

1	2
3	4
5	6
7	8

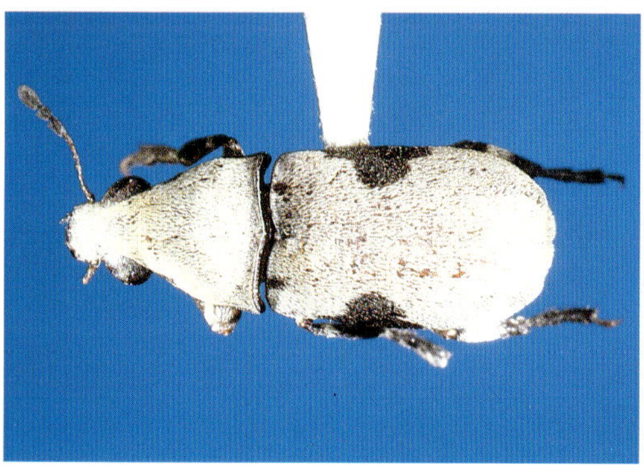

Plate 13 Anthribidae

1, 2 *Exillis australis* Frieser
Cairns, Queensland
Female; 4.5 mm

3, 4 *Exillis australis* Frieser
Edge Hill, Cairns, Queensland
Male; 4.8 mm; compare antennal lengths of female.

5, 6 *Howeanthribus bufo* Zimmerman
Lagoon Road, Lord Howe Island
Paratype male; 5.4 mm. Note the short carina on ventral margin of metatibia.

7 *Leaoanthribus crucifer* (Lea)
Gayndah, Queensland
3.3 mm; from the holotype of *Epargemus bicolor* Lea, a synonym.

8 *Leaoanthribus crucifer* (Lea)
Mount Chalmers, Queensland
Holotype; 5 mm

1	2
3	4
5	6
7	8

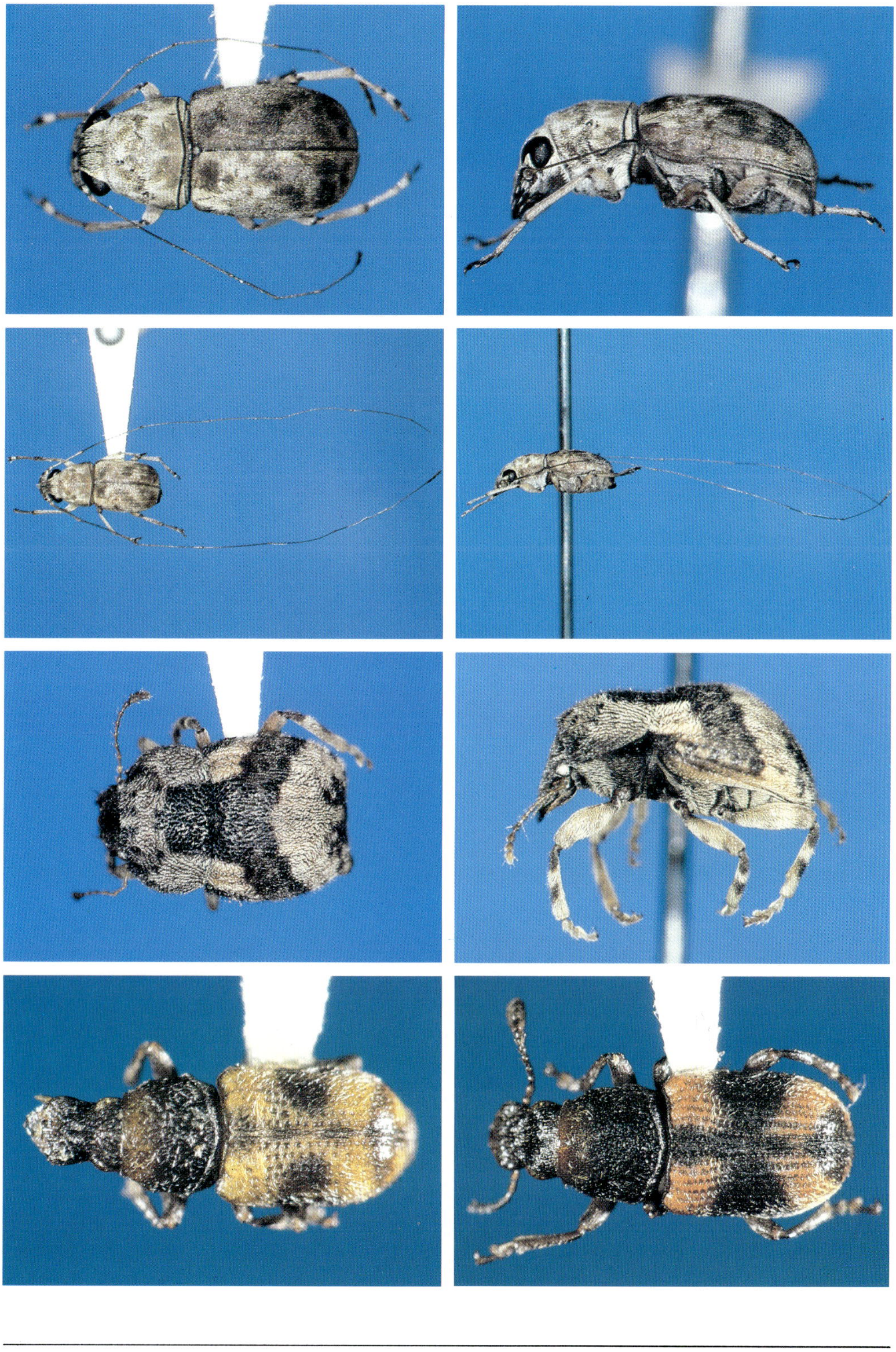

Plate 14 Anthribidae

1, 2 *Litocerus variegatus* Jordan
Iron Range, Queensland
Female; 7.5 mm; the colour pattern varies greatly.

3, 4 *Mecocerinopis amabilis* (Pascoe)
Iron Range, Queensland
Male; 11.5 mm; the male antennae are about twice the length of the body; the colour pattern is variable.

5, 6 *Notoecia reticulata* Blackburn
Cairns, Queensland
Male; 5 mm; see Plate 17:4 for face.

7, 8 *Phaulimia australiaca* (Jordan)
Iron Range, Queensland
Female; 5 mm; compared with holotype.

1	2
3	4
5	6
7	8

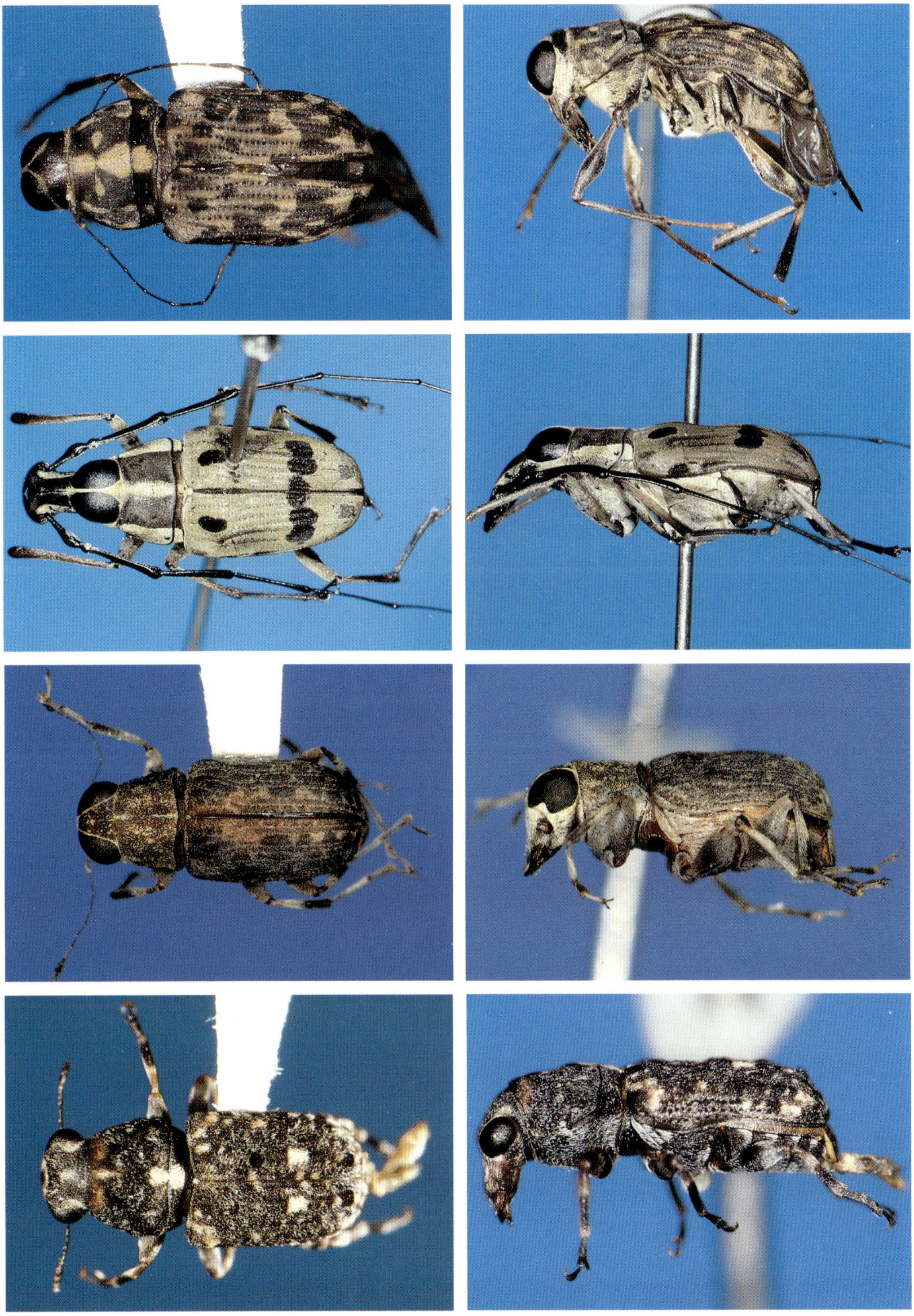

Plate 15 Anthribidae

1, 2 *Phloeobius gigas* (Fabricius)
Edge Hill, Cairns, Queensland
Male; 10.5 mm; see face on Plate 9:4.

3, 4 *Streneoderma planatum* Blackburn
Sydney, New South Wales
Male; 2.4 mm

5, 6 *Taburnus breviceps* Zimmerman
Wilton, New South Wales
Holotype female; 8.5 mm

7, 8 *Taburnus neglectus* Zimmerman
Black Mountain, Canberra, Australian Capital Territory
Holotype male; 15.5 mm

<table>
<tr><td>1</td><td>2</td></tr>
<tr><td>3</td><td>4</td></tr>
<tr><td>5</td><td>6</td></tr>
<tr><td>7</td><td>8</td></tr>
</table>

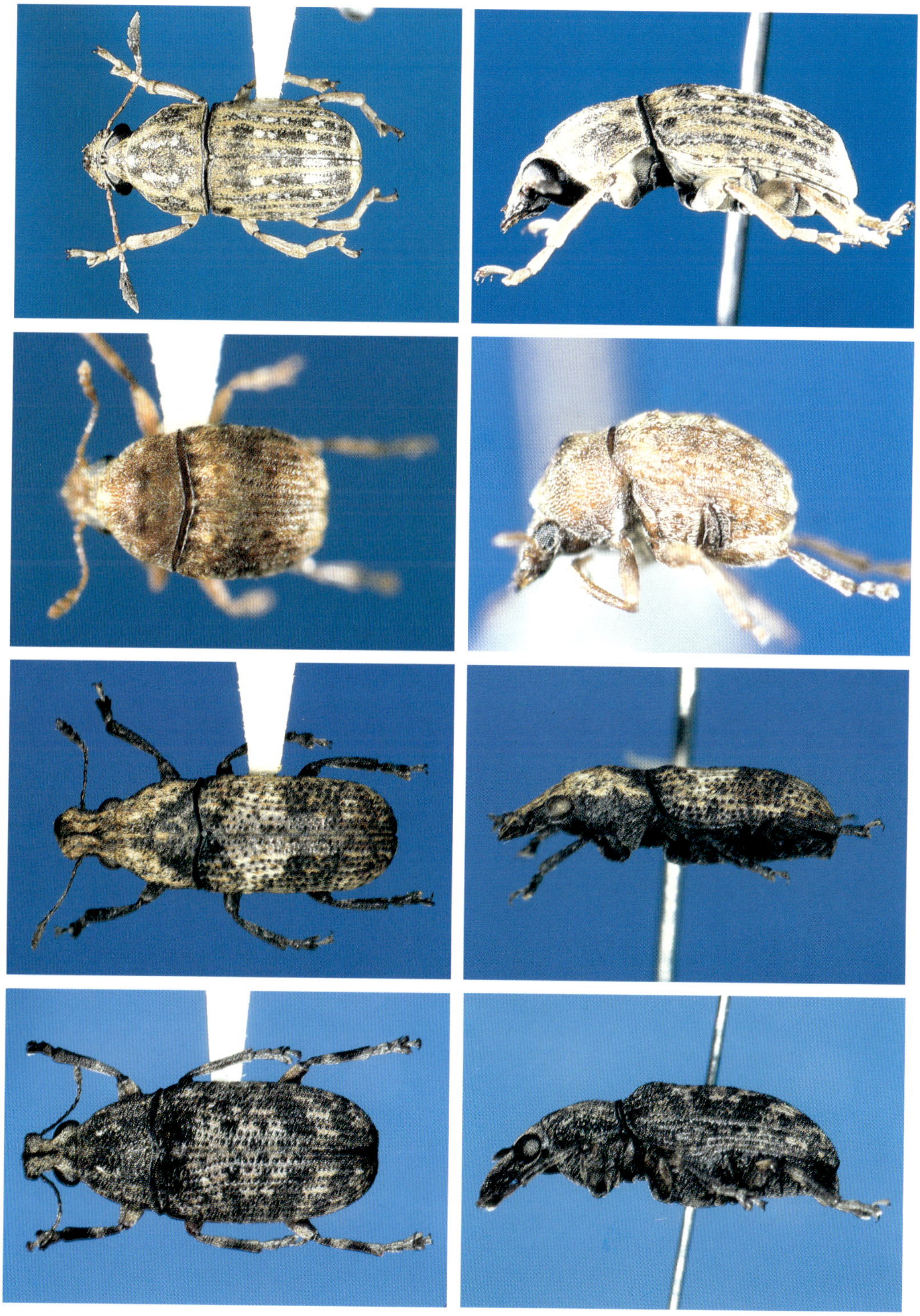

Plate 16 Anthribidae

1, 2 *Taburnus setosus* Zimmerman
Mackay, Queensland
Holotype male; 17.5 mm

3, 4 *Taburnus tessellatus* Zimmerman
Belgrave, Victoria
Holotype female; 12 mm

5, 6 *Telala musiva* (Erichson)
Piccadilly Circus, Australian Capital Territory
Female; 9 mm; compared with lectotype.

7, 8 *Telala* new species 1
Illawarra, New South Wales
Female; 8.5 mm; antennae of male extend to near apex of elytra.

1	2
3	4
5	6
7	8

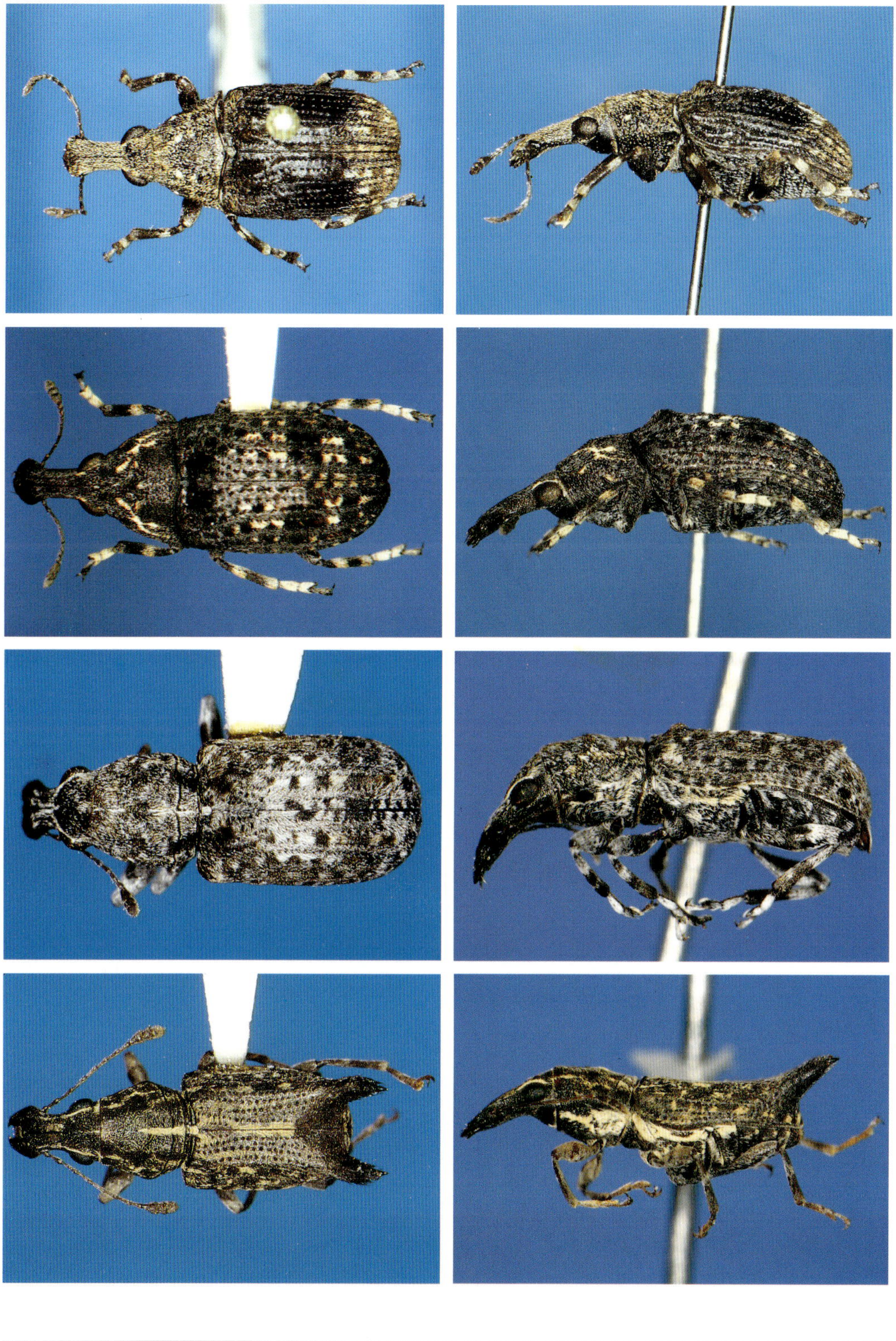

Plate 17 Anthribidae

1, 2 *Teratanthribus monstrosus* Zimmerman
Mount Superbus via Boonah, Queensland
Allotype female; 4.8 mm

3 The same, caudal view; a spider mimic?

4 Face of *Notoecia reticulata* Blackburn; compare Plate 14:5, 6.

5, 6 *Xynotropis micans* Blackburn
Near Weldborough, Tasmania
Male; 1.7 mm

7, 8 Genus "A"
Bunya Mountains, Queensland
Male; 5.3 mm

1	2
3	4
5	6
7	8

Plate 18 Anthribidae

1, 2 Genus "B"
Norfolk Island
Male; 4 mm. The antennae are 8 times the length of the body.

3, 4 Genus "C"
15 km west of Kiama, New South Wales
Pale form; female; 2.7 mm; compare following figures.

5, 6 Genus "C"
Royal National Park, 30 km south of Sydney, New South Wales
Dark form; male; 2.2 mm; compare foregoing figures.

7, 8 *Stenorhis* new species 1
Hugh Nelson Range, Queensland
Male; 1.6 mm

1	2
3	4
5	6
7	8

Plate 19 Anthribidae

1, 2 Genus "J"
Brindle Creek, north of Kyogle, New South Wales
Male; 2 mm

3, 4 Genus "K"
Near Kuranda, Queensland
Female; 1.8 mm

5, 6 Genus "L"
Near Lockerbie, north Queensland
Male; 7 mm; see next figure for face.

7 Face of previous specimen.

8 Face of *Telala* new species "1".

For more illustrations of Anthribidae, see Volume VI.

1	2
3	4
5	6
7	8

Plate 20 Belidae

1, 2 *Agathobelus bivittatus* Zimmerman
Marys Creek State Forest, Queensland
Holotype male; 9.3 mm
See Plate 49:7 for view of head.

3, 4 *Agnesiotis pilosula* Pascoe
Dividing Range, Victoria
Female; 10 mm

5, 6 *Apagobelus brevirostris* (Lea)
Yarraman, Queensland
Female; 10.5 mm
See Plate 49:5 for ventral view.

7, 8 *Araiobelus acicularis* (Pascoe)
Mount Barker, Western Australia
Female; 9.5 mm

1	2
3	4
5	6
7	8

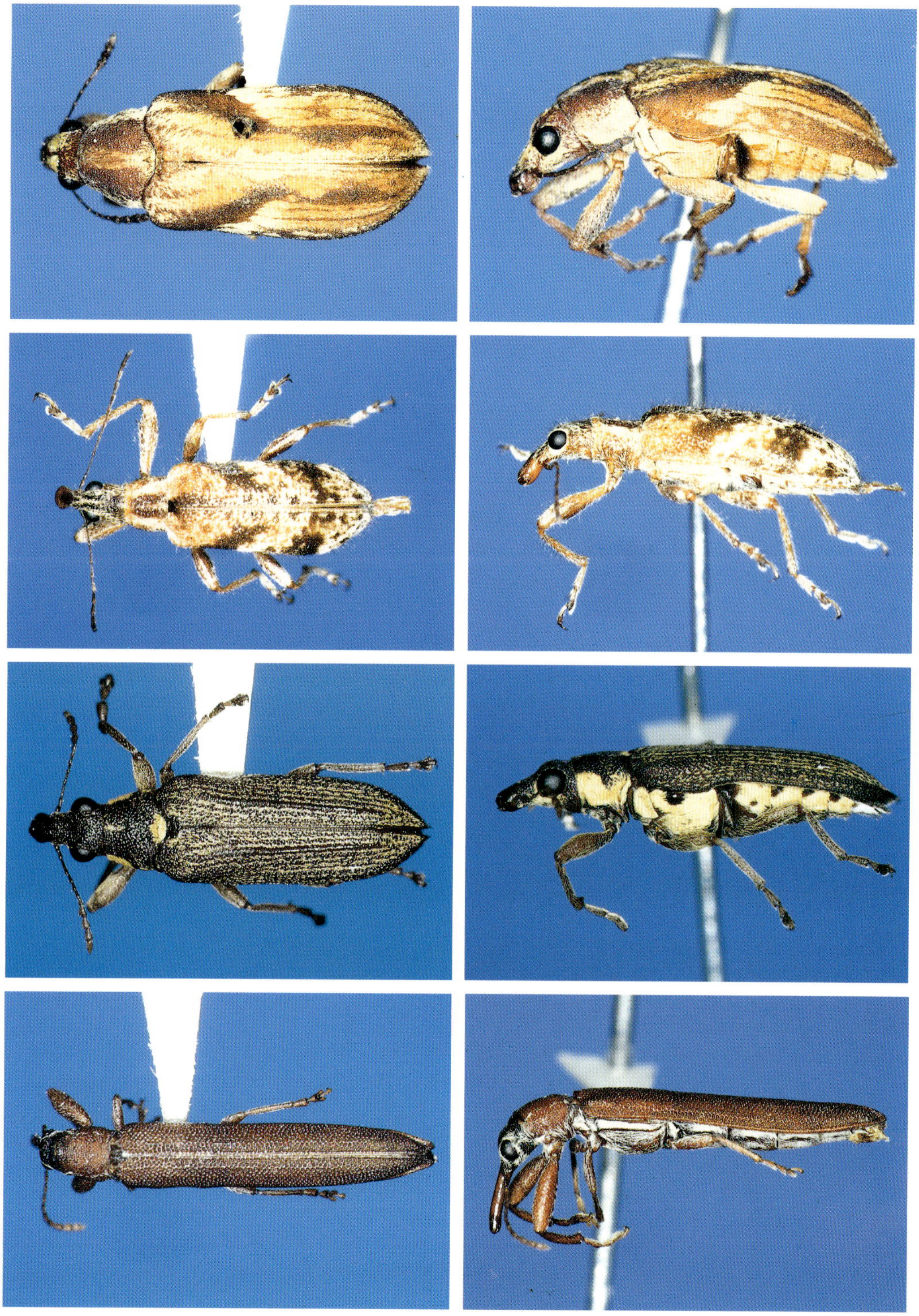

Plate 21 Belidae

1, 2 *Araiobelus filum* (Jekel)
North Sydney, New South Wales
Female; 9.5 mm; compared with holotype.

3, 4 *Araiobelus floccosus* (Lea)
Beverley, Western Australia
Paratype female; 9.5 mm

5, 6 *Araiobelus maculipennis* (Lea)
National Park, Queensland
Paratype male; 8.5 mm

7, 8 *Arhinobelus agathophagus* Zimmerman
Gallagowan State Forest, Queensland
Holotype female; 5.4 mm

1	2
3	4
5	6
7	8

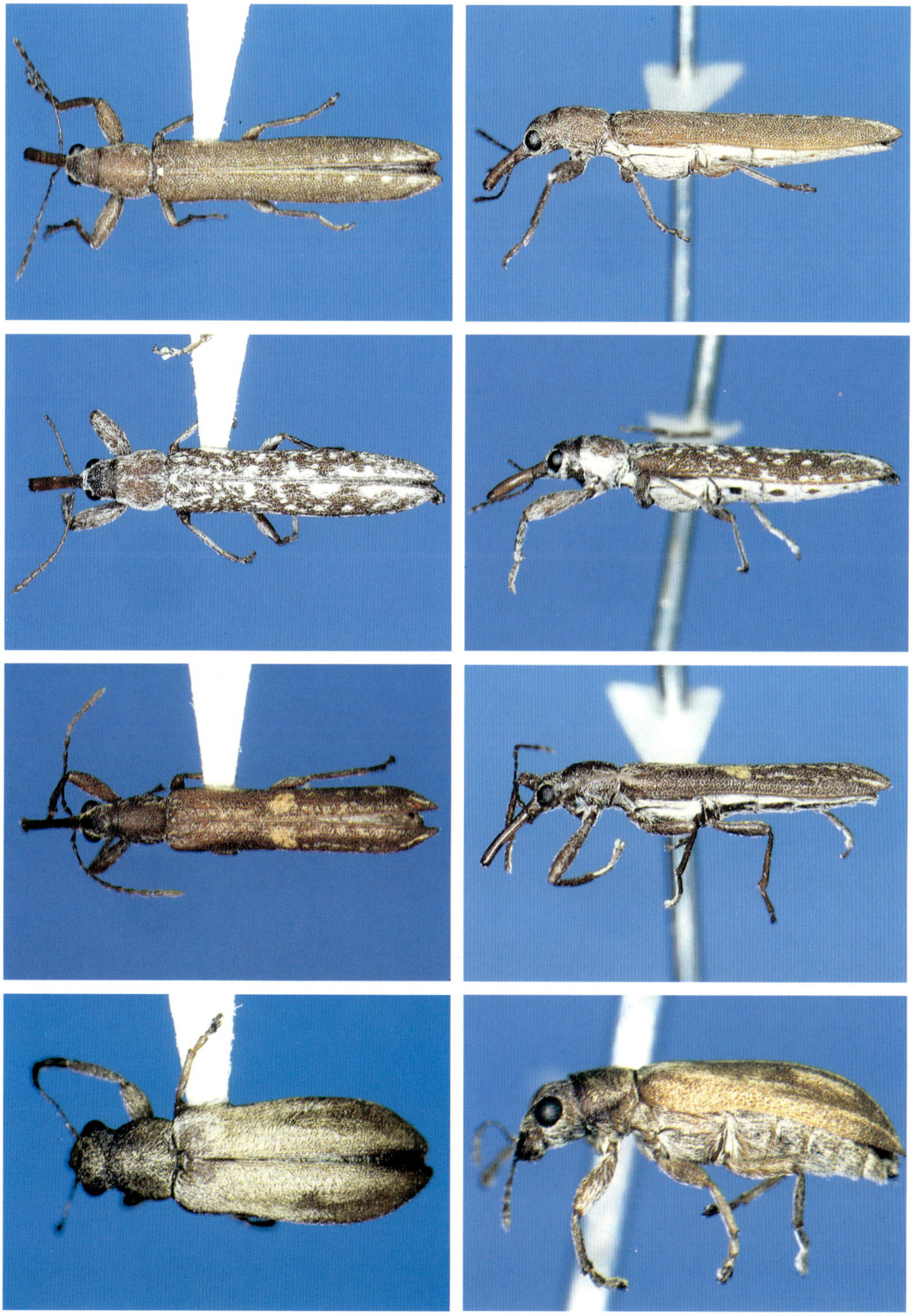

Plate 22 Belidae

1, 2 *Basiliobelus alveatus* Zimmerman
Brisbane, Queensland
Holotype female; 17 mm

3, 4 *Basiliobelus flavovittatus* Zimmerman
Cascade, New South Wales
Holotype female; 12.5 mm

5, 6 *Basiliobelus lepidus* Zimmerman
Beauty Spot 5, Imbil State Forest, Queensland
Allotype female; 7.3 mm

7, 8 *Brachybelus undulatus* Zimmerman
Imbil, Queensland
Paratype female; 13.5 mm

1	2
3	4
5	6
7	8

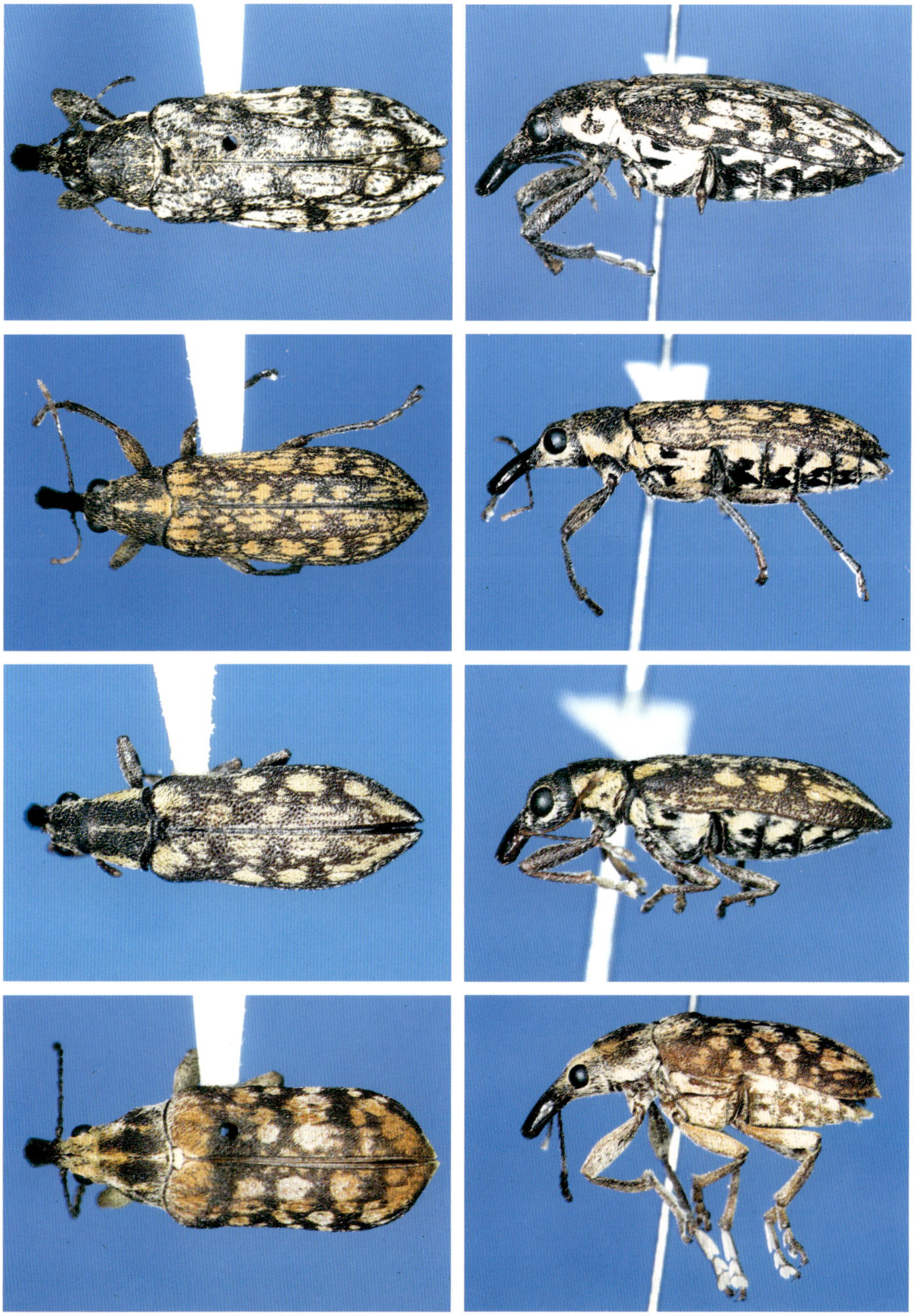

Plate 23 Belidae

1, 2 *Cyrotyphus blandus* (Faust)
Oakey Creek, Tidbinbilla, Australian Capital Territory
Female; 14.5 mm

3, 4 *Cyrotyphus fascicularis* Pascoe
Juranda Rockhole, 106 km south of Balladonia, Western Australia
Female; 19 mm; compared with syntype.

5, 6 *Cyrotyphus* new species 1
Monarto, South Australia
Male; 17.5 mm

7, 8 *Cyrotyphus variegatus* Lea
Mount Wellington, Tasmania
Female; 14 mm; compared with holotype.

1	2
3	4
5	6
7	8

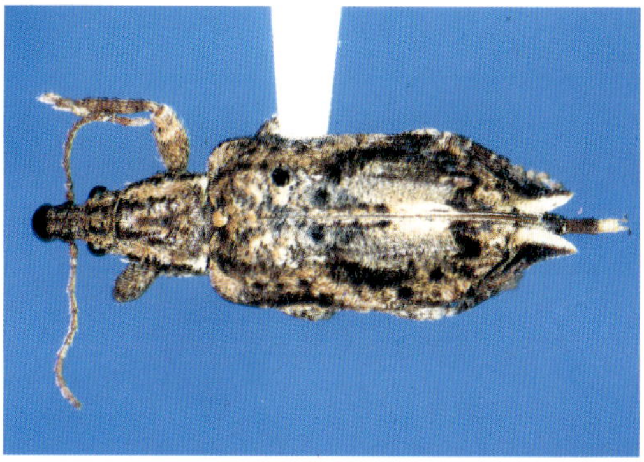

Plate 24 Belidae

1, 2 *Cyrotyphus vestitus* (Pascoe)
Styx River, New South Wales
Female; 9 mm

3, 4 *Cyrotyphus* new species 2
Dorrigo, New South Wales
Male; 12 mm

5, 6 *Habrobelus maculatus* Zimmerman
Benarkin Logging Area, State Forest 283, Queensland
Holotype male; 9.8 mm

7, 8 *Isacantha dermestiventris* (Boisduval)
Eidsvold, Queensland
Female; 11.5 mm

1	2
3	4
5	6
7	8

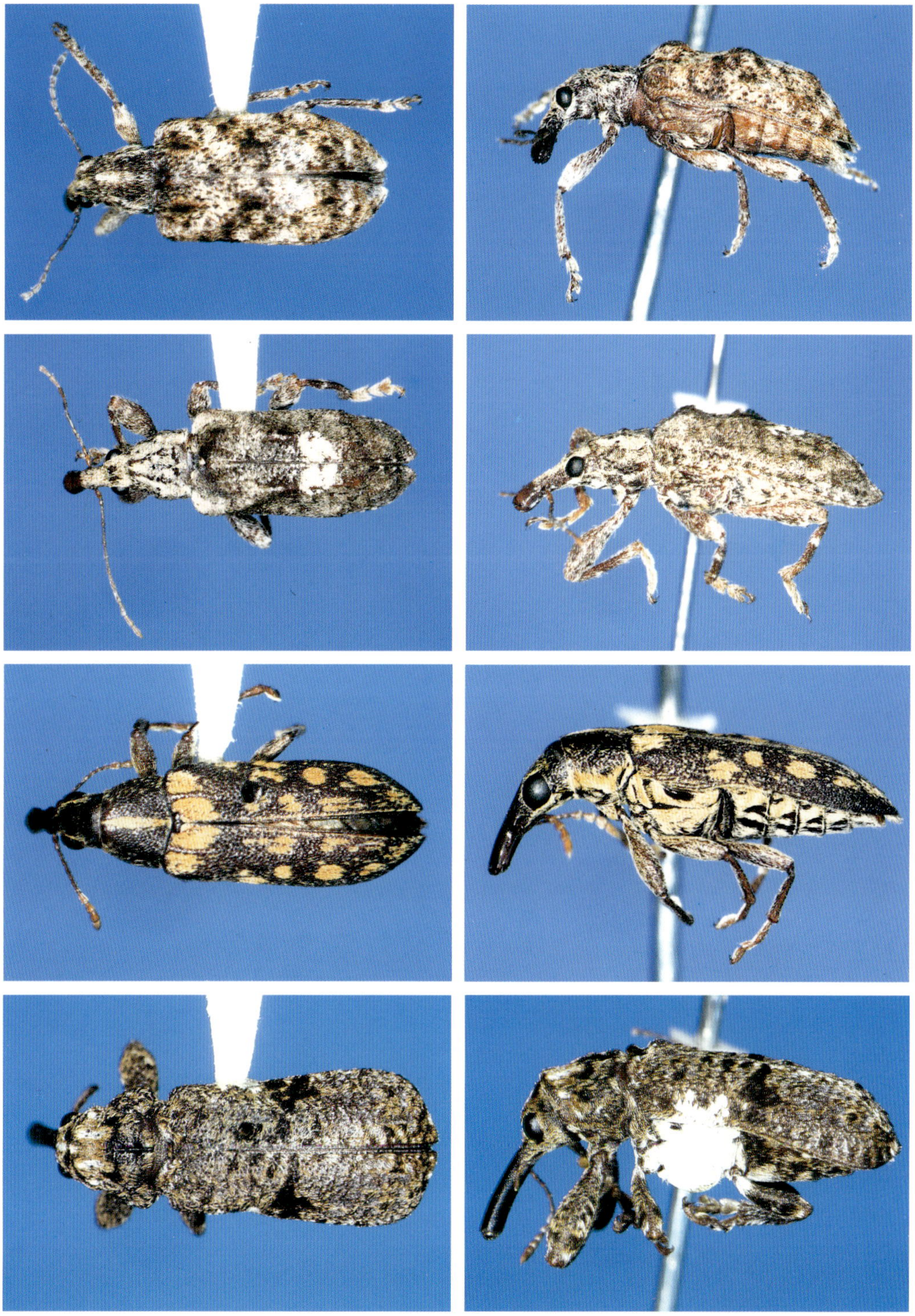

Plate 25 Belidae

1, 2 *Isacantha dermestiventris* variety *albicollis* Lea
Fletcher, Queensland
Female; 13.5 mm

3, 4 *Isacantha dermestiventris* (Boisduval), variety
Canberra, Australian Capital Territory
Male; 14 mm

5, 6 *Isacantha inculta* Olliff
Lord Howe Island
Female; 19.5 mm

7, 8 *Isacantha interrupta* (Lea)
North Queensland
Holotype male; 12 mm

1	2
3	4
5	6
7	8

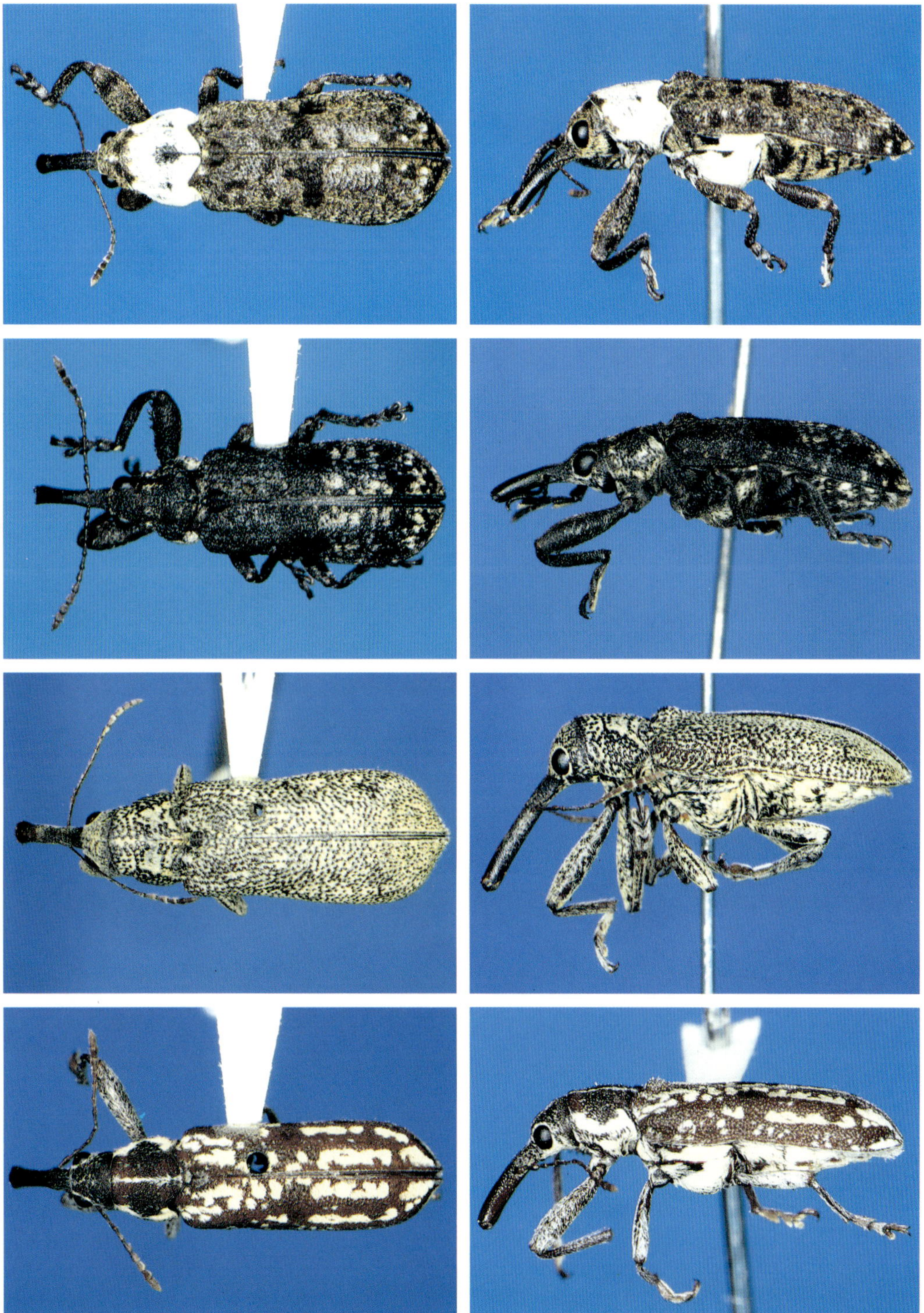

Plate 26 Belidae

1, 2 *Isacantha serrata* Lea
Eidsvold, Queensland
Female; 18 mm

3, 4 *Isacantha* new species 1
Tone Bridge, Western Australia
Female; 15.5 mm

5, 6 *Isacantha* new species 2
41 km north of Euston, New South Wales
Male; 15 mm

7, 8 *Isacantha* new species 3
Victoria
Female; 13.5 mm

1	2
3	4
5	6
7	8

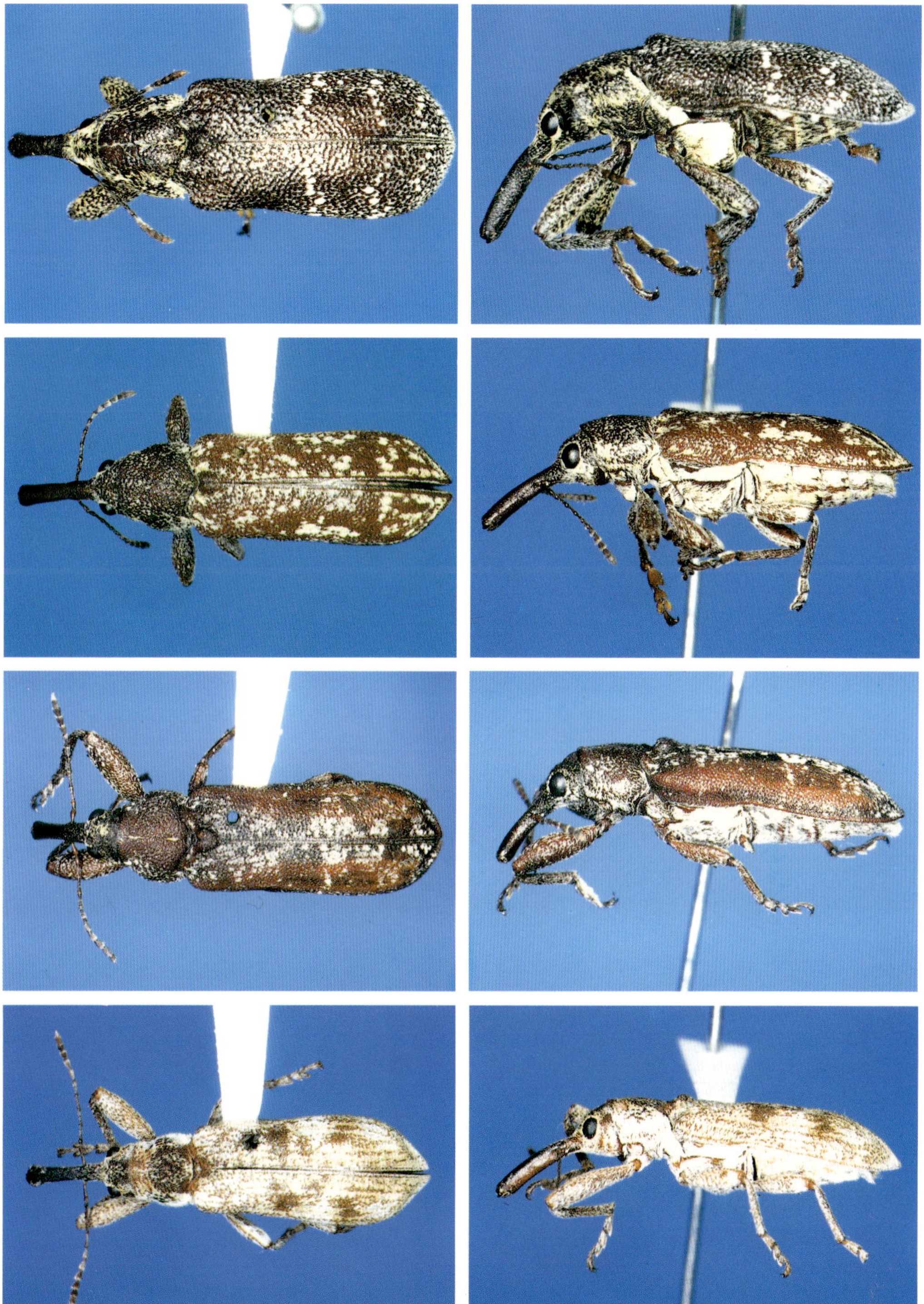

Plate 27 Belidae

1, 2 *Isacanthodes ganglionicus* (Pascoe)
Uriarra Road, Canberra, Australian Capital Territory
Male; 15 mm

3, 4 *Isacanthodes monilis* (Newman)
Uriarra Road, Canberra, Australian Capital Territory
Male; 11.5 mm

5, 6 *Isacanthodes* new species 1
Mount Walsh National Park, 5 km south by east of Biggenden, Queensland
Female; 14 mm

7, 8 *Isacanthodes* new species 2
Crystal Falls, Paluma Range Road, north Queensland
Female; 14 mm

1	2
3	4
5	6
7	8

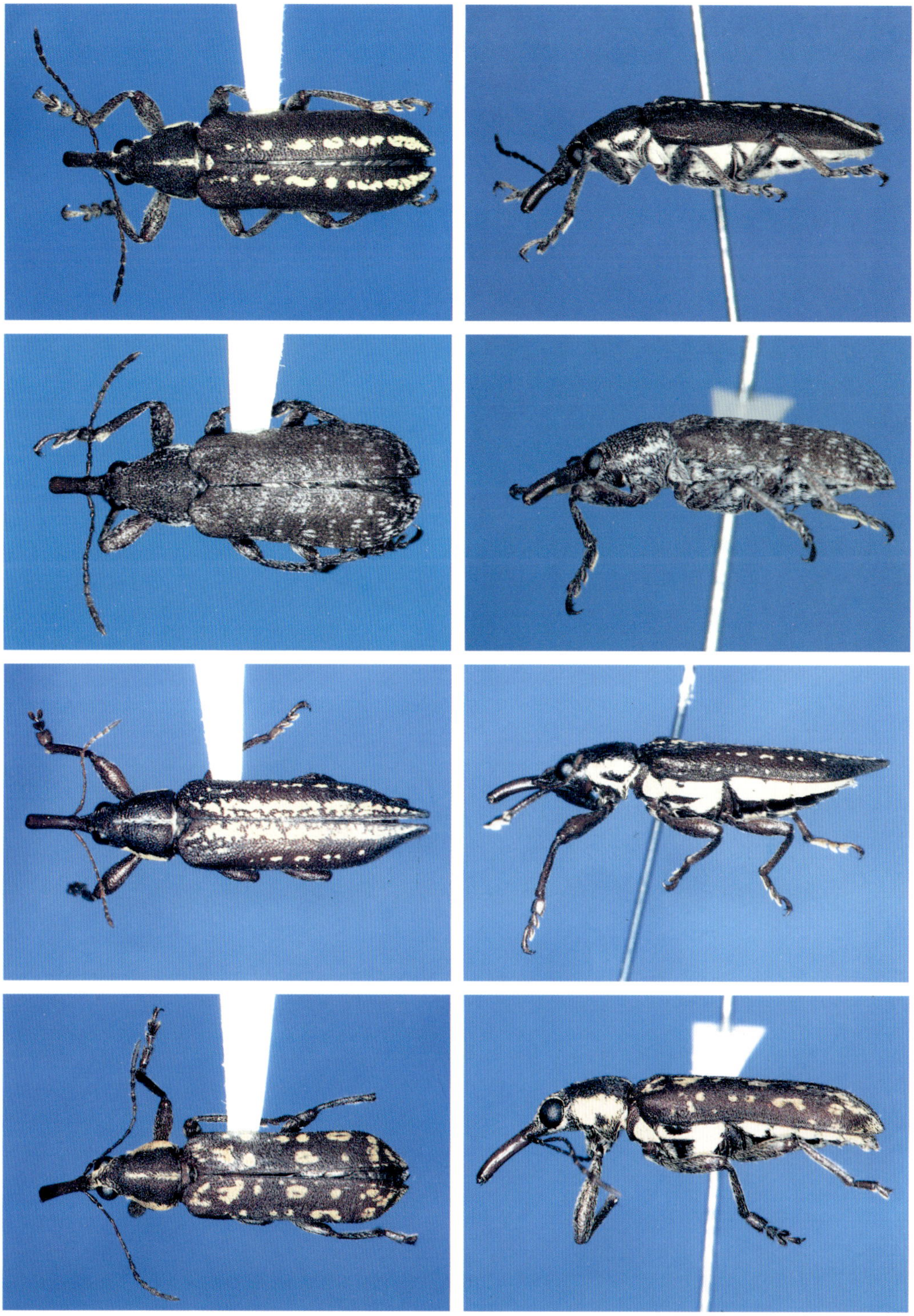

Plate 28 Belidae

1, 2 *Isacanthodes* new species 3
Bathurst or Melville Island, Northern Territory
Female; 13.5 mm

3, 4 *Isacanthodes* new species 4
Toolbrunup, Stirling Ranges, Western Australia
Male; 11.5 mm

5, 6 *Isacanthodes* new species 5
Israelite Bay, Western Australia
Female; 15 mm

7, 8 *Leptobelus linearis* (Pascoe)
King George Sound, Western Australia
Female; 10 mm

1	2
3	4
5	6
7	8

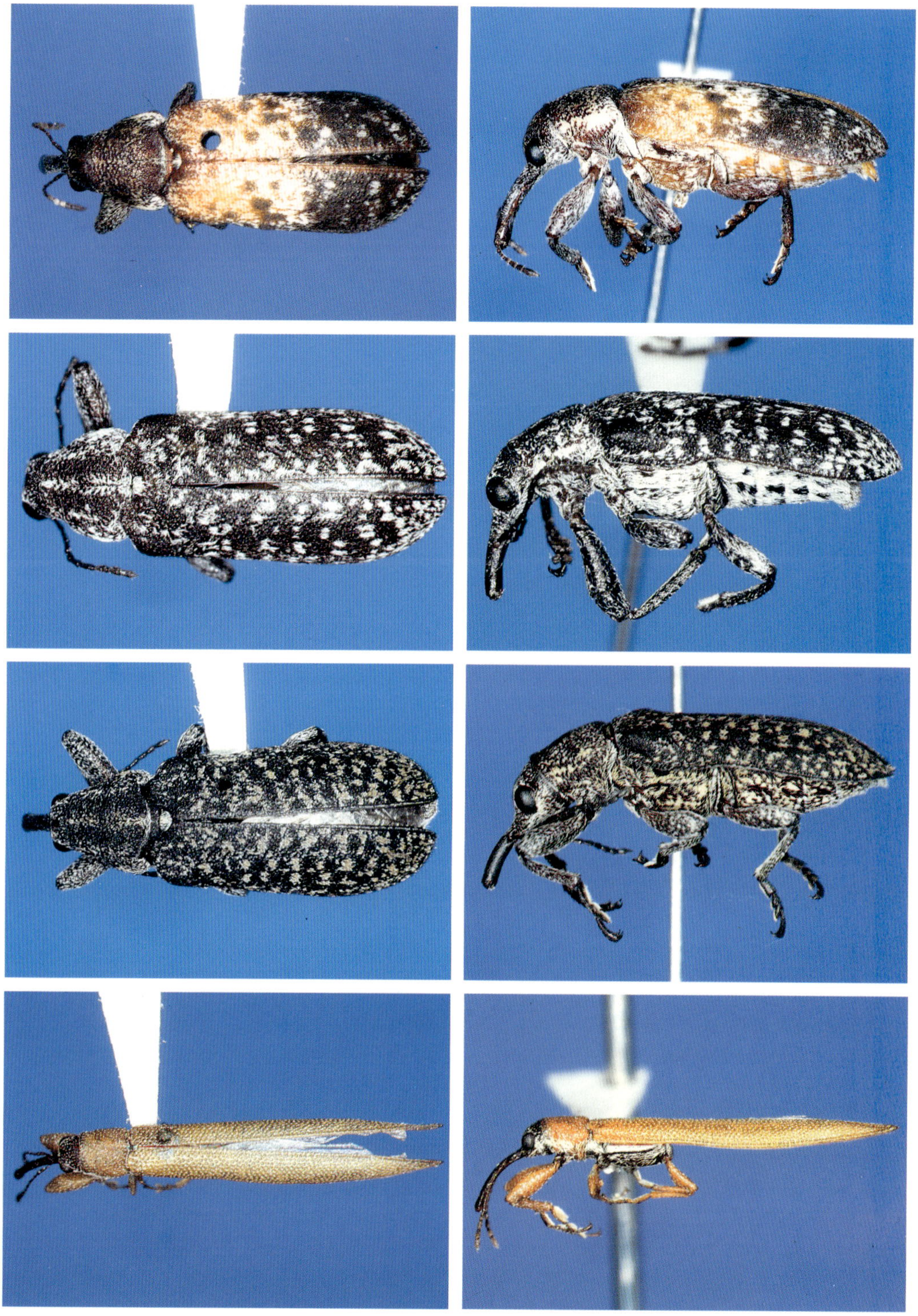

Plate 29 Belidae

1, 2 *Leptobelus tibialis* (Blackburn)
Lucindale, South Australia
Female, compared with holotype; 12.5 mm

3, 4 *Macrobelus insignis* Lea
Burnett River, Bundaberg, Queensland
Female; 16.5 mm

5, 6 *Pachybelus tuberculatus* Zimmerman
Cedar Creek, Queensland
Holotype female; 10.5 mm

7, 8 *Pachyura australis* Hope
Wilsons Promontory, Victoria
Female; 13 mm

1	2
3	4
5	6
7	8

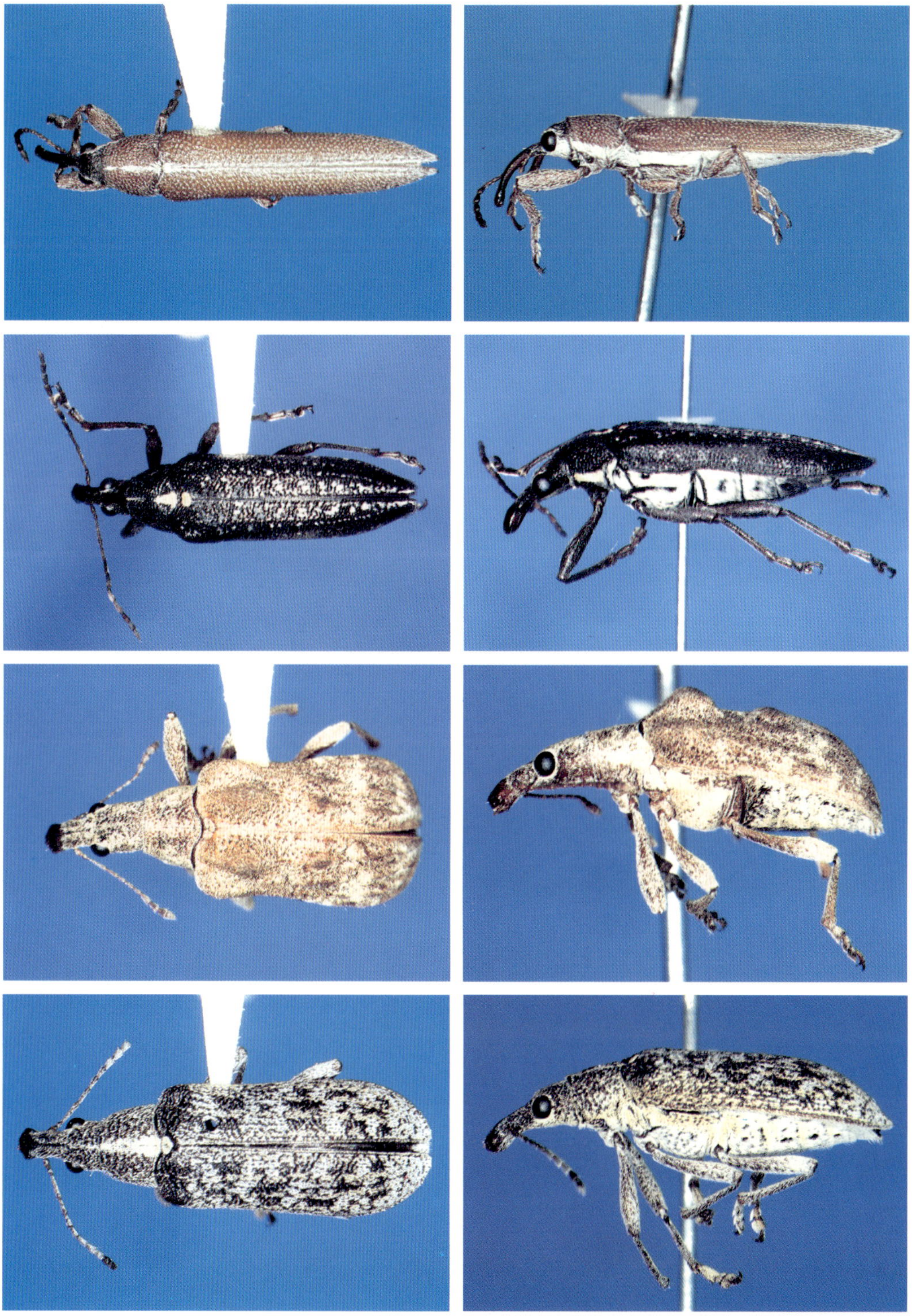

Plate 29A Belidae

1, 2 *Rhinotia acaciae* (Lea)
Geraldton, Western Australia
Holotype female; 10 mm

3, 4 *Rhinotia acanthoptera* (Lea)
Bunya Mountains, Queensland
Female; 14.5 mm

5, 6 *Rhinotia acrobela* (Olliff)
Smoking Tree Ridge, Lord Howe Island
Female, compared with holotype; 10 mm

7, 8 *Rhinotia acutipennis* (Lea)
Gosford, New South Wales
Holotype female; 11.5 mm

1	2
3	4
5	6
7	8

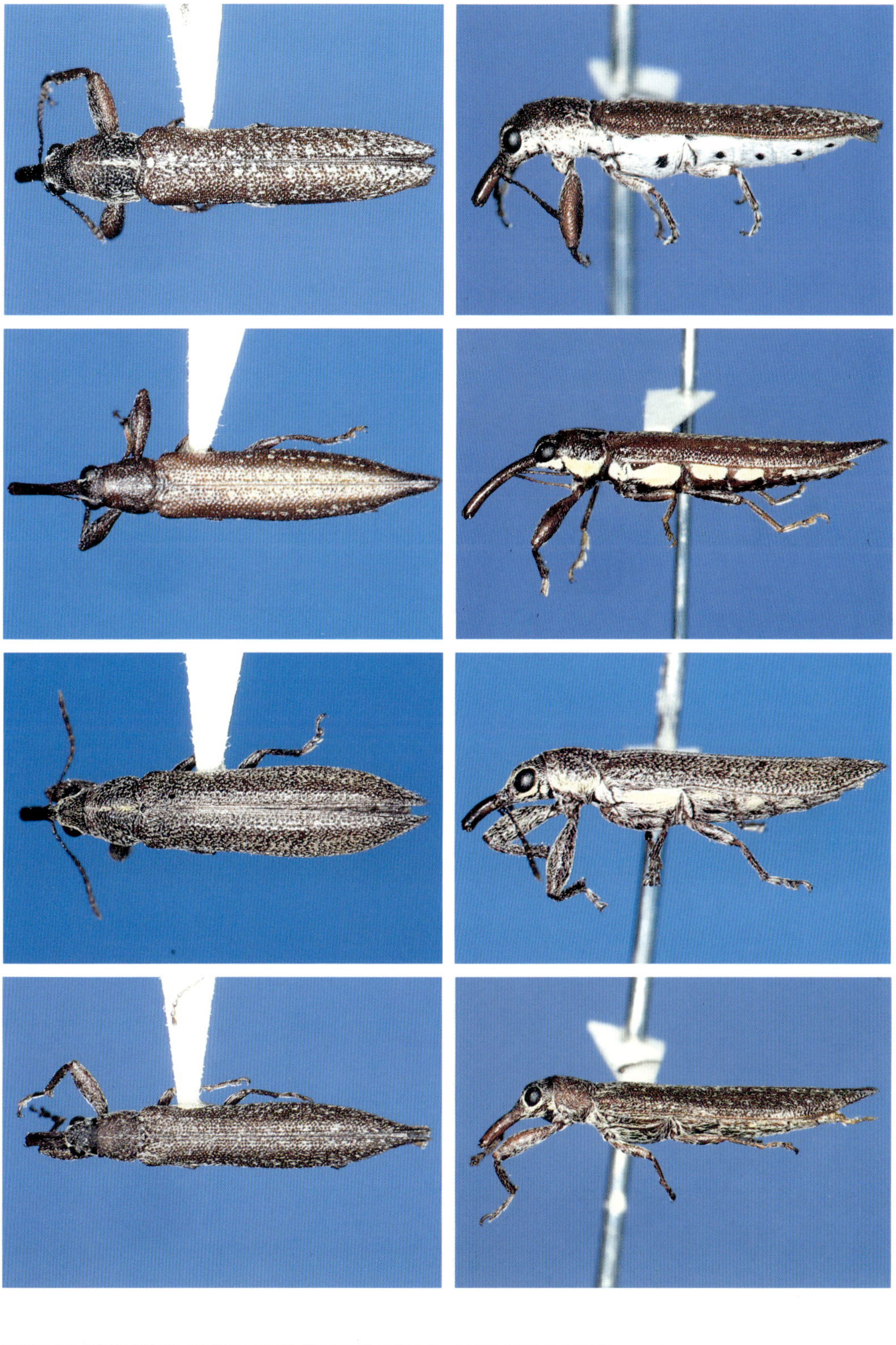

Plate 30 Belidae

1, 2 *Rhinotia adelaidae* (Blackburn)
Red Hill, Canberra, Australian Capital Territory
Female; 15 mm

3, 4 *Rhinotia angustata* (Lea)
Mindarie, South Australia
Paratype male; 14 mm

5, 6 *Rhinotia angustula* (Germar)
Fraser Range, Western Australia
Female; 14 mm

7, 8 *Rhinotia apicalis* Zimmerman
Brock Creek, Burnside, Northern Territory
Paratype female; 14.5 mm

1	2
3	4
5	6
7	8

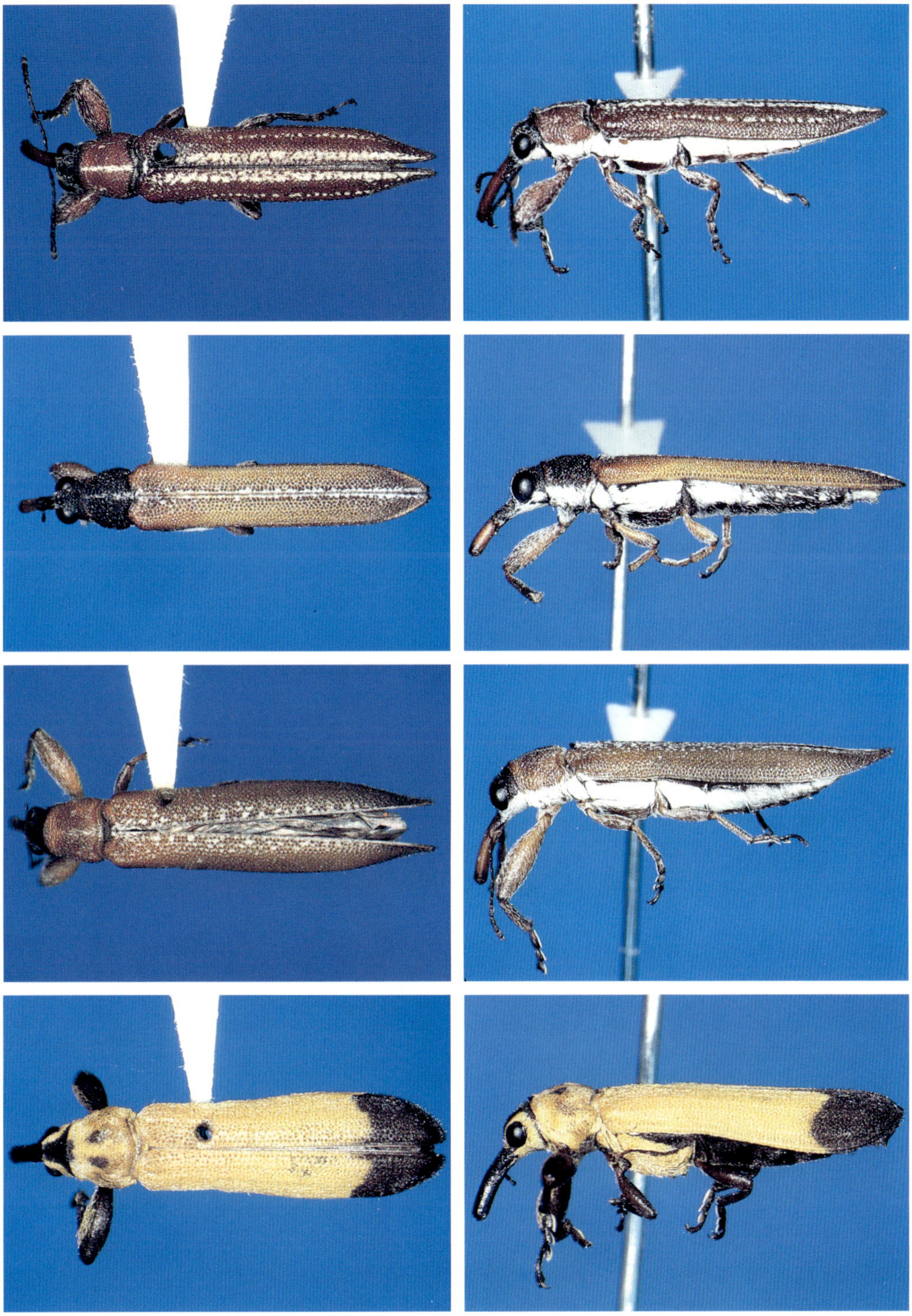

Plate 31 Belidae

1, 2 *Rhinotia bidentata* (Donovan)
Canberra, Australian Capital Territory
Female; 22.5 mm

3, 4 *Rhinotia bimaculata* (Pascoe)
Dorrigo, New South Wales
Male; 9 mm

5, 6 *Rhinotia brevipes* (Lea)
Cape York, Queensland
Holotype female; 13.5 mm

7, 8 *Rhinotia brunnea* (Guérin-Méneville)
Black Mountain, Canberra, Australian Capital Territory
Female; 19 mm

1	2
3	4
5	6
7	8

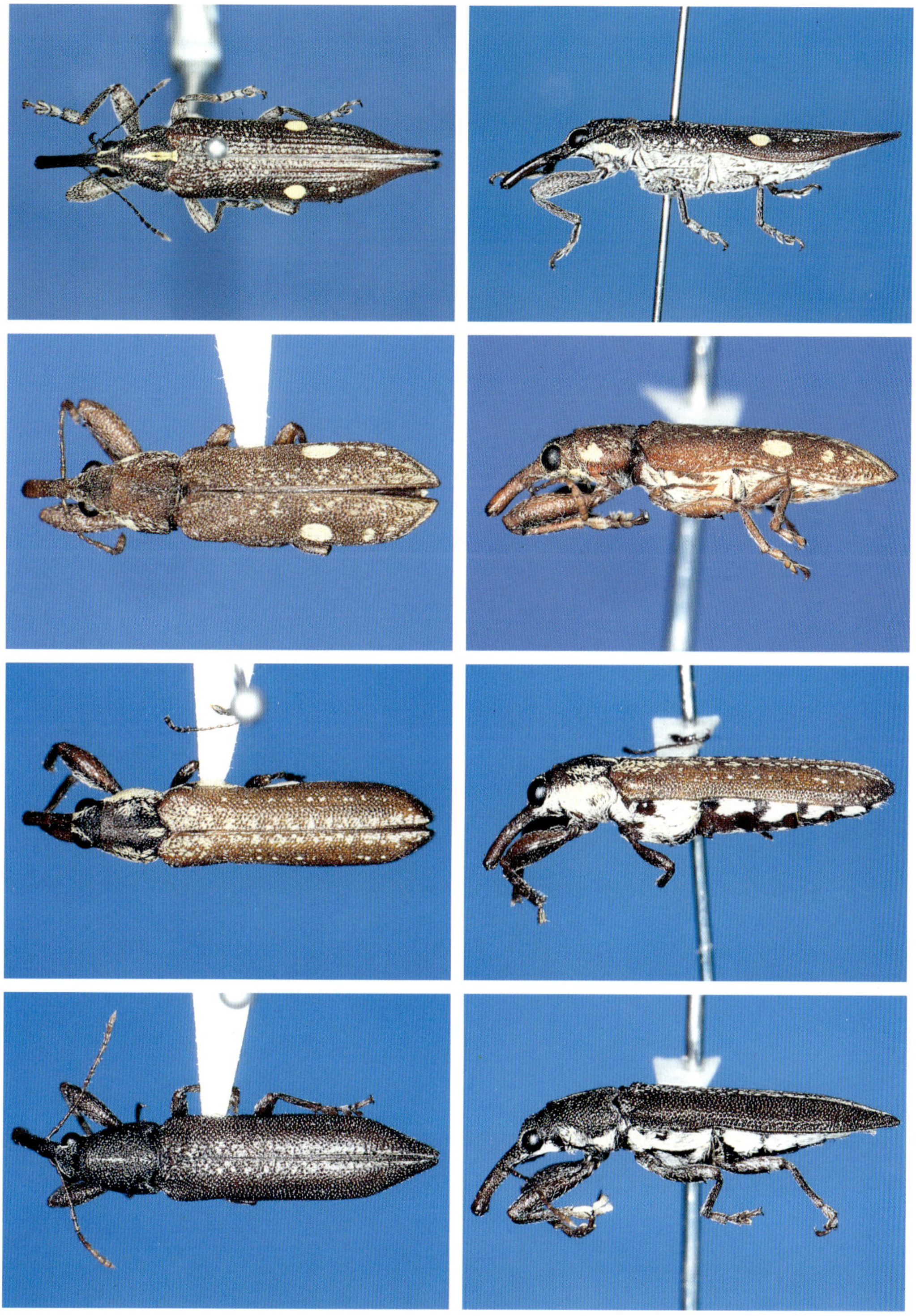

Plate 32 Belidae

1,2 *Rhinotia* new species near *brunnea*
Black Mountain, Canberra, Australian Capital Territory
Female; 16.5 mm

3, 4 *Rhinotia centralis* Pascoe
Port Lincoln, South Australia
Female; 19 mm

5, 6 *Rhinotia cristata* (Lea)
Port Lincoln, South Australia
Holotype male; 14 mm

7, 8 *Rhinotia cruenta* Pascoe
Geraldton, Western Australia
Female; 18 mm

1	2
3	4
5	6
7	8

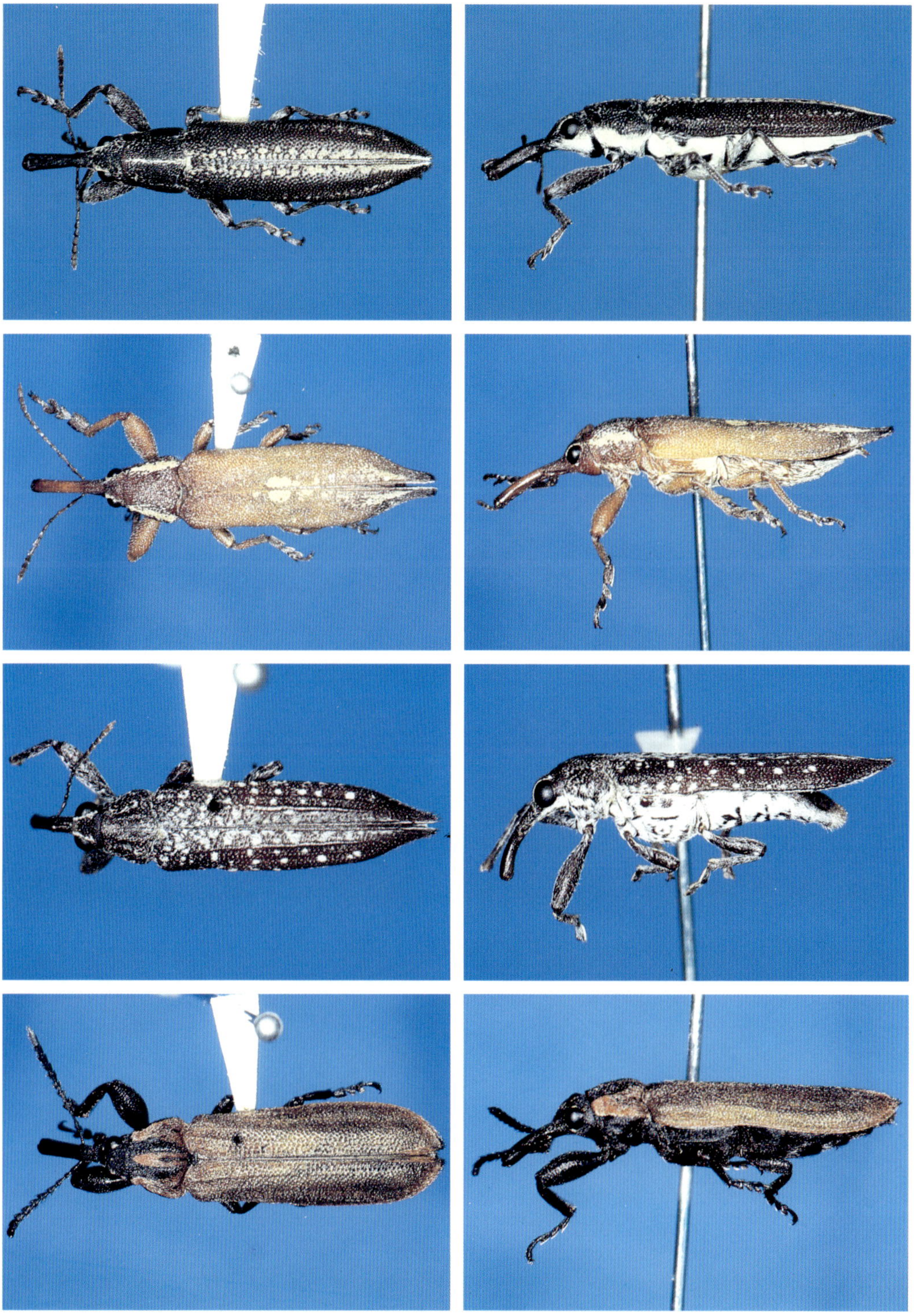

Plate 33 Belidae

1, 2 *Rhinotia cylindrica* (Lea)
Lake Austin, South Australia
Paratype female; 12.5 mm

3, 4 *Rhinotia difficilis* (Blackburn)
Fraser Range, Western Australia
Paratype female; 9.5 mm

5, 6 *Rhinotia divisa* (Pascoe)
Coen River, Queensland
Female; 12 mm

7, 8 *Rhinotia edentula* (Lea)
Braidwood, New South Wales
Paratype female, 15.5 mm

1	2
3	4
5	6
7	8

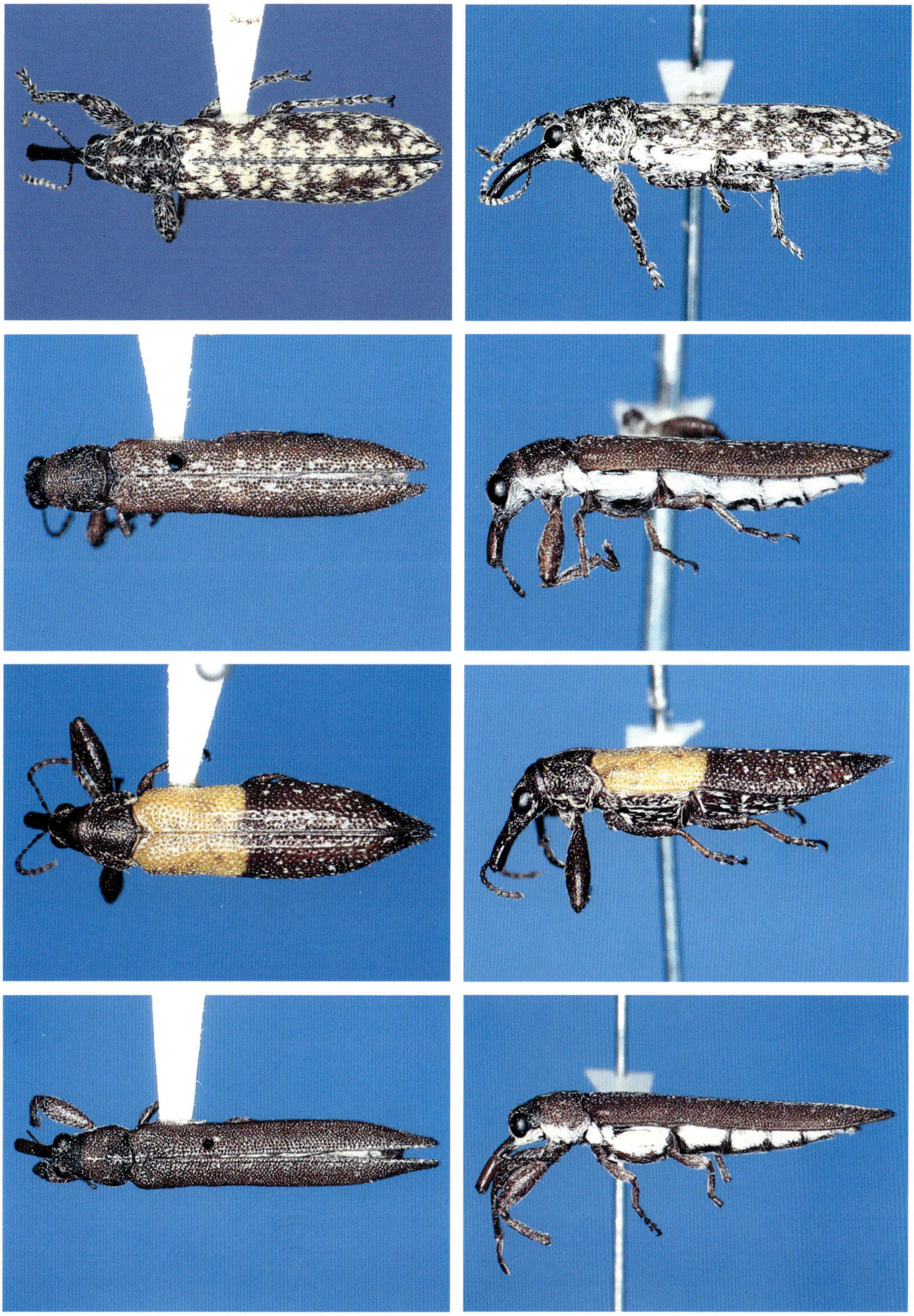

Plate 34 Belidae

1, 2 *Rhinotia exigua* (Pascoe)
Gayndah, Queensland
Female; 10 mm

3, 4 *Rhinotia exilis* (Lea)
Mittagong, New South Wales
Holotype female; 8.5 mm

5, 6 *Rhinotia filiformis* (Germar)
Canberra, Australian Capital Territory
Male; 11.5 mm

7, 8 *Rhinotia flindersi* (Blackburn)
Lake Hattah, Victoria
Female; 12 mm

1	2
3	4
5	6
7	8

Plate 35 Belidae

1, 2 *Rhinotia frater* (Blackburn)
Fraser Range, Western Australia
Holotype male; 14.5 mm

3, 4 *Rhinotia granicollis* (Lea)
"Australia"
Paratype female; 9.5 mm

5, 6 *Rhinotia haemoptera* Kirby
Canberra, Australian Capital Territory
Female; 21.5 mm
This species is a mimic of the following:

7, 8 *Metriorrhynchus rhipidius* (Macleay), Lycidae
Murrumbateman, New South Wales
Female; 15 mm
Lycids are the models for several mimetic Belidae.

1	2
3	4
5	6
7	8

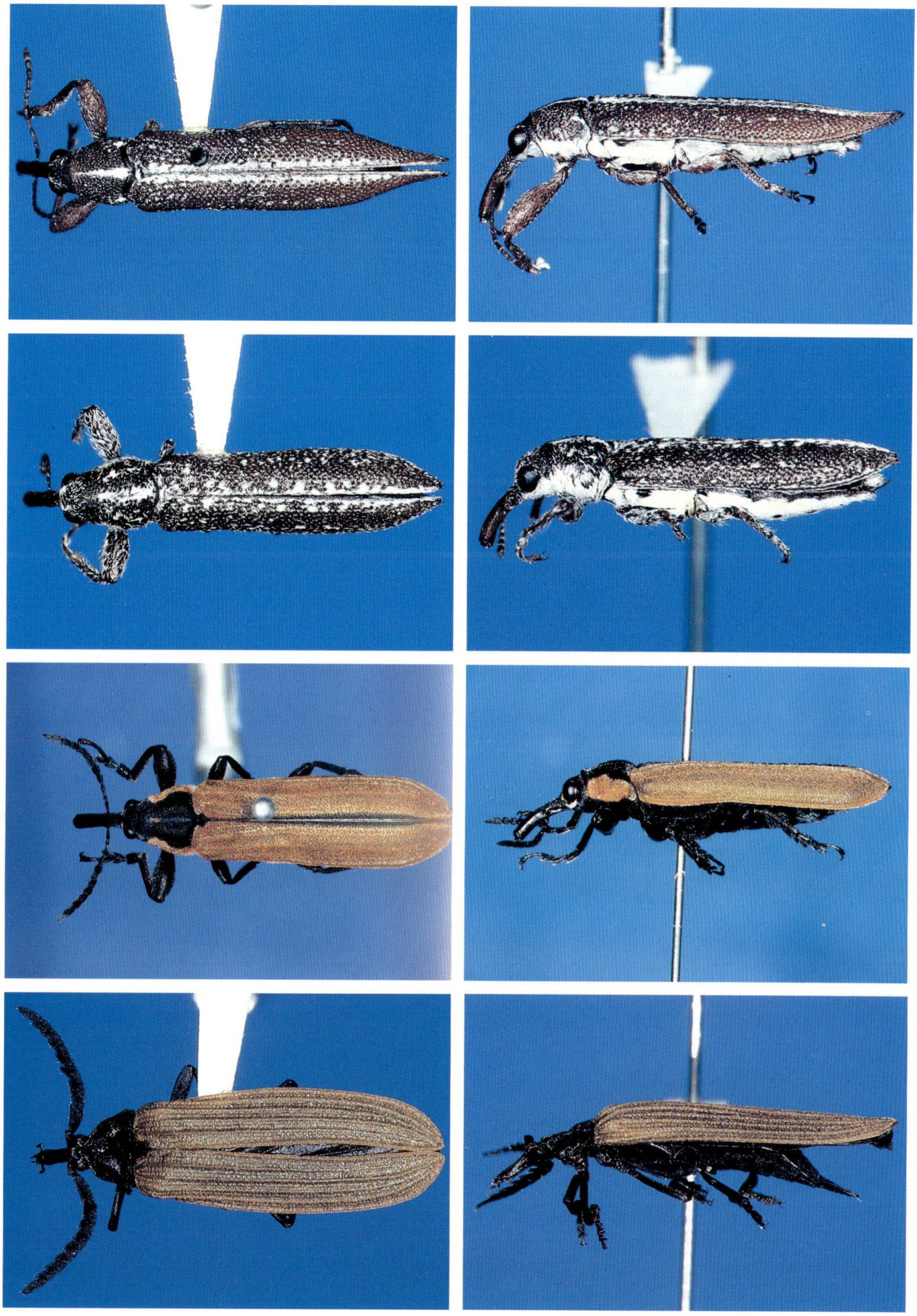

Plate 36 Belidae

1, 2 *Rhinotia halmaturina* (Lea)
Kangaroo Island, South Australia
Holotype male; 9 mm

3, 4 *Rhinotia helmsi* (Blackburn)
Fraser Range, Western Australia
Allotype male; 15.5 mm

5, 6 *Rhinotia hemisticta* (Germar)
Port Lincoln, South Australia
Male; 16 mm

7, 8 *Rhinotia inconstans* (Lea)
Werialda, New South Wales
Holotype female; 10 mm

1	2
3	4
5	6
7	8

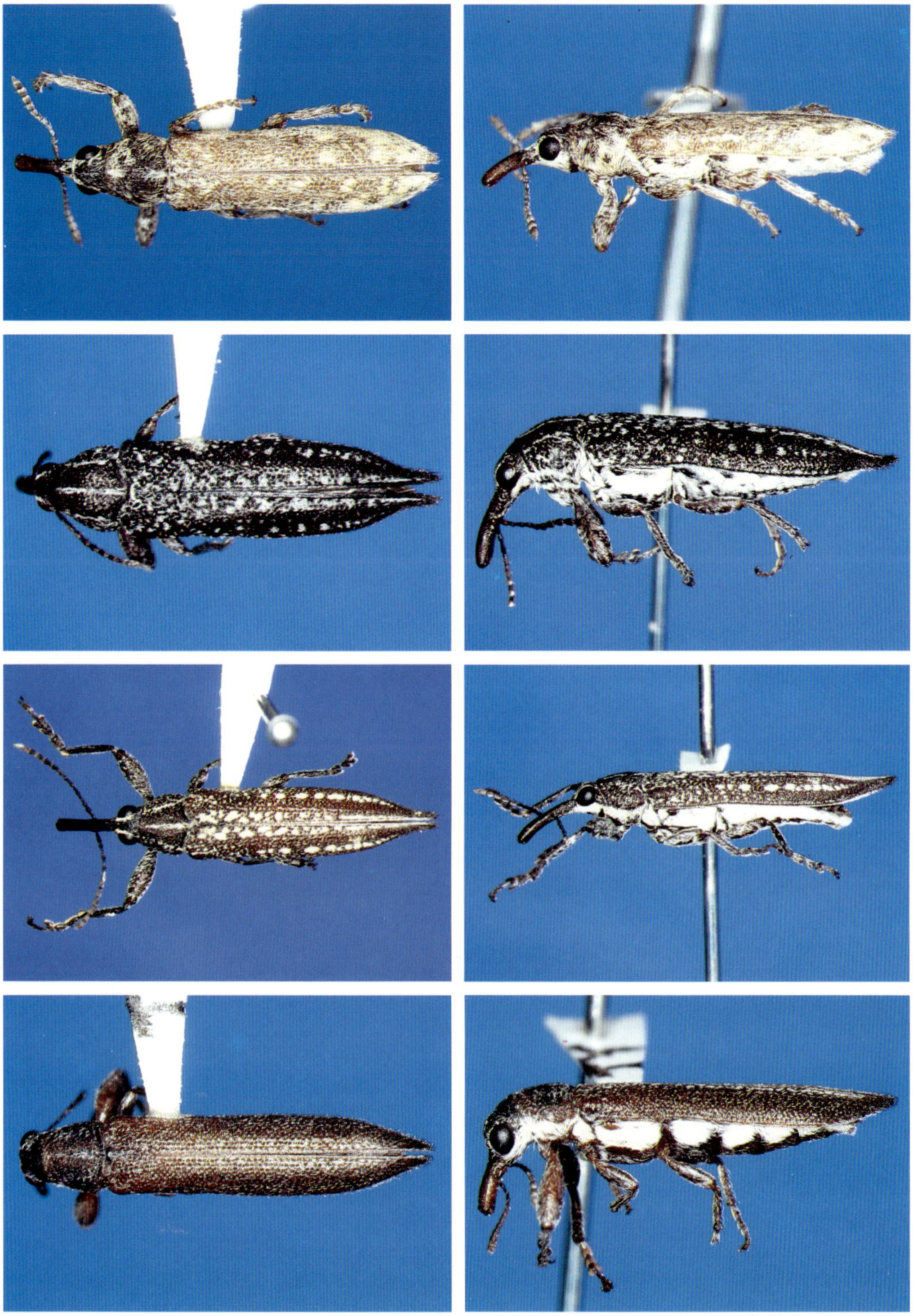

Plate 37 Belidae

1, 2 *Rhinotia insipida* (Blackburn)
Northern Territory
Holotype female; 14.5 mm

3, 4 *Rhinotia irrorata* (Jekel)
Leigh Creek, South Australia
Male; 9.5 mm

5, 6 *Rhinotia lacustris* (Lea)
Lake Callabonna, South Australia
Holotype female; 12.5 mm

7, 8 *Rhinotia lineata* (Donovan)
Mount Woowoonga, southwest of Childers, Queensland
Male; 15.5 mm

1	2
3	4
5	6
7	8

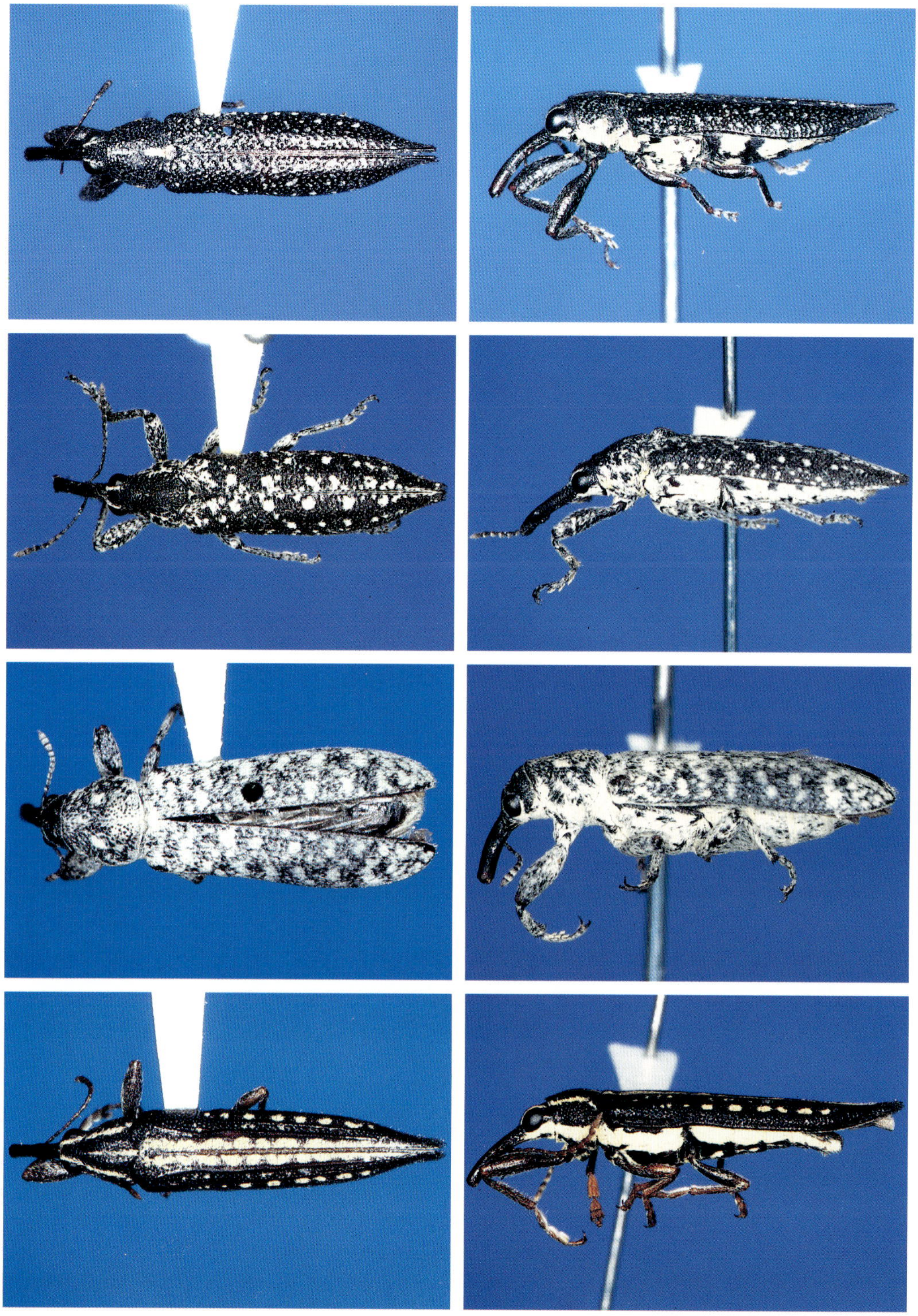

Plate 38 Belidae

1, 2 *Rhinotia marginella* Boheman
Sydney, New South Wales
Male; 19 mm

3, 4 *Rhinotia melanocephala* (Boheman)
Peterborough, South Australia
Male; 14.5 mm

5, 6 *Rhinotia mimica* (Lea)
Parachilna, South Australia
Female; 15 mm

7, 8 *Rhinotia multimaculata* (Lea)
Quorn, South Australia
Holotype male; 11.5 mm

1	2
3	4
5	6
7	8

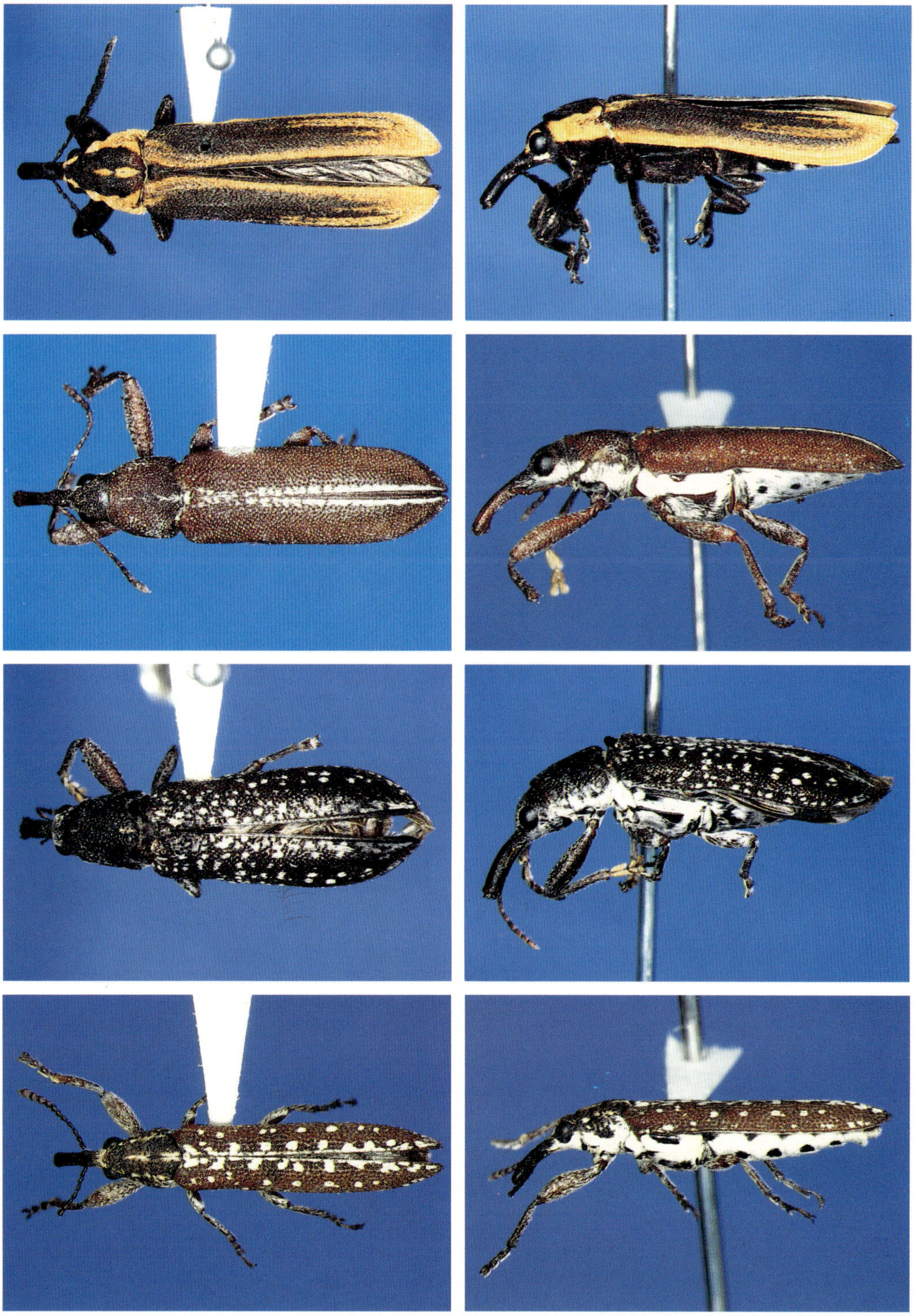

Plate 39 Belidae

1, 2 *Rhinotia niveopilosa* (Lea)
North Queensland
Male; 13 mm

3, 4 *Rhinotia orthodoxa* (Lea)
Mount Lofty, South Australia
Holotype female; 11.5 mm

5, 6 *Rhinotia parva* Lea
Tweed River, New South Wales
Female; 8.5 mm

7, 8 *Rhinotia perplexa* (Blackburn)
Western Australia
The type of the Western Australian form; male; 14 mm.

1	2
3	4
5	6
7	8

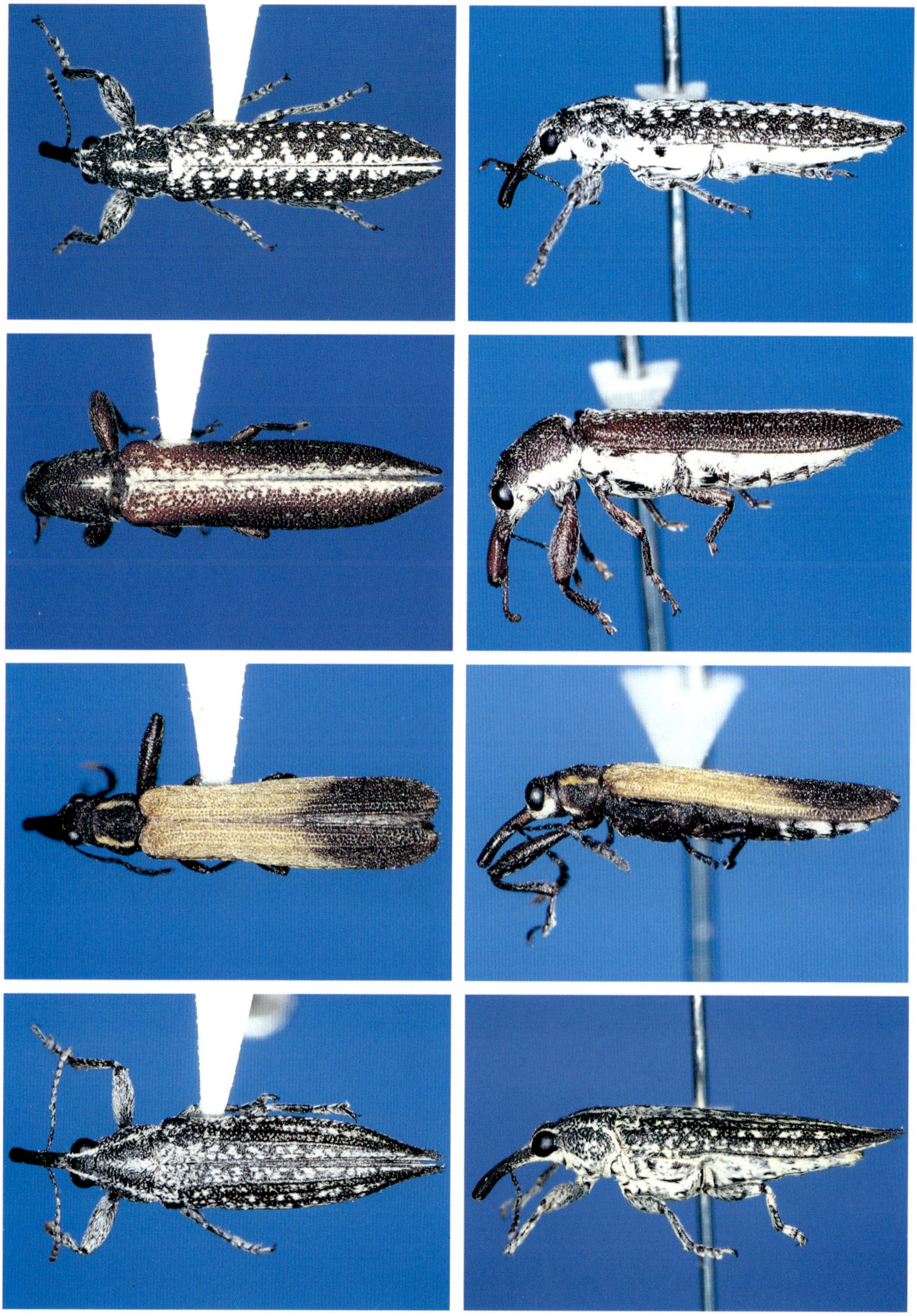

Plate 40 Belidae

1, 2 *Rhinotia phoenicoptera* (Germar)
Canberra, Australian Capital Territory
Female; 19.5 mm

3, 4 *Rhinotia plagiata* (Pascoe)
Cairns District, Queensland
Female; 17 mm

5, 6 *Rhinotia podagrosa* (Lea)
Murray Bridge, South Australia
Female; 17 mm

7, 8 *Rhinotia povera* (Lea)
Mundel Island, South Australia
Female; 15 mm

1	2
3	4
5	6
7	8

Plate 41 Belidae

1, 2 *Rhinotia princeps* Zimmerman
Marloo Station, Wurarga, Western Australia
Paratype female; 13.5 mm
See Plate 47:4 for ventral view.

3, 4 *Rhinotia pruinosa* Pascoe
South Australia
Female; 16 mm

5, 6 *Rhinotia pudica* (Lea)
Forest Reefs, New South Wales
Female; 13 mm

7, 8 *Rhinotia pulverulenta* (Lea)
Gayndah, Queensland
Female; 16 mm

1	2
3	4
5	6
7	8

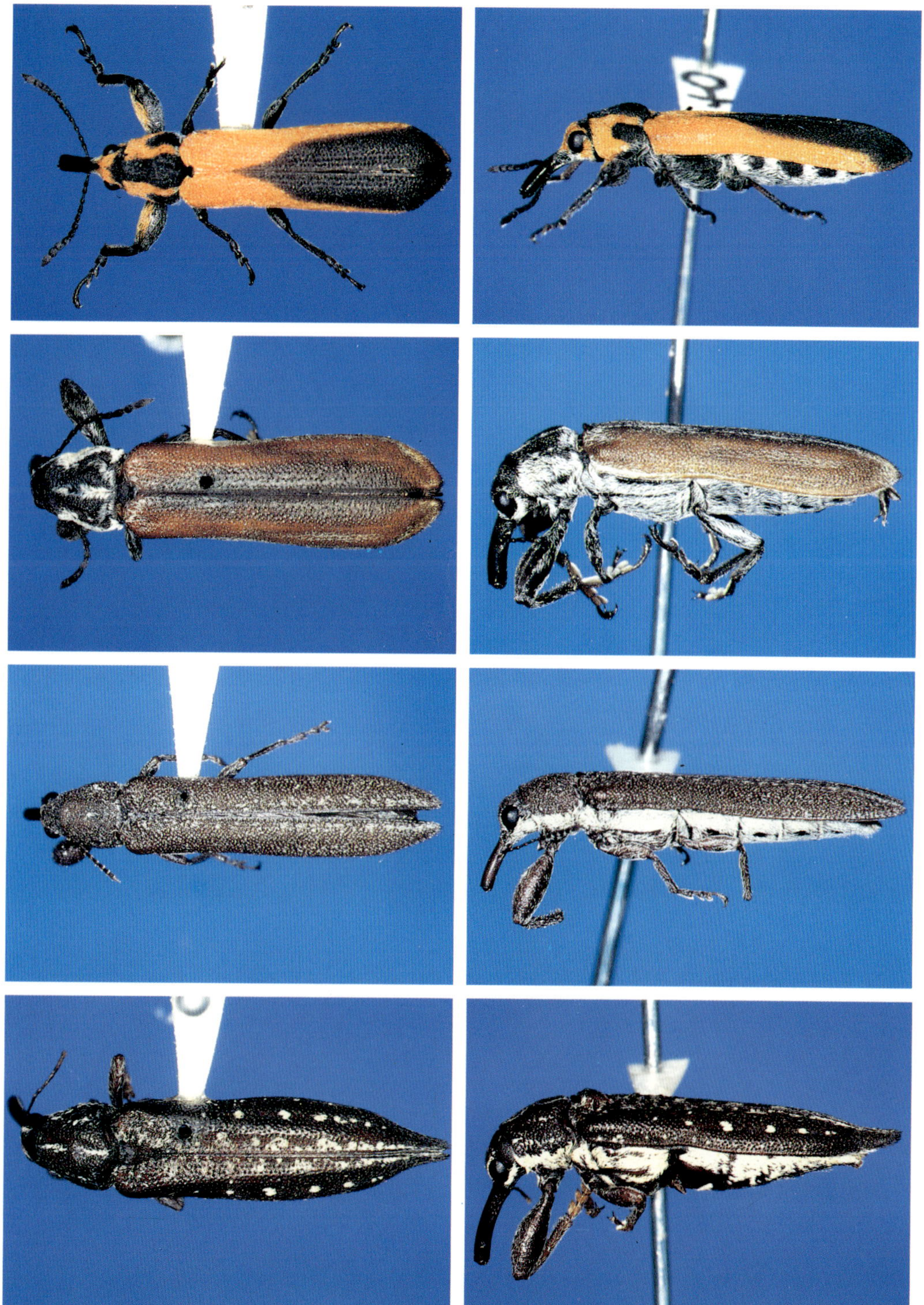

Plate 42 Belidae

1, 2 *Rhinotia ruficornis* (Lea)
Near Salisbury, New South Wales
Female; 18 mm
This species has often been mistakenly identified as *semipunctata*; compare 7 and 8 below.

3, 4 *Rhinotia regalis* (Blackburn)
Eyres Peninsula, South Australia
Paratype male; 19 mm

5, 6 *Rhinotia scalaris* (Germar)
South Australia
Male; 18 mm

7, 8 *Rhinotia semipunctata* (Fabricius)
40 Mile Scrub National Park, via Garnet, Queensland
Male, compared with holotype; 12 mm

1	2
3	4
5	6
7	8

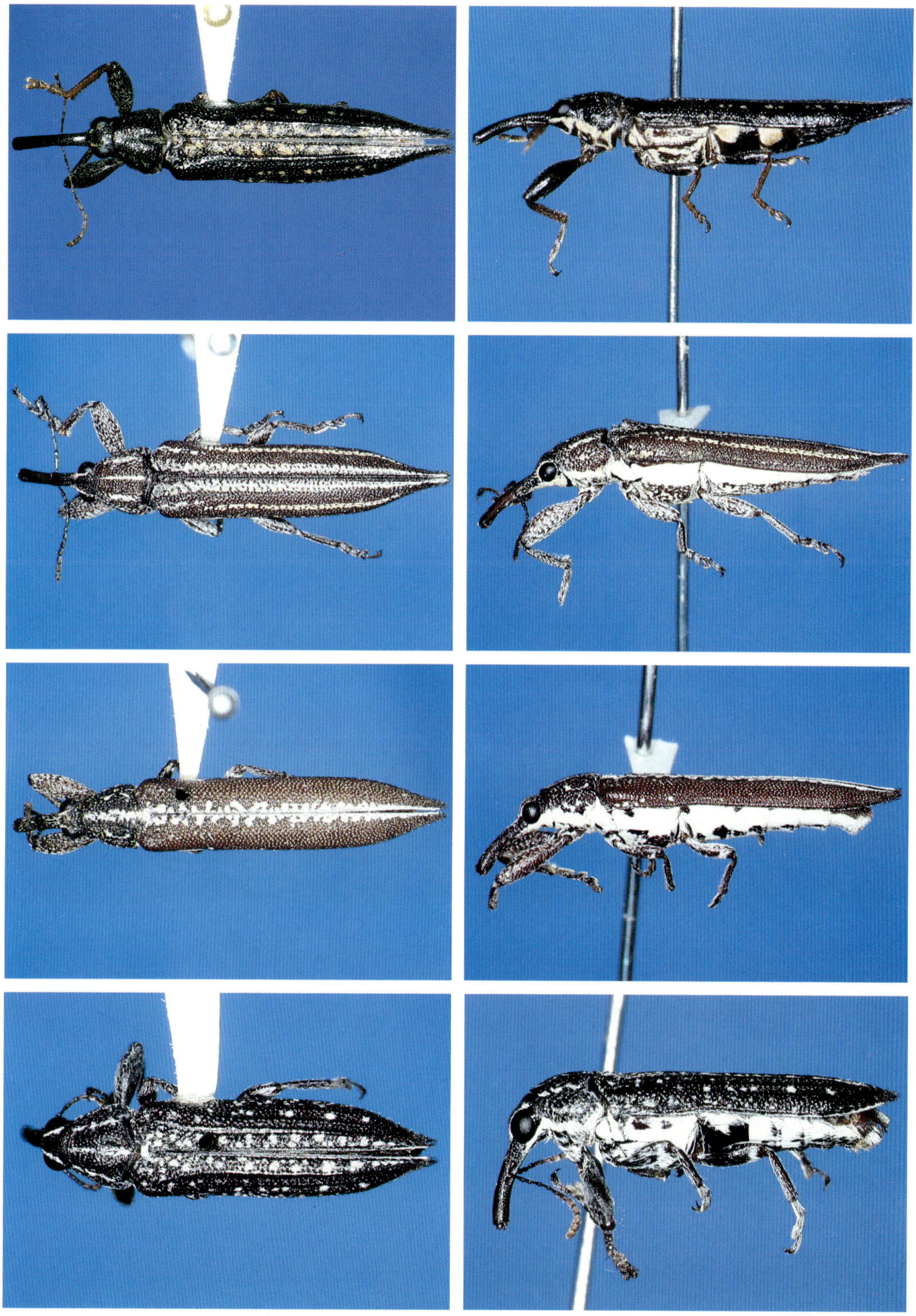

Plate 43 Belidae

1, 2 *Rhinotia serpens* (Pascoe)
Western Australia
Paratype male; 12 mm

3, 4 *Rhinotia simplicipennis* (Lea)
Williams River, New South Wales
Male; 10 mm

5, 6 *Rhinotia sparsus* Germar
Fraser Range, Western Australia
Male; 12.5 mm

7,8 *Rhinotia subparallela* (Jekel)
Macleay River, New South Wales
Female; 12.5 mm

1	2
3	4
5	6
7	8

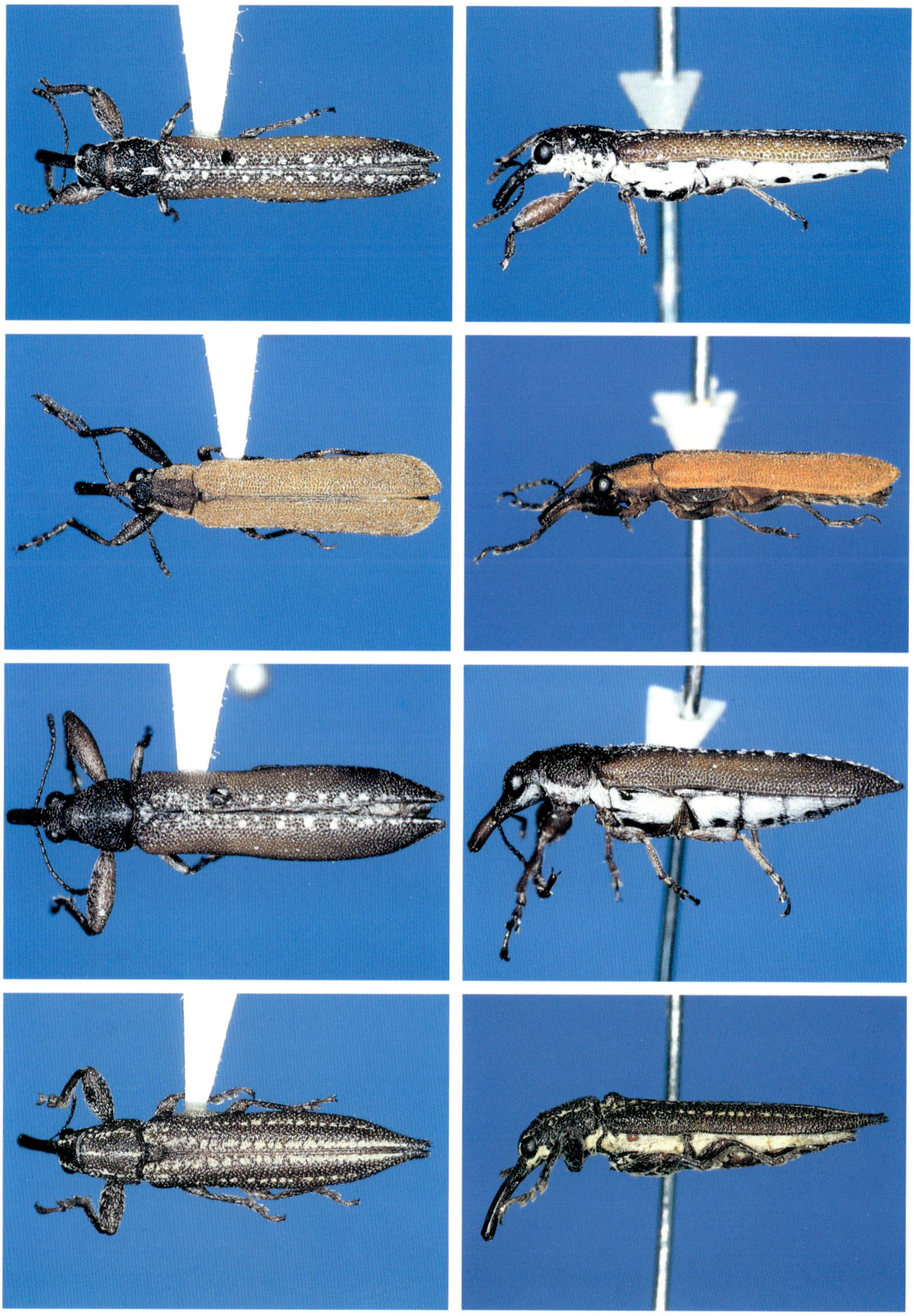

Plate 44 Belidae

1, 2 *Rhinotia subsuturalis* (Lea)
Cairns, Queensland
Paratype female; 16 mm

3, 4 *Rhinotia suturalis* (Boisduval)
Canberra, Australian Capital Territory
Male; 17.5 mm

5, 6 *Rhinotia tenuis* (Lea)
Galston, New South Wales
Paratype female; 10 mm

7, 8 *Rhinotia trilinealba* (lea)
Cue, Western Australia
Holotype male; 17 mm

1	2
3	4
5	6
7	8

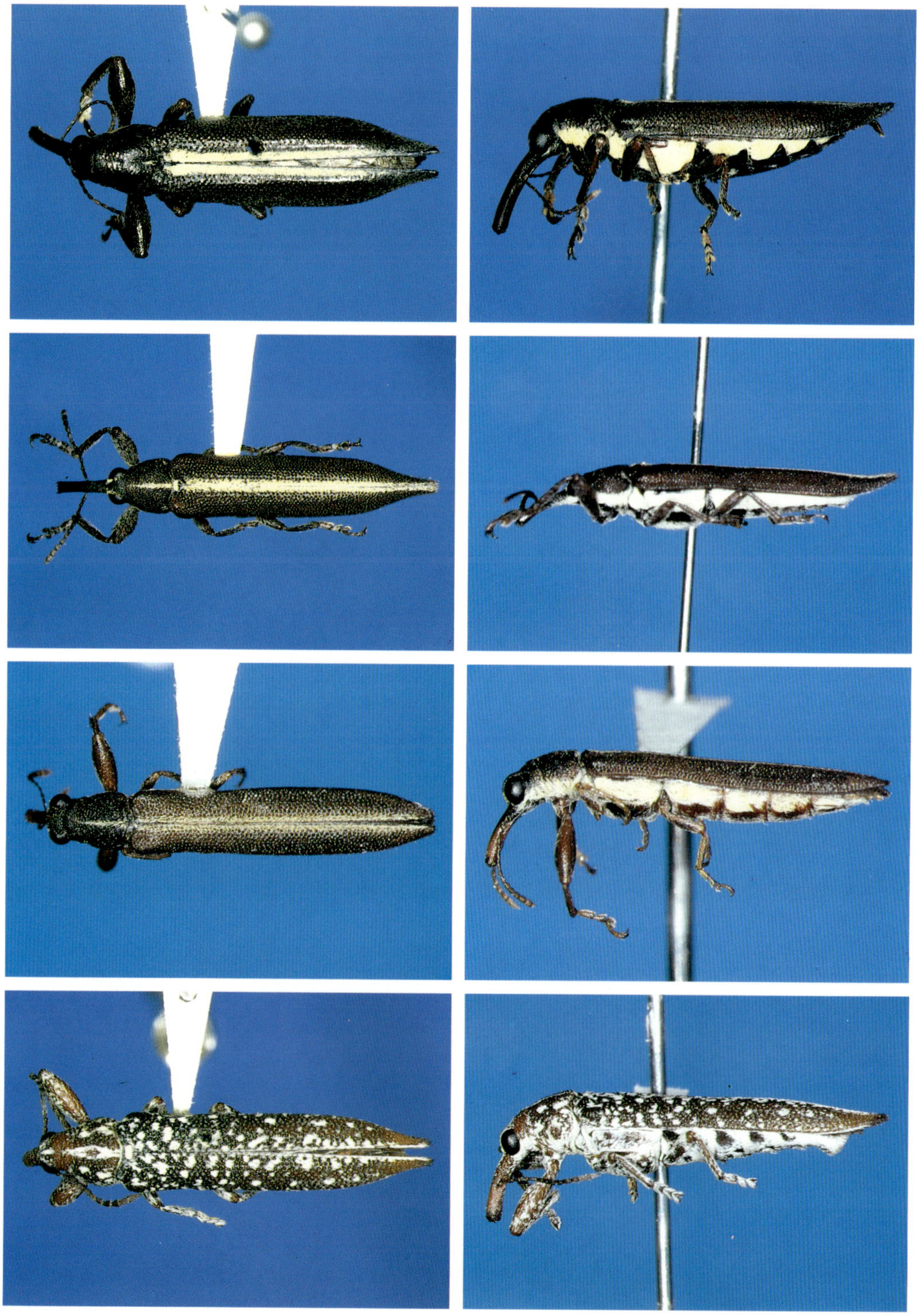

Plate 45 Belidae

1, 2 *Rhinotia ursa* (Lea)
Hay, New South Wales
Holotype male; 9 mm

3, 4 *Rhinotia variabilis* (Lea)
Lucindale, South Australia
Holotype male; 9 mm

5, 6 *Rhinotia venusta* Pascoe
Rockhampton, Queensland
Colour form; female; 16.5 mm

7, 8 *Rhinotia venusta* Pascoe
Wondai, southeast Queensland
Colour form; female; 18 mm; (dark spot on right side of pronotal disc is an abrasion).

1	2
3	4
5	6
7	8

Plate 46 Belidae

1, 2 *Rhinotia vetusta* (Pascoe)
Geraldton, Western Australia
Male, compared with holotype; 10 mm

3, 4 *Rhinotia villosa* (Lea)
Cue, Western Australia
Paratype male; 10 mm

5, 6 *Rhinotia* new species 1
Lismore, New South Wales
Female; 10.5 mm

7, 8 *Rhinotia* new species 2
Broken Hill, New South Wales
Male; 11 mm

1	2
3	4
5	6
7	8

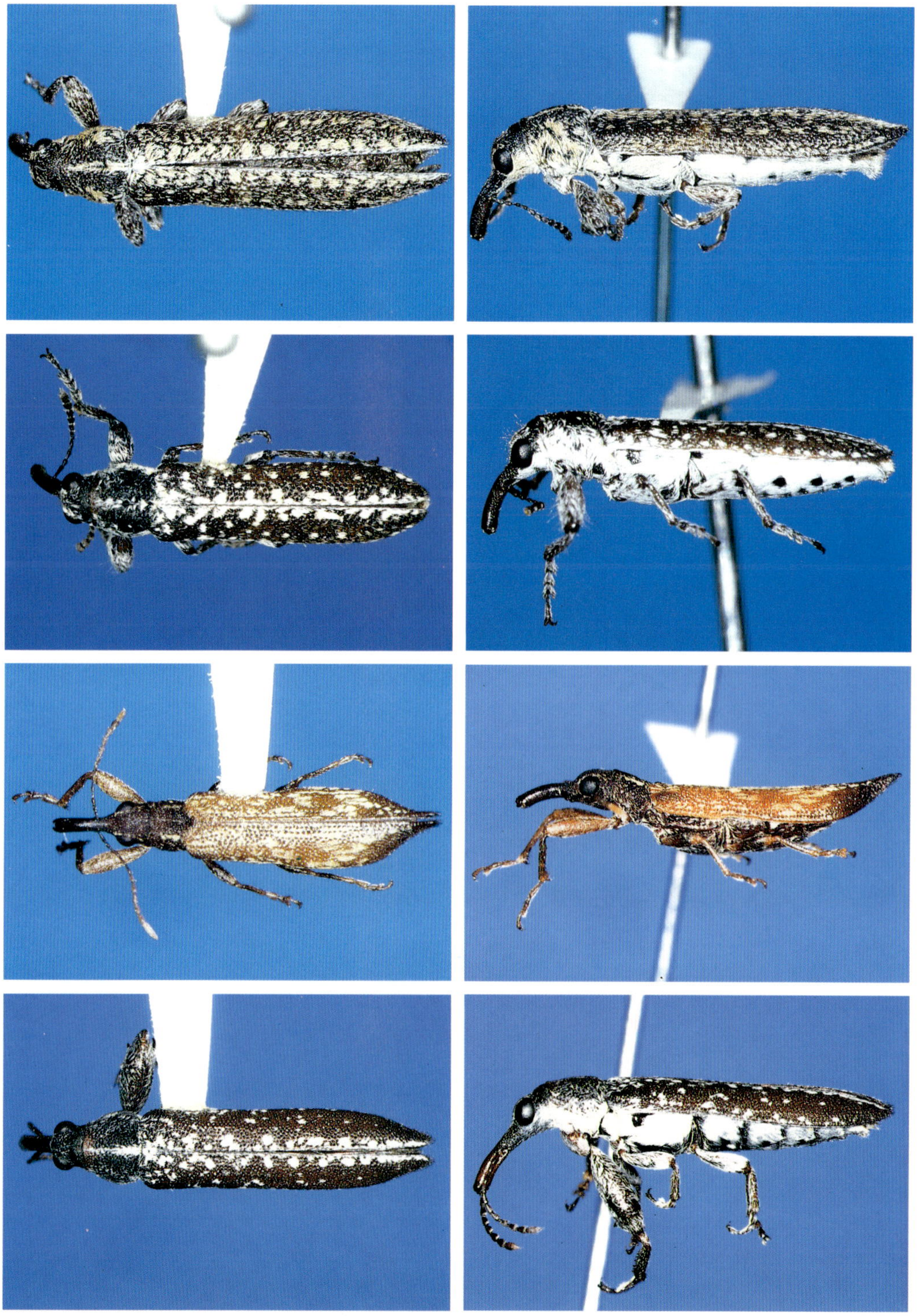

Plate 47 Belidae

1, 2 *Rhinotia* new species 3
Dorrigo, New South Wales
Male; 11 mm

3 *Rhinotia cruenta* Pascoe
Culham, Western Australia
Adult and larva cut from *Loranthus* mistletoe growing on *Acacia acuminata*. Collected and photographed by Peter McMillan.

4 *Rhinotia princeps* Zimmerman
Ventral view of the female illustrated on Plate 41:1,2.

5, 6 *Sphinctobelus ater* Zimmerman
Near Fitzroy Falls, New South Wales
Holotype female; 4.8 mm

7, 8 *Sphinctobelus cinereus* (Blanchard)
4 km west of Molong, New South Wales
Female; 6.8 mm

1	2
3	4
5	6
7	8

Plate 48 Belidae

1, 2 *Sphinctobelus niger* Zimmerman
Cunangra, Queensland
Holotype female; 5.2 mm

3, 4 *Sphinctobelus pyriatrus* (Lea)
Dorrigo, New South Wales
Female; 6 mm

5, 6 *Sphinctobelus quadrimaculatus* (Lea)
Warrumbungle National Park, New South Wales
Female; 7 mm

7, 8 *Spinctobelus rufibecus* Zimmerman
Mount Mee, Queensland
Paratype female; 6 mm

1	2
3	4
5	6
7	8

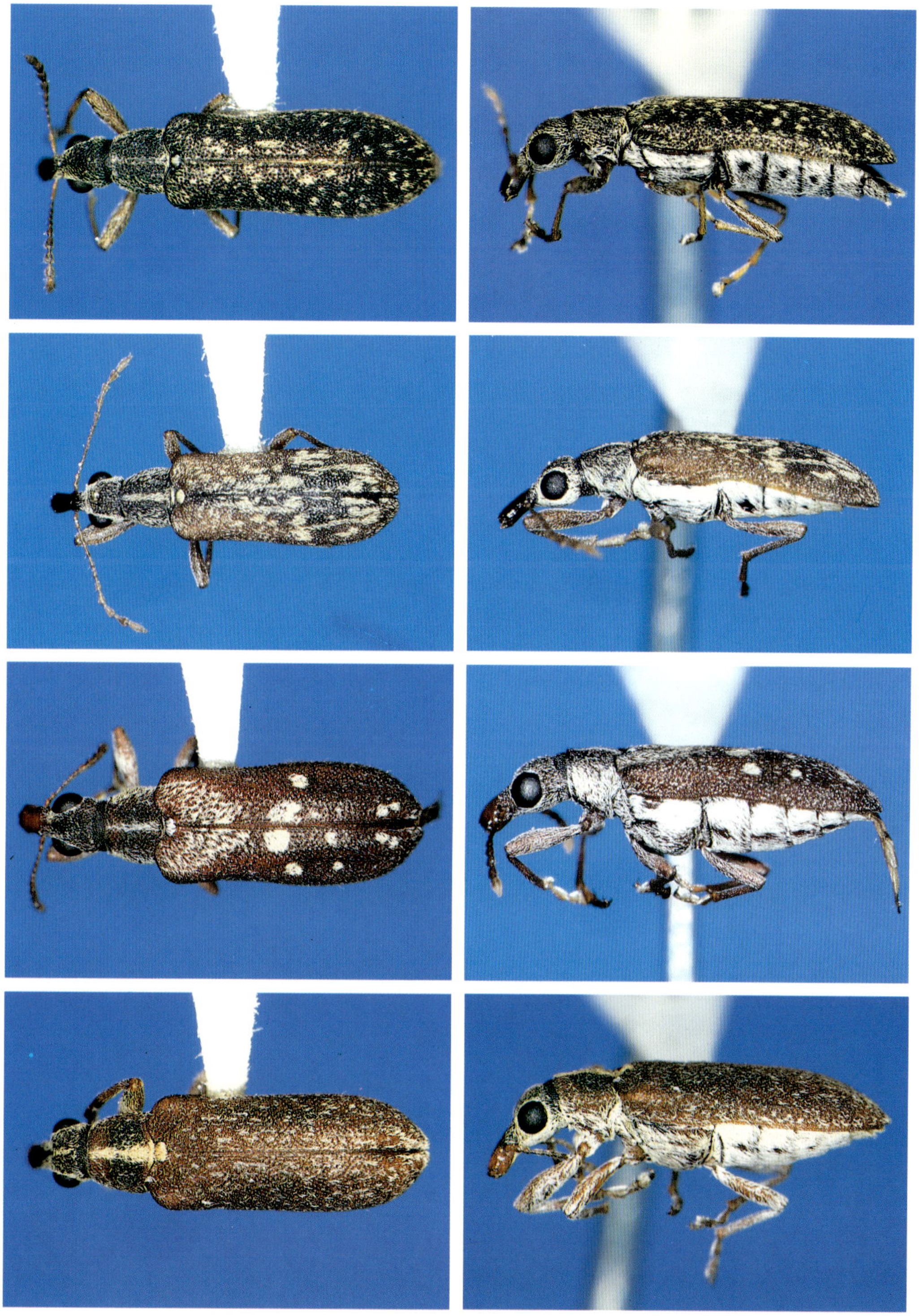

Plate 49 Belidae

1, 2 *Rhinotiodes spinipennis* (Lacordaire)
Ravensbourne, near Crows Nest, Queensland
Female; 14 mm

3, 4 *Rhinotiodes spinipennis* colour form *bella* (Lea)
Port Macquarie, New South Wales
Male; 12.5 mm

5 *Apagobelus brevirostris* (Lea)
Ventral view of specimen illustrated on Plate 20:5, 6. Note the well-defined gula.

6 *Basiliobelus flavovittatus* Zimmerman
Ventral view of specimen illustrated on Plate 22:3, 4.

7 *Agathobelus bivittatus* Zimmerman
Head and rostrum of specimen illustrated on Plate 20:1, 2.

8 *Pachybelus tuberculatus* Zimmerman
Head and rostrum of specimen illustrated on Plate 29:5, 6.

1	2
3	4
5	6
7	8

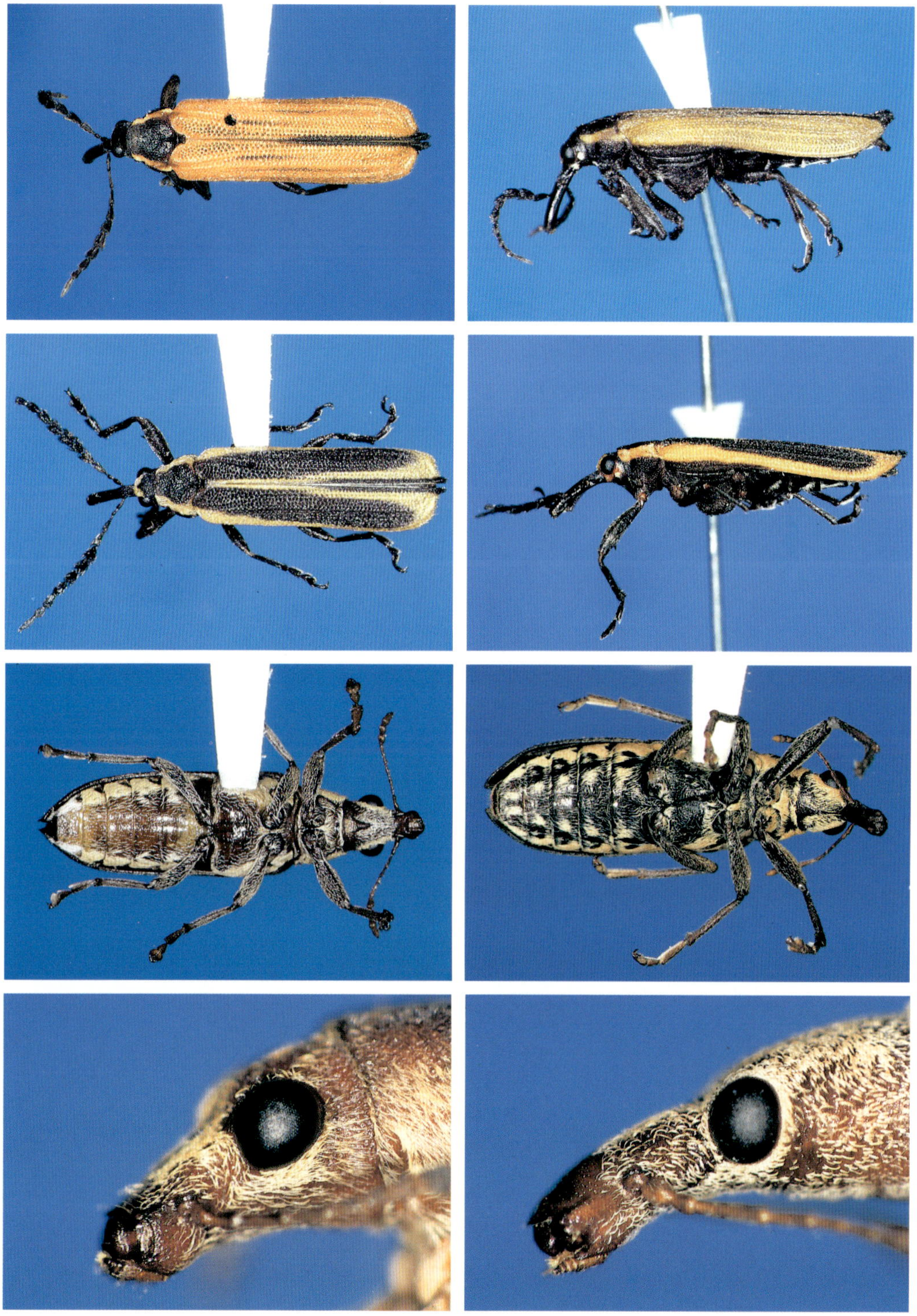

Plate 50 Nemonychidae

1, 2 *Aragomacer leai* Kuschel
Cainable, Queensland
Male; 5 mm

3, 4 *Aragomacer uniformis* Kuschel
Cainable, Queensland
Male; 5.7 mm

5, 6 *Basiliogeus prasinus* Kuschel
Bunya Mountains, Queensland
Female; 4.2 mm

7, 8 *Basiliorhinus araucariae* Kuschel
Bunya Mountains, Queensland
Female; 5.3 mm

1	2
3	4
5	6
7	8

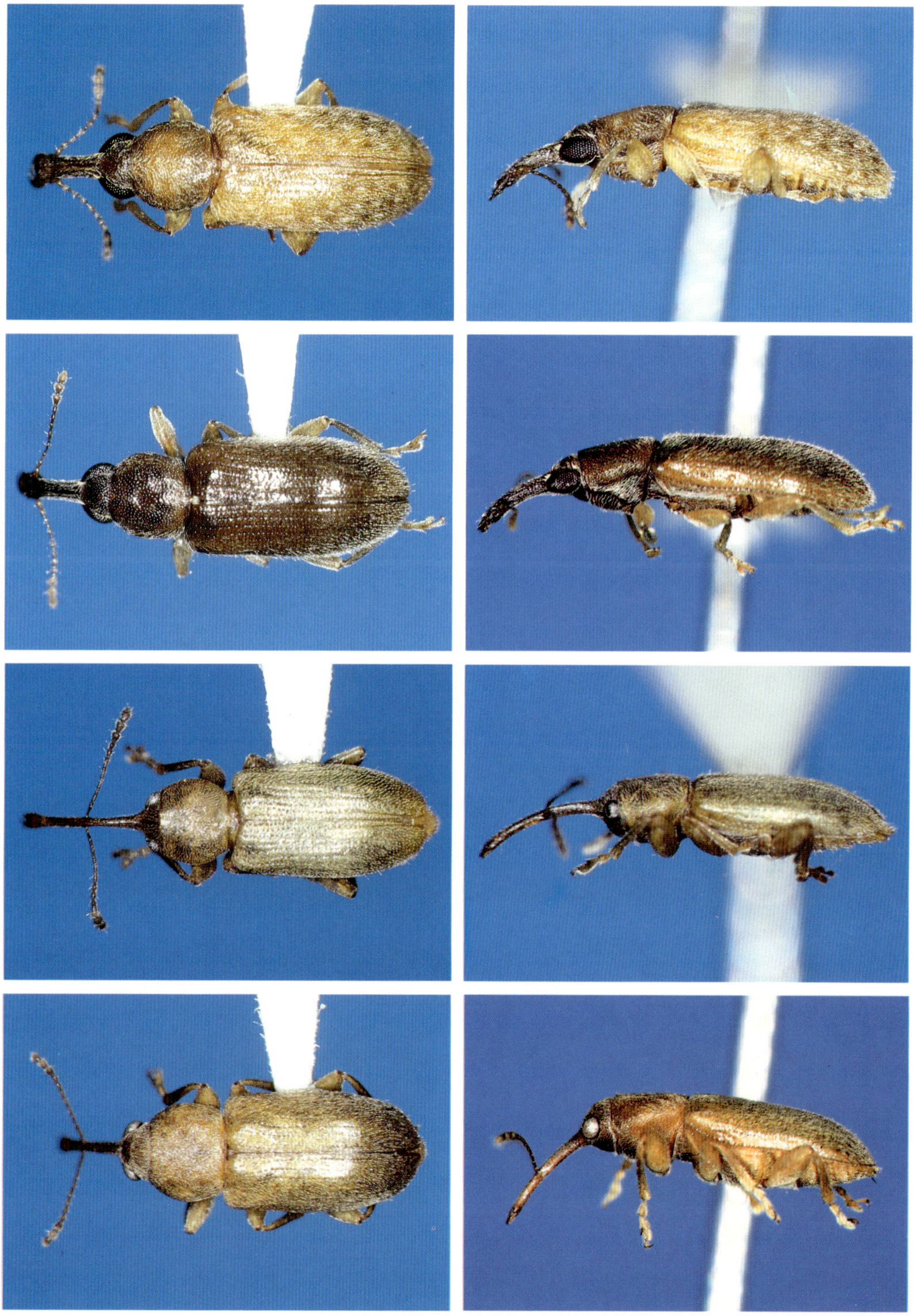

Plate 51 Nemonychidae

1, 2 *Bunyaeus eutactae* Kuschel
Cainable, Queensland
Male; 2.7 mm

3, 4 *Bunyaeus monteithi* Kuschel
Bunya Mountains, Queensland
Male; 3.4 mm

5, 6 *Eutactobius puellus* Kuschel
Cainable, Queensland
Female; 2.7 mm

7, 8 *Notomacer australis* (Lea)
Kuranda State Forest, Queensland
Female; 5 mm

1	2
3	4
5	6
7	8

Plate 52 Nemonychidae

1, 2 *Notomacer brittoni* Kuschel
Bellenden Ker Range, Queensland
Paratype female; 4.1 mm

3, 4 *Notomacer eximus* Kuschel
Cainable, Queensland
Paratype female; 4.6 mm

5, 6 *Notomacer reginae* Kuschel
Cainable, Queensland
Female; 4.3 mm

7, 8 *Notomacer zimmermani* Kuschel
Cainable, Queensland
Holotype male; 3.8 mm

1	2
3	4
5	6
7	8

Plate 53 Nemonychidae – Rhynchitidae

Nemonychidae

1, 2 *Pogomacer deceptus* Kuschel
Bellenden Ker Range, Queensland
Holotype male; 2.5 mm

Rhynchitidae

Note the superficial resemblance to Nemonychidae.

3, 4 *Auletobius anthracinus* (Lea)
Peterborough, South Australia
Paratype female; 2.5 mm

5, 6 *Auletobius brevirostris* (Lea)
Newport, New South Wales
Compared with holotype; 2.7 mm

7, 8 *Auletobius calceatus* (Pascoe)
Albany, Western Australia
Female; 4.3 mm

1	2
3	4
5	6
7	8

Plate 54 Rhynchitidae

1, 2 *Auletobius eucalypti* (Lea)
Swan River, Western Australia
Paratype female; 2.2 mm

3, 4 *Auletobius filirostris* (Pascoe)
Swan River, Western Australia
Female, compared with holotype; 3.2 mm

5, 6 *Auletobius monticola* Voss
Tamborine Mountain, Queensland
Female; 2.8 mm; determined by Voss

7, 8 *Auletobius obscurus* (Lea)
Port Lincoln, South Australia
Lectotype male; 2.6 mm

1	2
3	4
5	6
7	8

Plate 55 Rhynchitidae – Caridae

Rhynchitidae

1, 2 *Auletobius postscutellaris* (Lea)
North Queensland
Holotype female; 3.8 mm

3, 4 *Auletobius varicollis* (Lea)
Mount Wellington, Tasmania
Paratype male; 3.1 mm

Caridae

5, 6 *Car condensatus* Blackburn
Near Cocklebiddy, Western Australia
Female?; 6.2 mm

7, 8 *Car intermedius* Lea
Bribie Island, Queensland
Holotype male?; 3.4 mm

1	2
3	4
5	6
7	8

Plate 56 Caridae – Attelabidae

Caridae

1,2 *Car pini* Lea
Near Queanbeyan, New South Wales
Male, compared with holotype; 3.3 mm

See Volume VI for additional Caridae.

Attelabidae

See Volume VI for *Euops affinis* Voss.

3, 4 *Euops bakewelli* Jekel
Mount Tamborine, Queensland
Male; 2.2 mm

5, 6 *Euops clavigera* Pascoe
Queensland
Male; 3.1 mm

7, 8 *Euops contacta* Lea
Babinda, Queensland
Holotype male; 3.3 mm

1	2
3	4
5	6
7	8

Plate 57 Attelabidae

1, 2 *Euops corrugata* Lea
New South Wales
Paratype male; 2.2 mm

3, 4 *Euops coxalis* Lea
Dorrigo, New South Wales
Paratype female; 3.7 mm

5, 6 *Euops effulgens* Lea
Quorn, South Australia
Female; 3.6 mm

7, 8 *Euops episternalis* Lea
Cairns District, Queensland
Paratype female; 2.6 mm

1	2
3	4
5	6
7	8

Plate 58 Attelabidae

1, 2 *Euops eucalypti* Pascoe
Near Jenolan Caves, New south Wales
Male, 2.9 mm

3, 4 *Euops falcata* Guérin-Méneville
Lucindale, South Australia
Male; 7.5 mm

5, 6 *Euops flavomaculata* Lea
Cairns, Queensland
Paratype female; 2.9 mm

7, 8 *Euops impuncticollis* Lea
North Queensland
Male; 2.9 mm

1	2
3	4
5	6
7	8

Plate 59 Attelabidae

1, 2 *Euops leai* Voss
Gosford, New South Wales
Lectotype female; 2.5 mm

3, 4 *Euops micans* Lea
Stanthorpe, Queensland
Male; 3.3 mm

5, 6 *Euops parvoarmata* Lea
Mount Tamborine, Queensland
Holotype female; 3.7 mm

7, 8 *Euops quadrifasciculata* Lea
Queensland
Paratype male; 2.3 mm

1	2
3	4
5	6
7	8

Plate 60 Attelabidae

1, 2 *Euops rudis* Lea
Ben Lomond, New South Wales
Lectotype female; 6.3 mm

3, 4 *Euops strigiventris* Lea
Mount Tamborine, Queensland
Male, compared with holotype; 4.5 mm

5, 6 *Euops suturalis* Lea
Brisbane, Queensland
Female; 2.9 mm

7, 8 *Euops tuberculatus* Lea
Nanango District, Queensland
Female, compared with holotype; 3.2 mm

1	2
3	4
5	6
7	8

Plate 61 Attelabidae – Brentidae

Attelabidae

1, 2 *Euops victoriensis* Blackburn
Blue Mountains, New South Wales
Female; 2.8 mm

Brentidae – Amorphocephalini

3, 4 *Ankleineella australis* (Lacordaire)
Fortescue River, Hamersley Range, northern Western Australia
Male; 12 mm

5, 6 Female of the same, from the same locality: 13 mm

7, 8 *Ankleineella mniszechi* (Power)
Sir Graham Moore Island, Western Australia
Male; 7.5 mm

For enlarged details of heads of Brentidae, see Volume II.

1	2
3	4
5	6
7	8

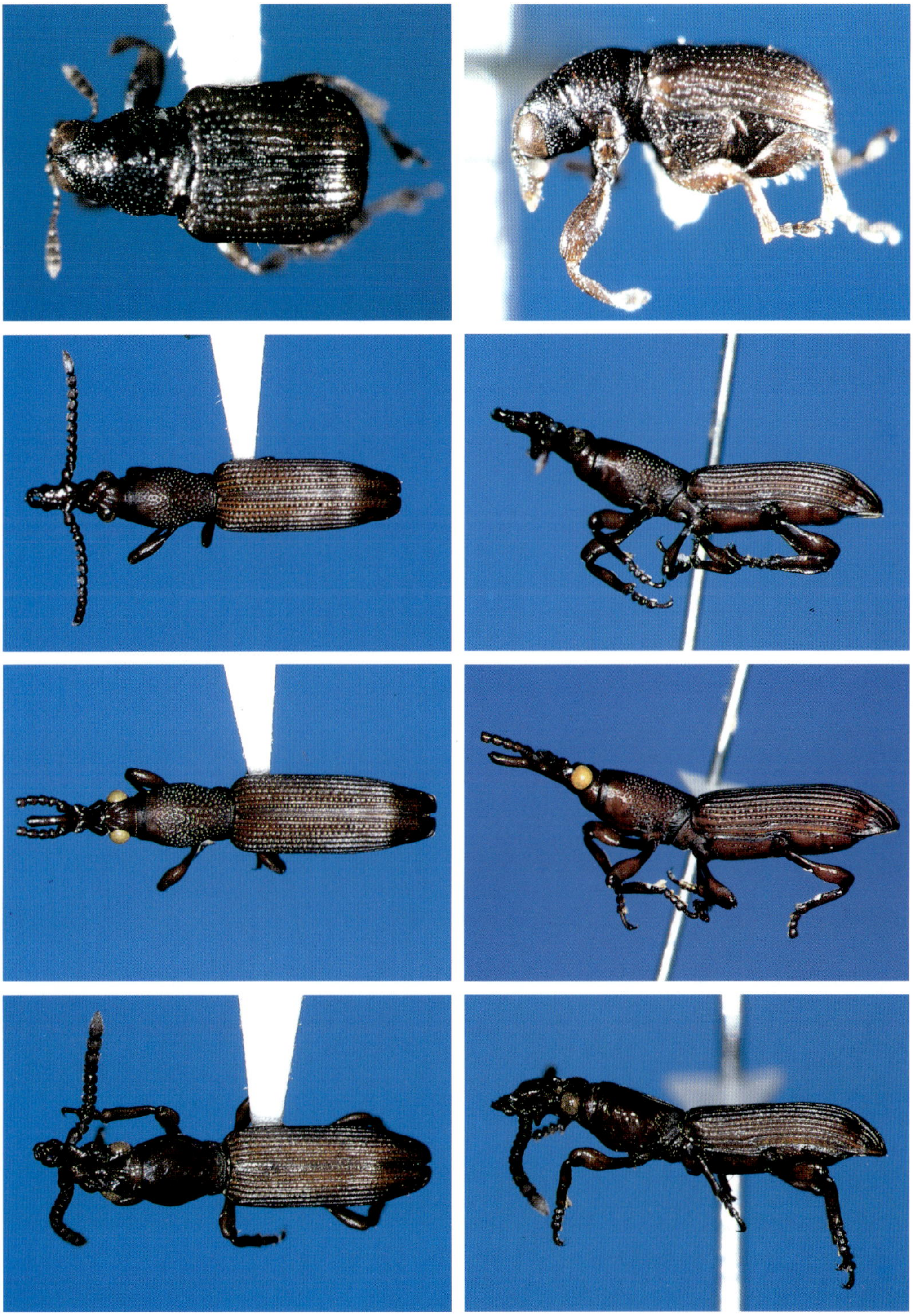

Plate 62 Brentidae – Amorphocephalini

1, 2 *Ankleineella mniszechi* (Power)
Sir Graham Moore Island, Western Australia
Female; 9.5 mm

3, 4 *Ankleineella sulcicollis* (Power)
Marloo Station, Wuraga, Western Australia
Male; 11 mm

5, 6 The same, female; 9 mm; from the same locality

7, 8 *Cordus demarzi* Damoiseau
Groote Eylandt, Northern Territory
Paratype female; 8 mm

1	2
3	4
5	6
7	8

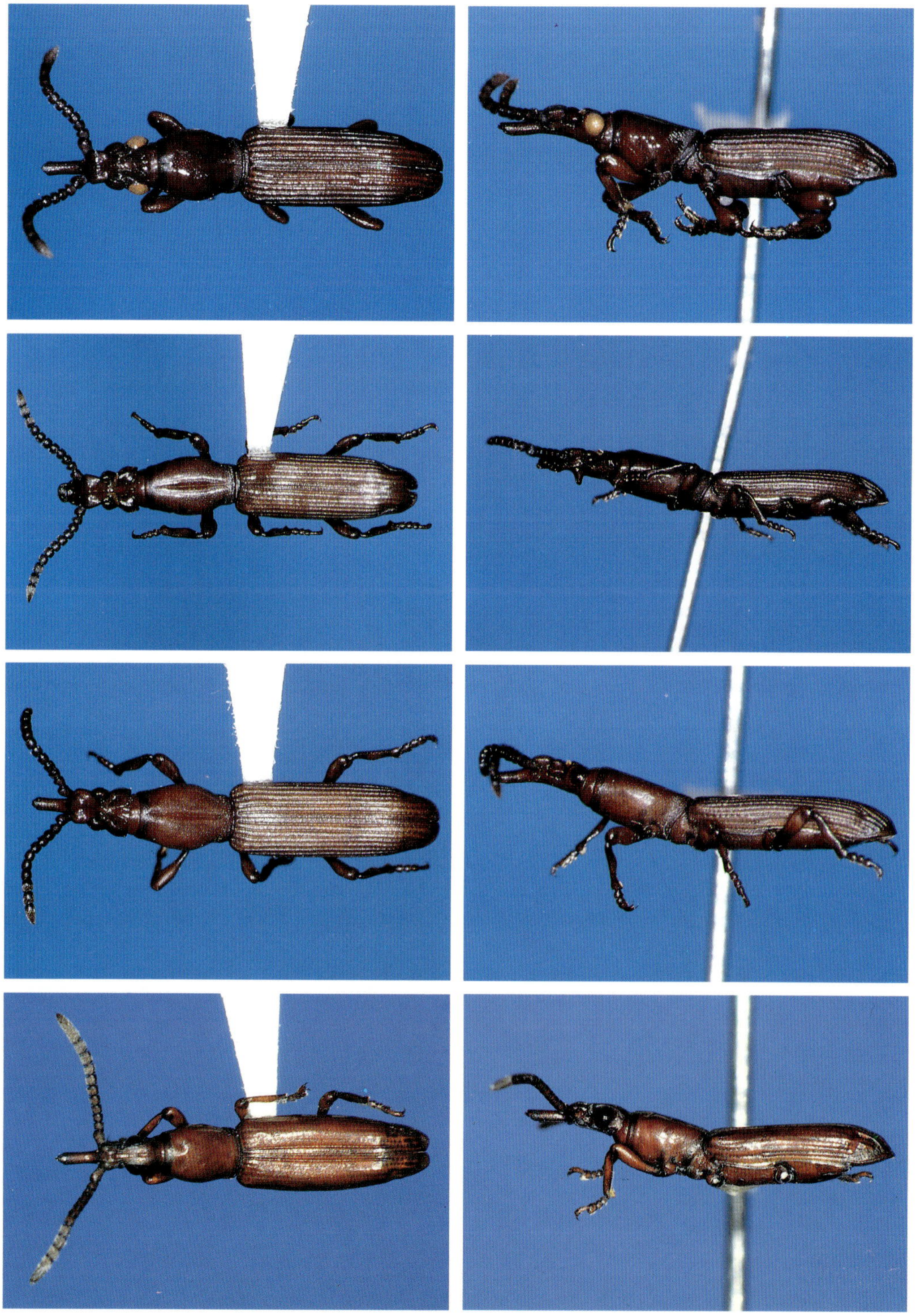

Plate 63 Brentidae – Amorphocephalini

1, 2 *Cordus festus* Damoiseau
Northwestern Western Australia
Holotype male; 11 mm

3, 4 *Cordus firmus* Damoiseau
Western Australia
Allotype female; 11 mm

5, 6 *Cordus ganglebaueri* Senna
Blackstone Range, Western Australia
Male; 8.4 mm

7, 8 The same, allotype female
Cunnamulla, Queensland
8.4 mm

1	2
3	4
5	6
7	8

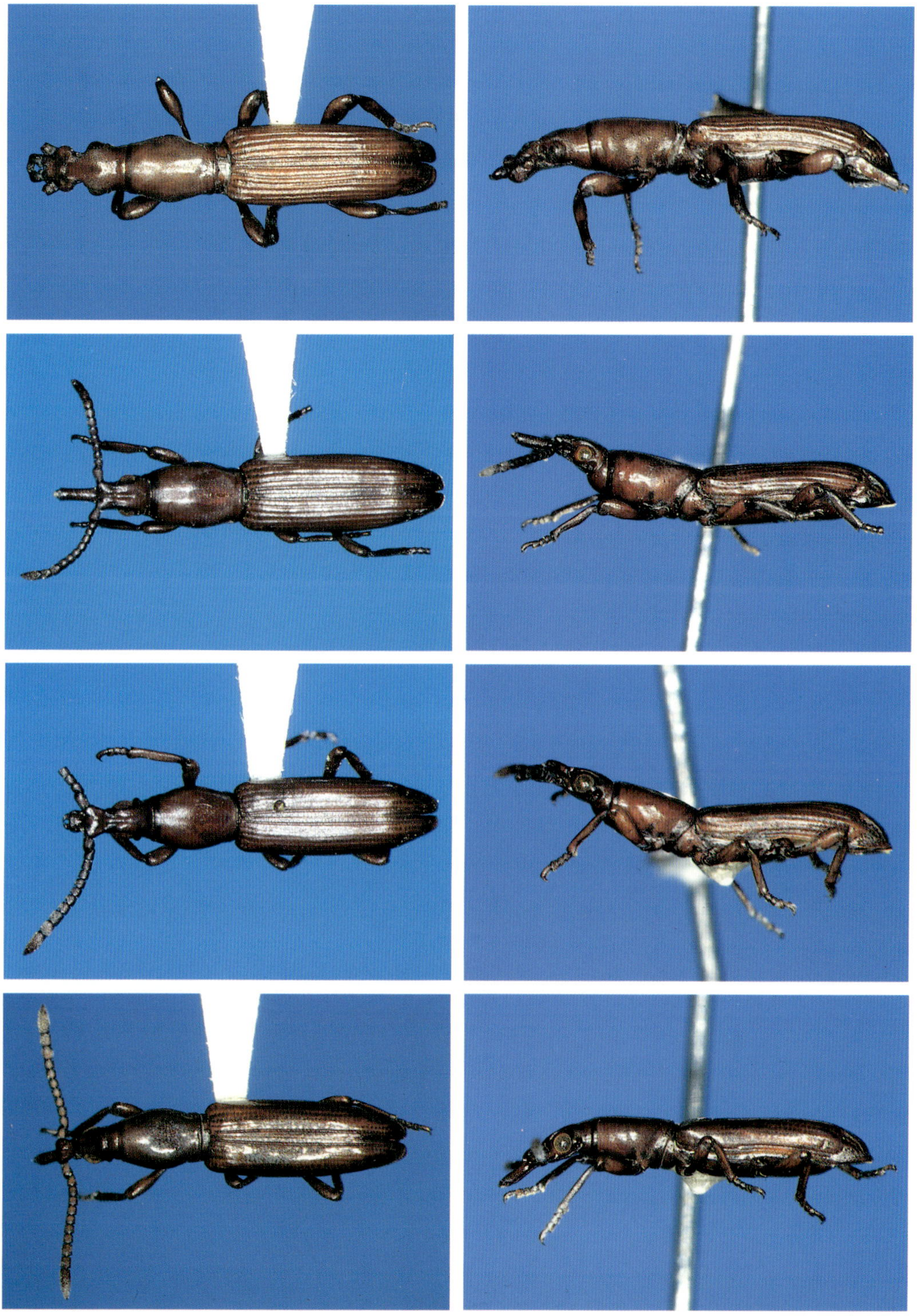

Plate 64 Brentidae – Amorphocephalini

1, 2 *Cordus hospes* Schoenherr
No locality data.
Male; 9.5 mm

3, 4 *Cordus laevis* Damoiseau
Swan River, Western Australia
Holotype male; 8 mm

5, 6 *Cordus* species
Ocean Grove, Victoria
Female; 9.5 mm
This is the specimen designated by Damoiseau as a paratype of *laevis* but which probably is not that species. See Volume II for discussion.

7, 8 *Cordus occidentalis* Damoiseau
Gilgai, Western Australia
Paratype male; 6.5 mm

1	2
3	4
5	6
7	8

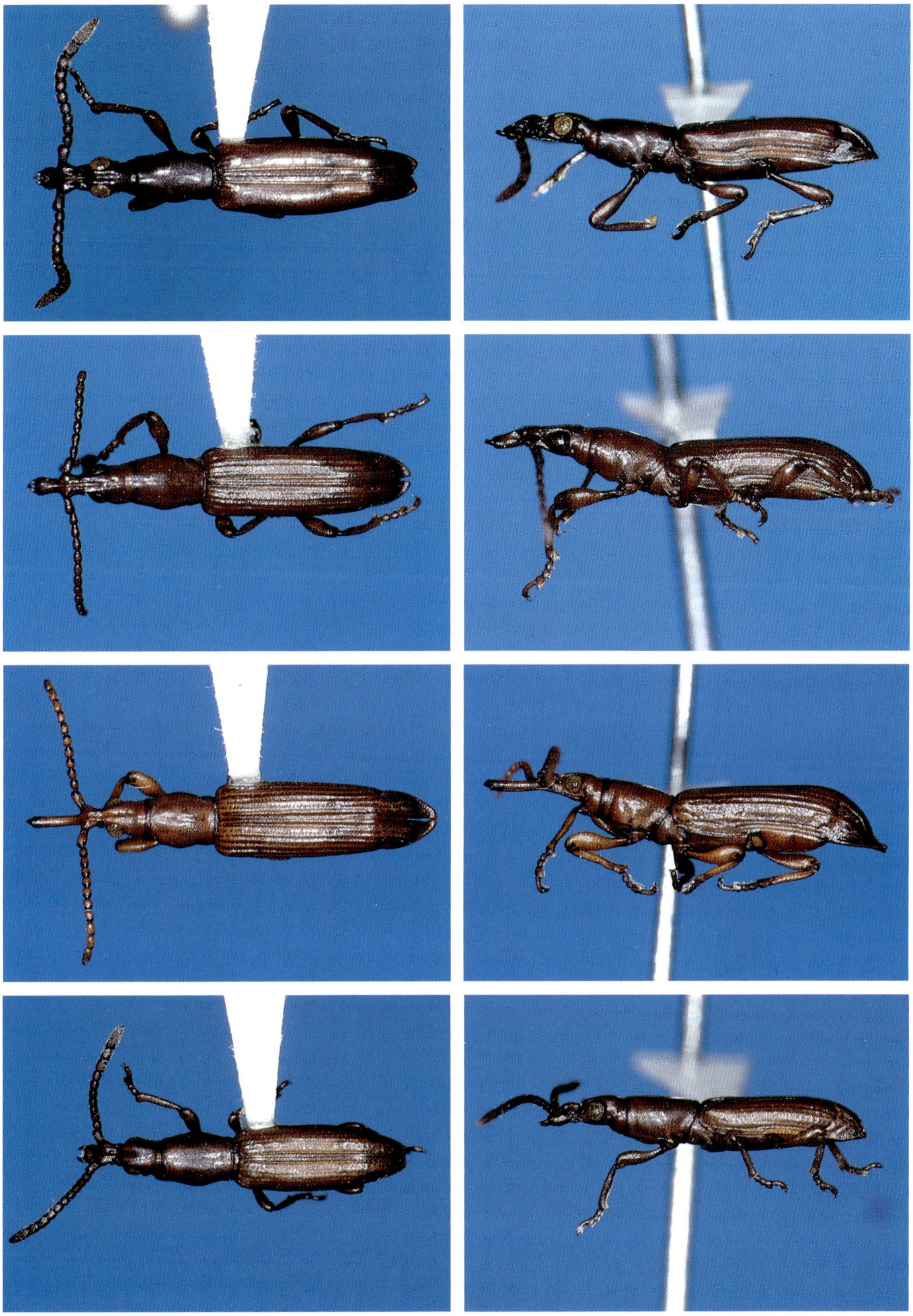

Plate 65 Brentidae – Amorphocephalini

1, 2 *Cordus pascoei* Power
Adelaide, South Australia
Female; 9.5 mm

3, 4 *Cordus queenslandicus* Senna
Bowen, Queensland
Male; 8.2 mm

5, 6 *Cordus queenslandicus* Senna
No field data.
Allotype female; 8.5 mm

7, 8 *Cordus recticornis* Damoiseau
Fortescue River, northwestern Western Australia
Holotype male; 8.6 mm

1	2
3	4
5	6
7	8

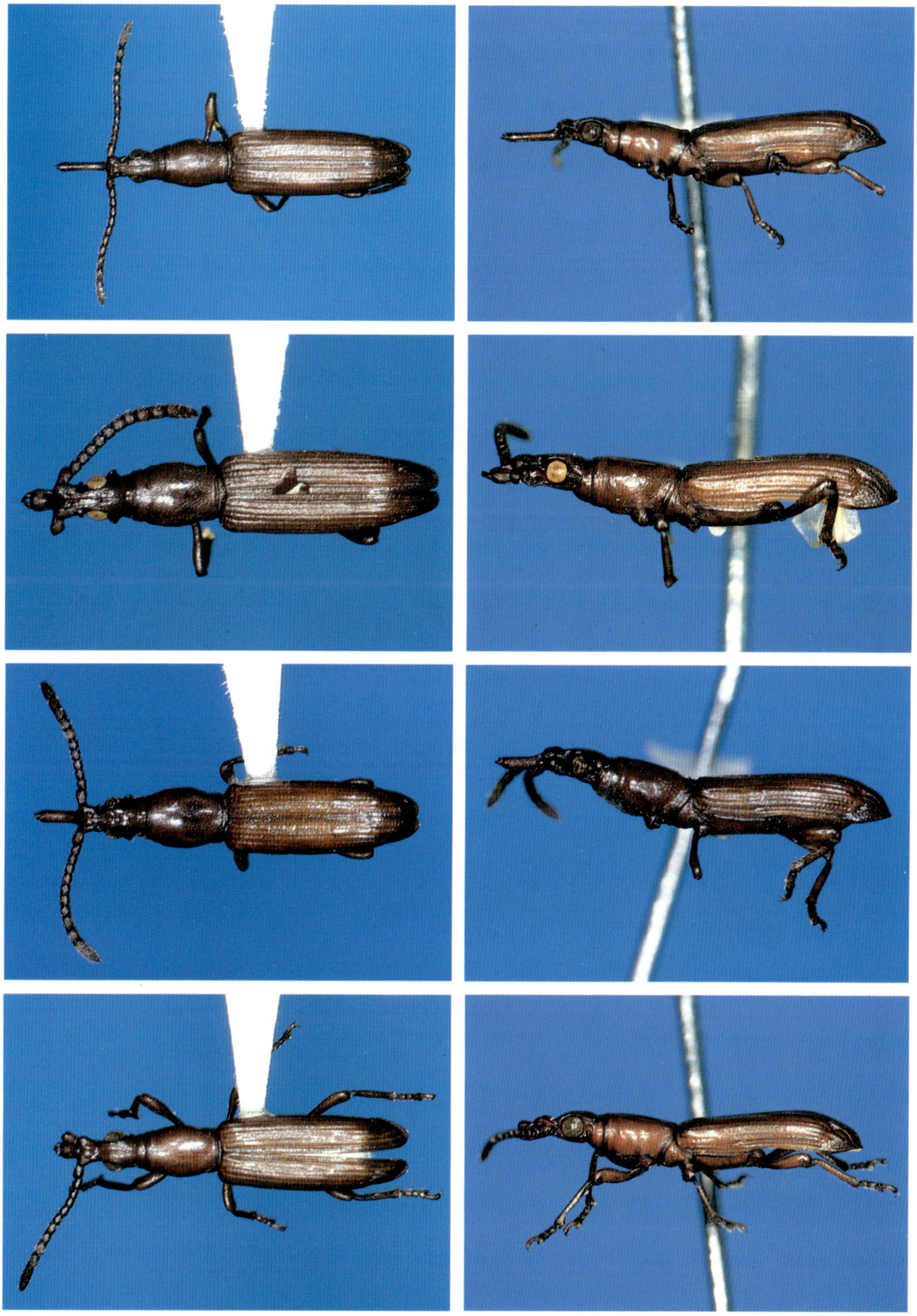

Plate 66 Brentidae – Amorphocephalini

1, 2 *Cordus recticornis* Damoiseau
Fortescue River, Hamersley Range, Western Australia
Allotype female; 7 mm

3, 4 *Cordus rostralis* Damoiseau
Atherton, Queensland
Paratype male; 4.6 mm

5, 6 *Cordus ultimus* Damoiseau
Iron Range, north Queensland
Holotype male; 6.9 mm

7, 8 *Cordus vermiculatus* Damoiseau
Daly River, Northern Territory
Female; 8 mm

1	2
3	4
5	6
7	8

Plate 67 Brentidae – Amorphocephalini

1, 2 *Cordus vicinus* Senna
Fortescue River, Hamersley Range, northwest Western Australia
Female; 7.3 mm

3, 4 *Cordus* new species 1
19-32 km south-southwest of Kalumburu Mission, King Edward Range, northwestern Western Australia.
Female; 5.5 mm; found in nest of *Iridomyrmex conifera.*

5, 6 *Cordus* new species 2
Halftide, near Mackay, Queensland
Female; 6.7 mm

7, 8 *Cordus* new species 4
Yumbarra National Park, South Australia
Male; 12 mm

1	2
3	4
5	6
7	8

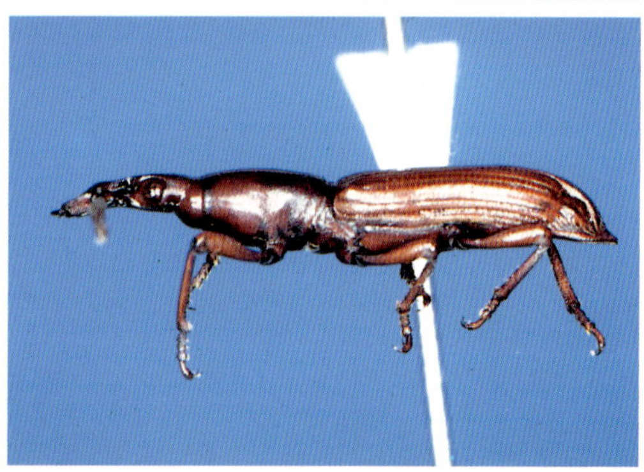

Plate 68 Brentidae – Amorphocephalini

1, 2 *Cordus* new species 5
Lockerbie, Queensland
Female; 12.5 mm

3, 4 *Cordus* new species 6
Bamaga, north Queensland
Male; 8 mm

5, 6 *Kleineella barbata* Kleine
Eubenangee, north Queensland
Male; 11 mm

7, 8 *Kleineella barbata* Kleine
Innisfail, New South Wales
Female; 10 mm

1	2
3	4
5	6
7	8

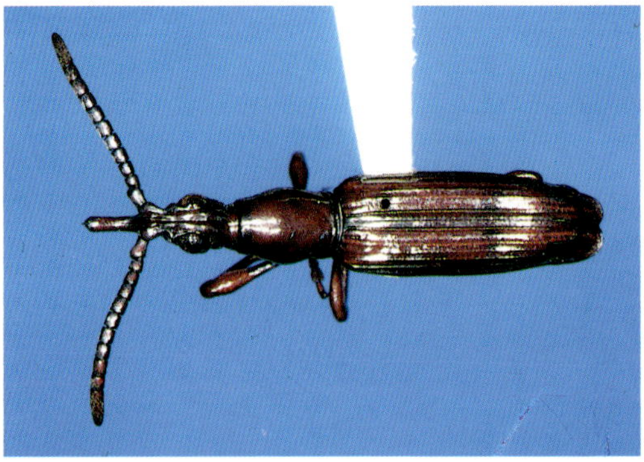

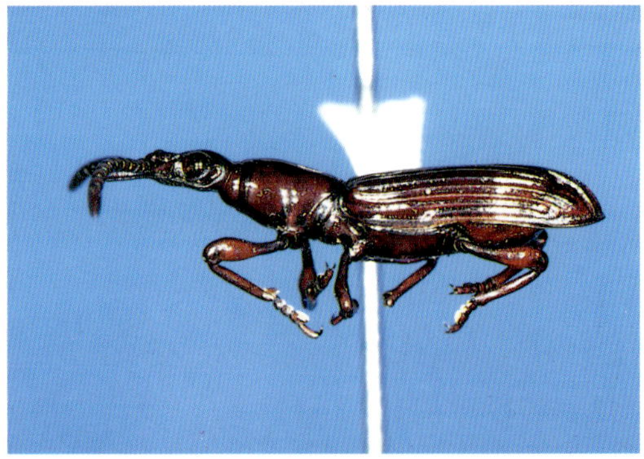

Plate 69 Brentidae – Arrhenodini, Cyphagogini

Arrhenodini

1, 2 *Baryrhynchus lineicollis* Power
Claudi River, Queensland
Female; 17 mm

3 *Ectocemus decemmaculatus* (Montrouzier)
Eubenangee, north Queensland
Male; 29 mm

4 The same species from the same locality; female; 25 mm

Cyphagogini

5, 6 *Allaeometrus breviceps* Senna
11 km east-northeast of Mount Tozer, Queensland
Female; 7. 5 mm

7, 8 *Callipareius plannitarsus* (Perroud)
Dunbulla Forest Reserve near Yungaburra, Queensland
Female; 10.5 mm

1	2
3	4
5	6
7	8

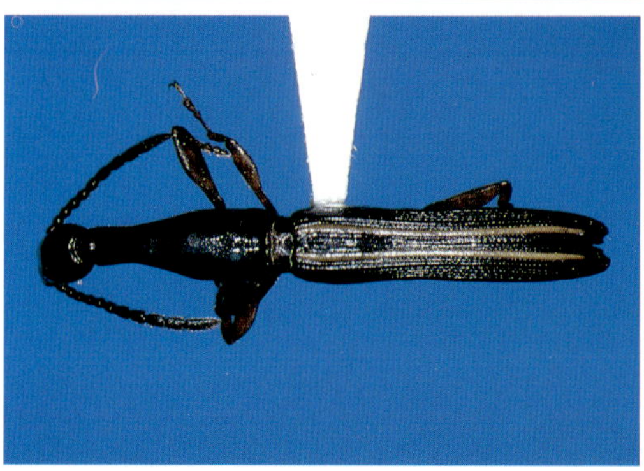

Plate 70 Brentidae – Cyphagogini

1, 2 *Catagogus diorymerus* (Lea)
Kuranda, Queensland
Male; 7.5 mm
Note the mites.

3, 4 *Cyphagogus bipunctatus* Senna
Mount Field National Park, Tasmania
Female; 7.5 mm

5, 6 *Cyphagogus crassitarsus* Damoiseau
Near Cairns, Queensland
Male; 6.5 mm

7, 8 *Cyphagogus delicatus* Lea
Mount Spec, north Queensland
Female; 6 mm

1	2
3	4
5	6
7	8

Plate 71 Brentidae – Cyphagogini

1, 2 *Cyphagogus modiglianii* Senna
Eubenangee, north Queensland
Male; 5.8 mm

3, 4 *Cyphagogus odewahni* Pascoe
Bindogundy, New South Wales
Male; 5.8 mm

5, 6 *Cyphagogus splendens* Kleine
Wareo, Finchhaven, New Guinea
Female; 111.5 mm

7, 8 *Isomorphus* species
11 km east-northeast of Mount Tozer, Queensland
Female; 3.1 mm

1	2
3	4
5	6
7	8

Plate 72 Brentidae – Cyphagogini, Hoplopisthiini

Cyphagogini

1,2 *Mesoderes guttatus* (Kleine)
Cairns, Queensland
Male; 6.3 mm

3 *Mesoderes guttatus* (Kleine)
11 km east-northeast of Mount Tozer, Queensland
Female; 3.7 mm

4 *Mesoderes guttatus* (Kleine)
West Claudi River, Iron Range, Queensland
Male, showing the extraordinary structure of the head. Compare female at 3.

5, 6 *Mesoderes sexnotatus* Senna
Malaya
Male, determined by Kleine; 6.7 mm

Hoplopisthiini

7, 8 *Carcinopisthius kolbei* Senna
Iron Range, Cape York, Queensland
Male; 6.7 mm

1	2
3	4
5	6
7	8

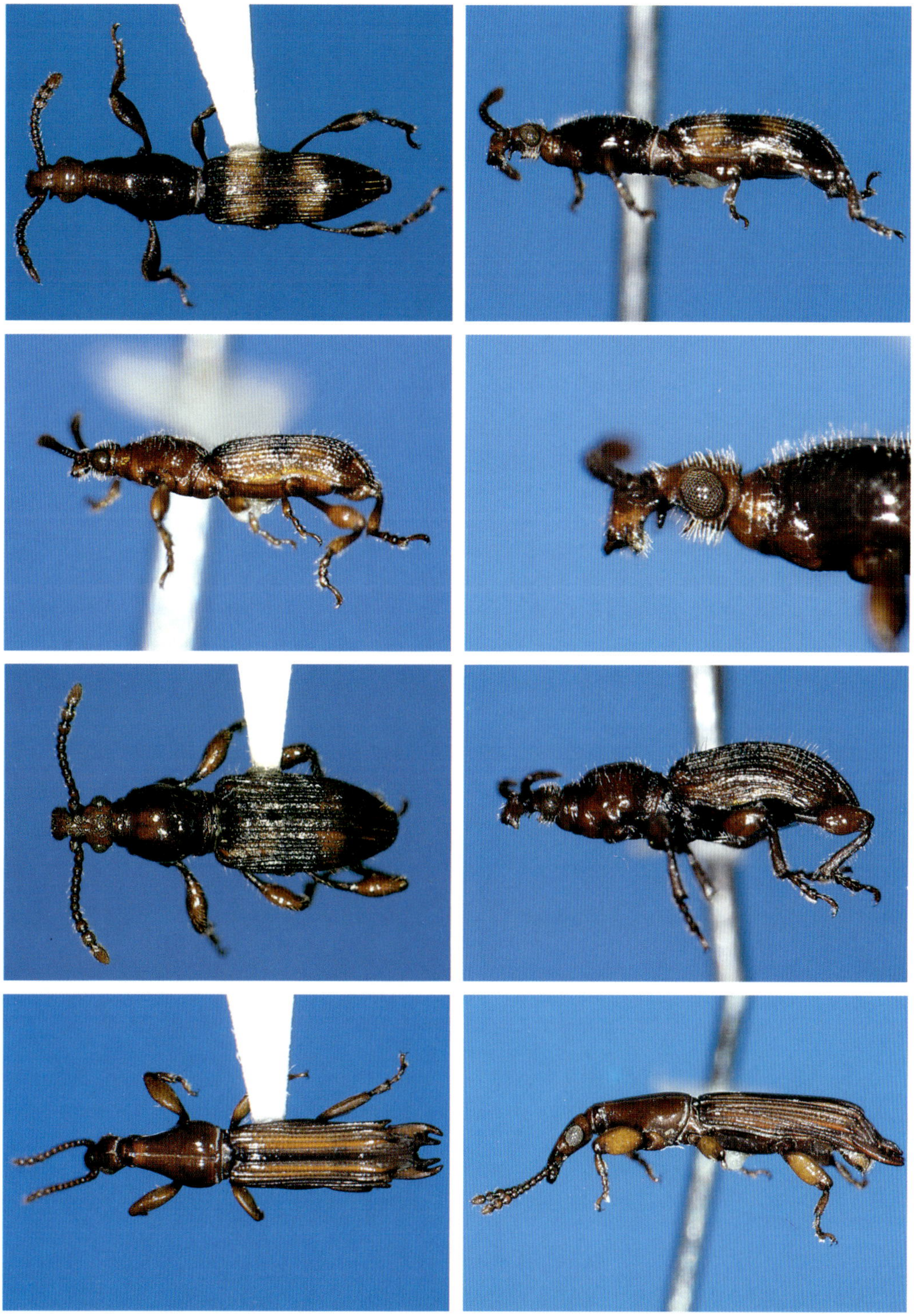

Plate 73 Brentidae – Trachelizini

Ceocephalina

1, 2 *Hormocerus compressitarsis* (Senna)
Selanger, Malaya
Male; 15.5 mm

3, 4 *Hormocerus fossulatus* Blackburn
Cairns, Queensland
Female; 15 mm

5, 6 *Uropteroides gestroi* (Senna)
Kuranda, Queensland
Female; 34.5 mm

Microtrachelizina

7, 8 *Microtrachelizus laevis* Damoiseau
Mount Glorious State Forest, Queensland
Male; 4.3 mm

1	2
3	4
5	6
7	8

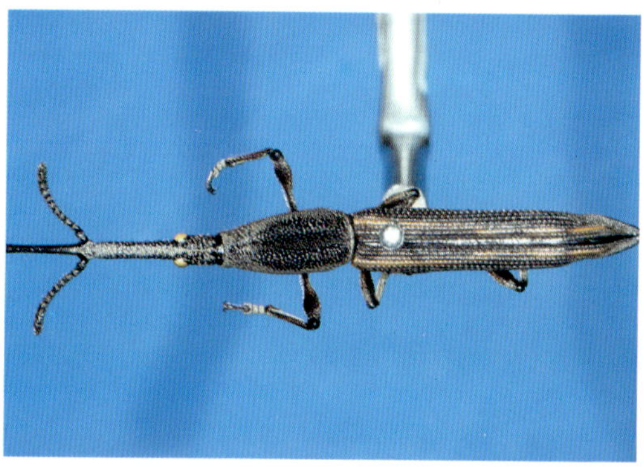

Plate 74 Brentidae – Trachelizini

1, 2 *Microtrachelizus* species 2
Rivertree, Queensland
Male; 4.5 mm

3, 4 *Microtrachelizus* species 3
Shiptons Flat, Queensland
Male; 3.9 mm

5, 6 *Microtrachelizus* species 4
Iron Range, Cape York, Queensland
Male; 5.6 mm

7, 8 *Microtrachelizus* species 5
Iron Range, Cape York, Queensland
Male; 4.9 mm

1	2
3	4
5	6
7	8

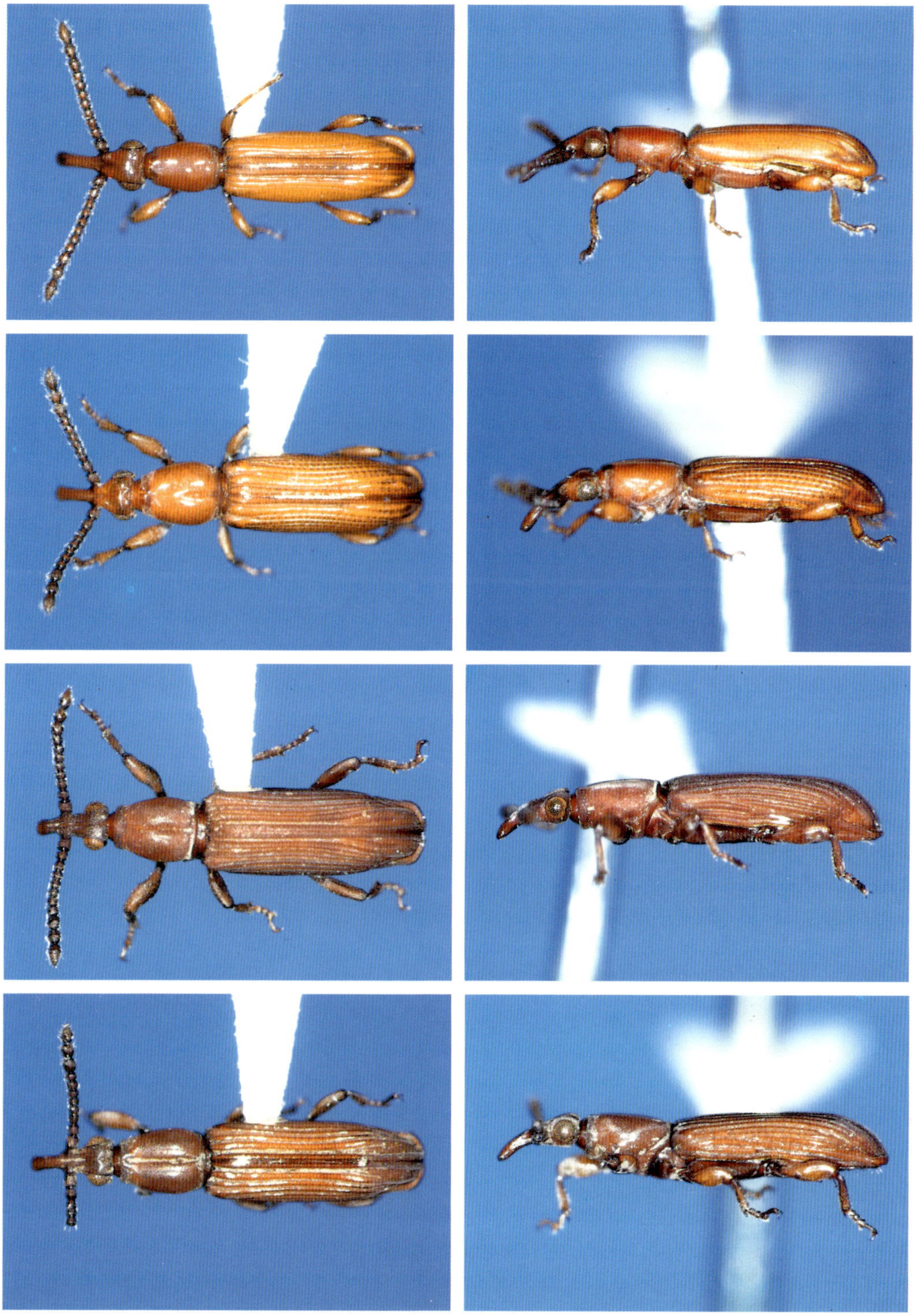

Plate 75 Brentidae – Trachelizini

Microtrachelizina

1, 2 *Microtrachelizus* species 6
Windsor Tableland, Queensland
Female; 4 mm

3, 4 *Microtrachelizus* species 7
Lilyvale, New South Wales
Male; 7.2 mm

Trachelizina

5, 6 *Euschizus alarius* Kleine
Queensland
Male; 14 mm; determined by Kleine

7, 8 *Euschizus dictatorius* Kleine
North Queensland
Male; 13 mm

1	2
3	4
5	6
7	8

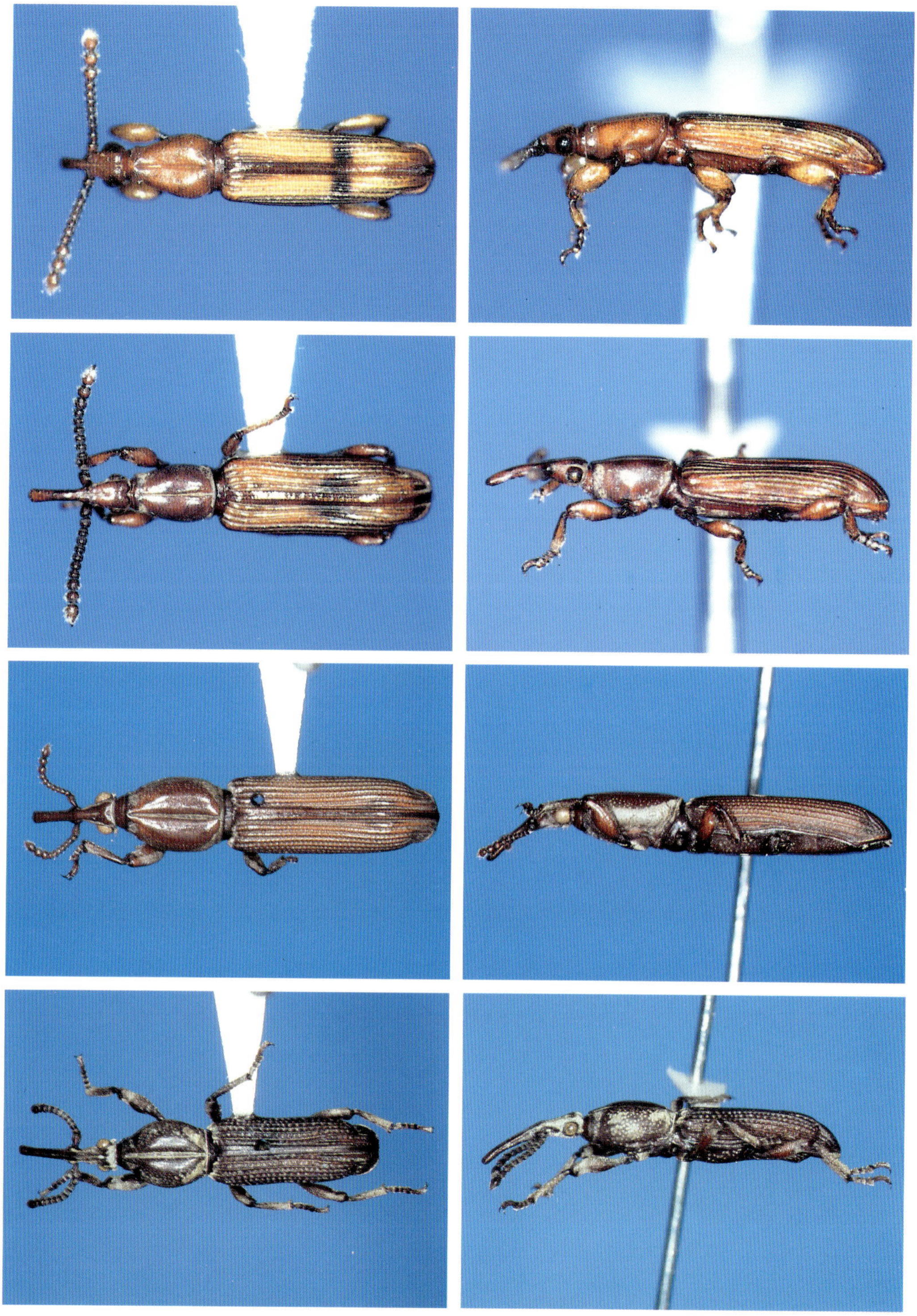

Plate 76 Brentidae – Trachelizini

1, 2 *Miolispa australiana* Senna
Cairns, Queensland
Female; 12 mm

3, 4 *Miolispa novaguineensis* (Guérin-Méneville)
Wau, Papua New Guinea
Male; 8.25 mm

5, 6 The same species with the same data.
Female; 8.50 mm

7, 8 Queensland genus A
Mount Lewis, north Queensland
Male; 9.5 mm

1	2
3	4
5	6
7	8

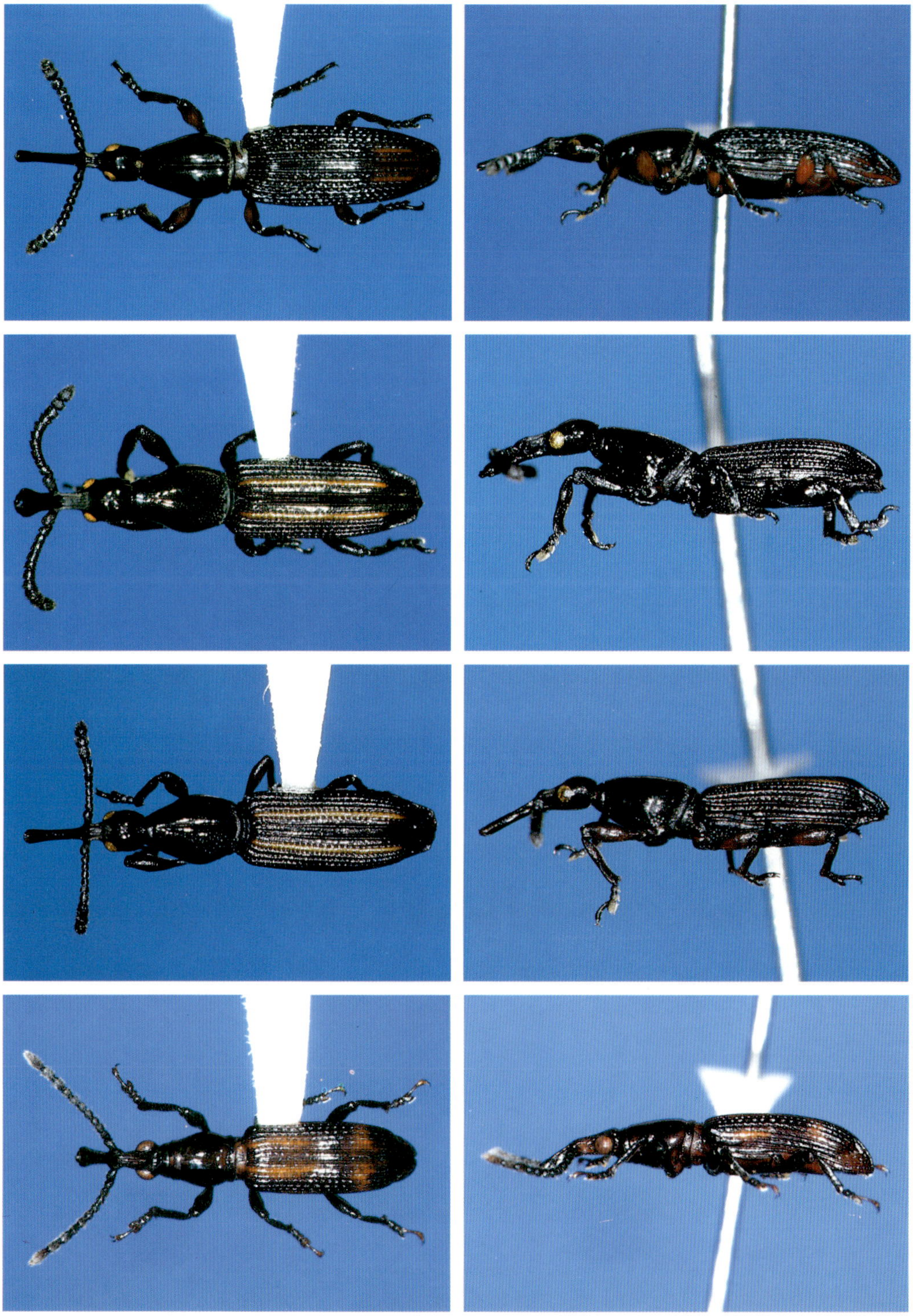

Plate 77 Brentidae

1, 2 Queensland genus B
Mount Glorious, Queensland
Female; 13.5 mm

3, 4 Queensland genus C
Kuranda, Queensland
Female; 10.1 mm

5, 6 Lord Howe Genus A
Lord Howe Island
Male; 6.9 mm

7, 8 *Araiorrhinus howittii* (Pascoe)
Windsor Tableland, north Queensland
Male; 4.2 mm

1	2
3	4
5	6
7	8

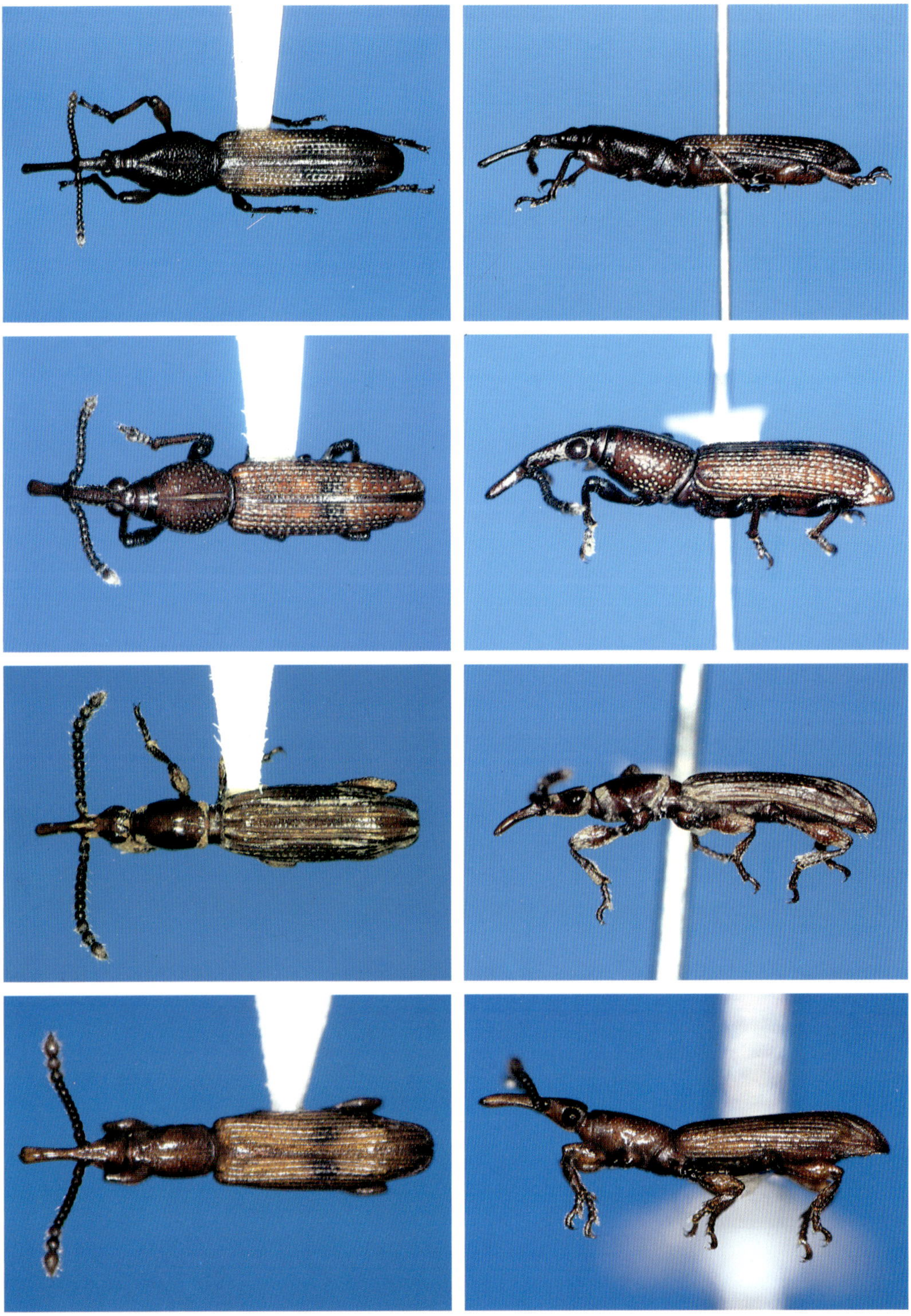

Plate 78 Eurhynchidae

1, 2 *Aporhina australis* (Heller)
Blue Mountains, New South Wales
Male; 13.5 mm

3, 4 *Aporhina splendida* (Blackburn)
Kulara, Queensland
Male; 10.5 mm

5, 6 *Eurhynchus acanthopterus* Boisduval
North Sydney, New South Wales
Male; 15.5 mm

7, 8 *Eurhynchus laevior* (Kirby)
Lake Tabournie, New South Wales
Male; 16.5 mm
Found *in copula* on *Persoonia* species (Proteaceae).

1	2
3	4
5	6
7	8

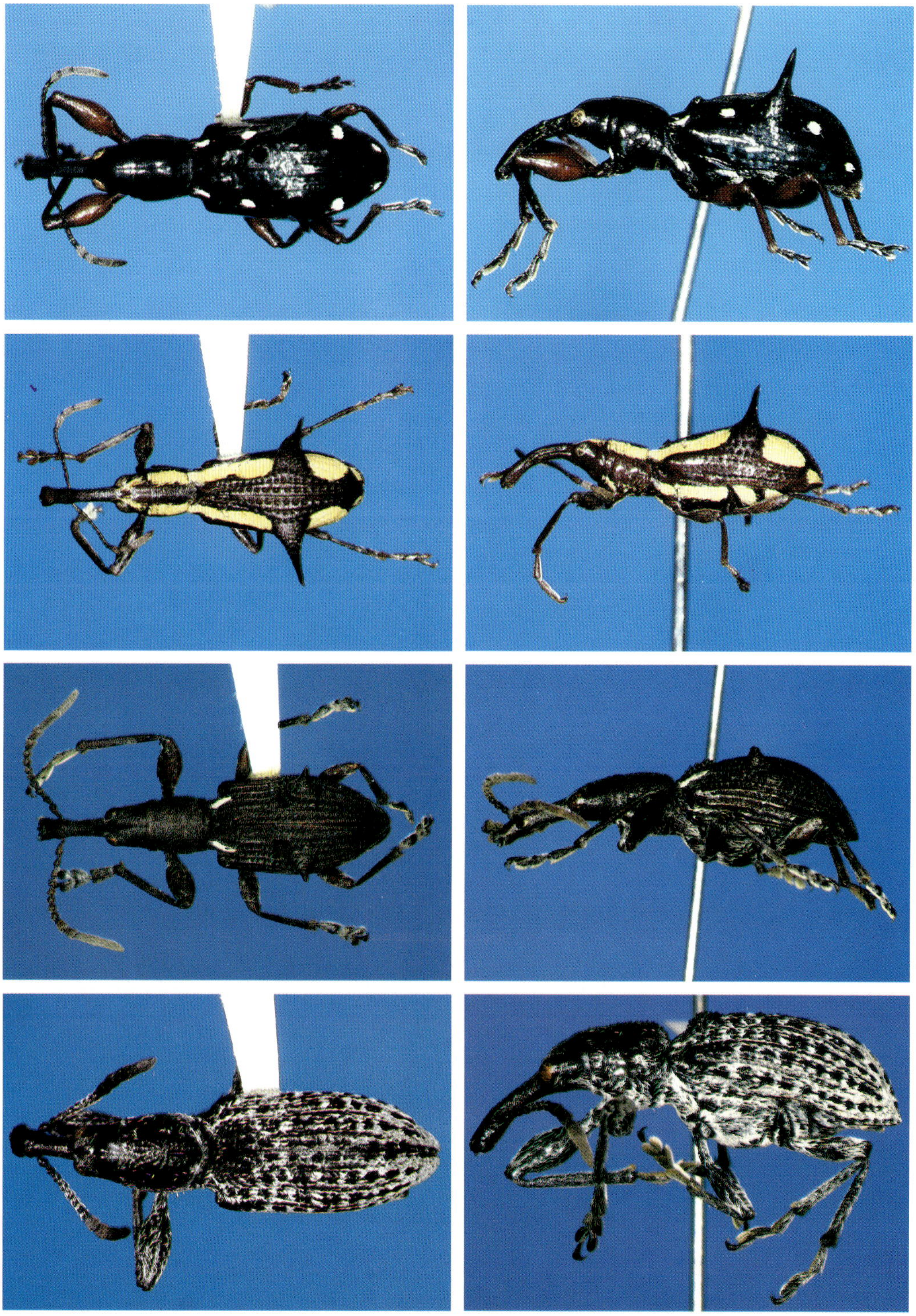

Plate 79 Eurhynchidae

1, 2 *Eurhynchus quadridens* Erichson
Ardrossan, South Australia
Female; 12 mm

3, 4 *Eurhynchus quadrinodosus* Erichson
Tasmania
Male, compared with holotype; 15 mm

5, 6 *Eurhynchus quadrituberculatus* Boheman
Blue Mountains, New South Wales
Male, compared with holotype; 21.5 mm

7, 8 *Eurhynchus scabrior* (Kirby)
Mount Corre, Australian Capital Territory
Male; 17.5 mm

1	2
3	4
5	6
7	8

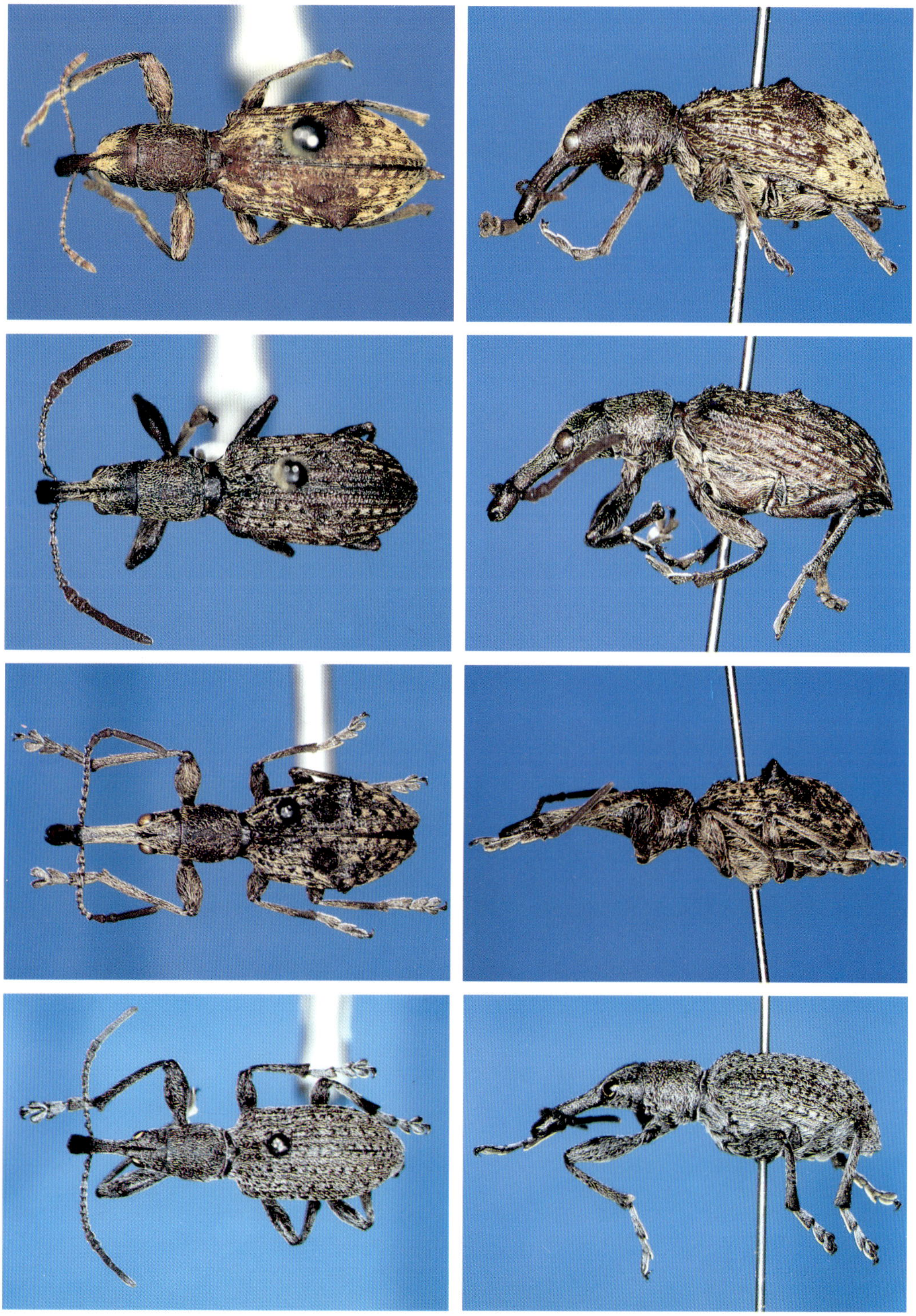

Plate 80 Apionidae

1, 2 *Apion aemulum* Lea
Chillagoe, Queensland
Female; 2.9 mm

3, 4 *Apion amabile* Lea
Tweed River, New South Wales
Holotype female; 4.2 mm

5, 6 *Apion argutulum* Lea
Dalmorton, New South Wales
Male; 2.5 mm

7, 8 *Apion astri* Lea
Murray River, New South Wales
Paratype male; 2.5 mm

Note: Some of the specimens of *Apion*, including types, available for photographs are abraded or imperfect. Some of the illustrations, therefore, may be misleading.

1	2
3	4
5	6
7	8

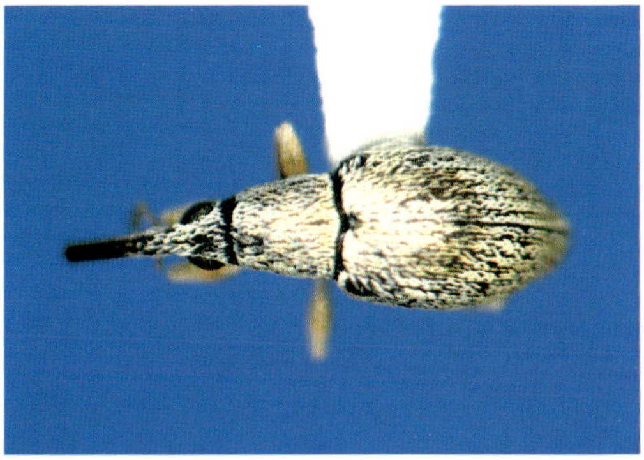

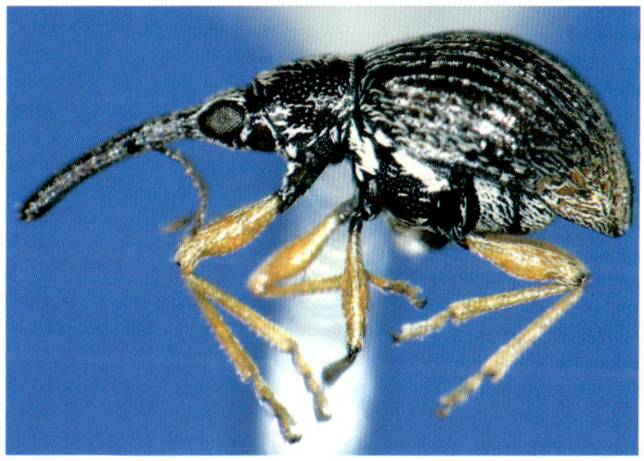

Plate 81 Apionidae

1, 2 *Apion basiinflatum* Lea
Cairns, Queensland
Paratype male; 2.4 mm

3, 4 *Apion binotatum* Lea
Swan River, Western Australia
Holotype male; 2.3 mm

5, 6 *Apion carpophagum* Lea
Rottnest Island, Western Australia
Paratype male; 2.75 mm

7, 8 *Apion comosum* Pascoe
Bunbury, Western Australia
Male; 2.3 mm; identified by Lea

1	2
3	4
5	6
7	8

Plate 82 Apionidae

1, 2 *Apion condensatum* Lea
Tamworth, New South Wales
Paratype male; 2.6 mm

3, 4 *Apion congestum* Lea
Chillagoe, Queensland
Holotype male; 3.2 mm

5, 6 *Apion congruum* Lea
Bunbury, Western Australia
Holotype female; 3.6 mm

7, 8 *Apion convexipenne* Lea
Townsville, Queensland
Holotype male; 4.7 mm

1	2
3	4
5	6
7	8

Plate 83 Apionidae

1, 2 *Apion foveicolle* Lea
Donnybrook, Western Australia
Holotype male; 2.1 mm

3, 4 *Apion fuscosuturale* Lea
Swan River, Western Australia
Holotype female; 2.1 mm

5, 6 *Apion hoblerae* Lea
Dalby, Queensland
Paratype female; 2.2 mm

7, 8 *Apion immundum* Lea
Galston, New South Wales
Paralectotype female; 2.7 mm

1	2
3	4
5	6
7	8

Plate 84 Apionidae

1, 2 *Apion integricolle* Lea
Albany, Western Australia
Paratype male; 2.6 mm

3, 4 *Apion macleayense* Lea
Macleay River, New South Wales
Holotype male; 2.9 mm

5, 6 *Apion melvillense* Lea
Melville Island, Northern Territory
Paratype male; 1.9 mm

7, 8 *Apion microscopicum* Lea
Frankford, Tasmania
Paratype female; 1.7 mm

1	2
3	4
5	6
7	8

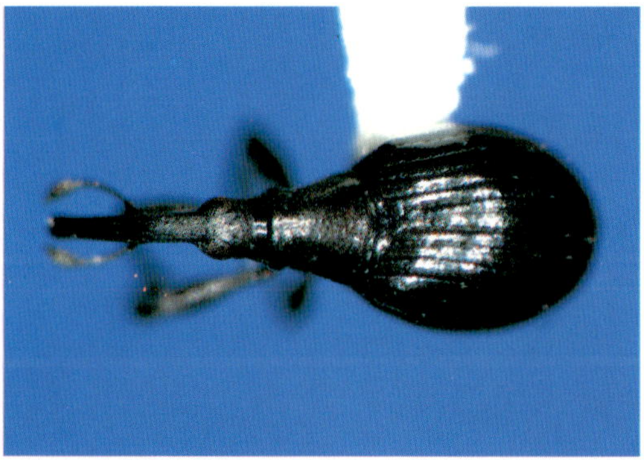

Plate 85 Apionidae

1, 2 *Apion nigrosuturale* Lea
Northwestern Western Australia
Paratype male; 2.0 mm

3, 4 *Apion nigroterminale* Lea
Sydney, New South Wales
Paratype female; 2.4 mm

5, 6 *Apion niveodispersum* Lea
Chillagoe, Queensland
Holotype male; 2.4 mm

7, 8 *Apion parvocastaneum* Lea
Murray Bridge, South Australia
Holotype male (abraded); 1.75 mm

1	2
3	4
5	6
7	8

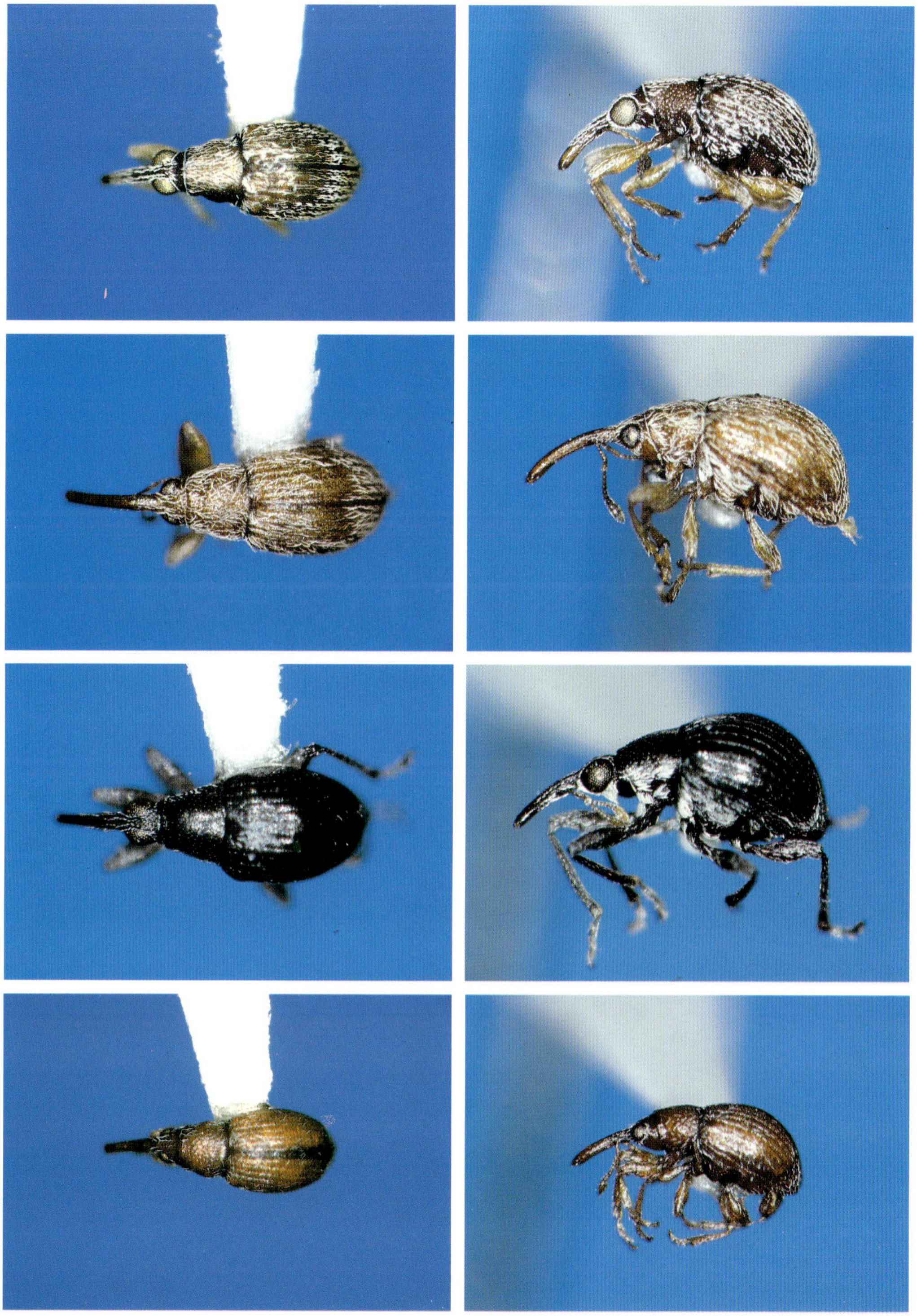

Plate 86 Apionidae

1, 2 *Apion philanthum* Lea
Behn River, northwest Western Australia
Paratype male; 2.0 mm

3, 4 *Apion pictipes* Lea
Cairns, Queensland
Holotype male; 2.1 mm

5, 6 *Apion pilistriatum* Lea
Mulgrave River, Queensland
Paratype male; 1.8 mm

7, 8 *Apion pudicum* Lea
Behn River, northwest Western Australia
Paratype female; 3.3 mm

1	2
3	4
5	6
7	8

Plate 87 Apionidae

1, 2 *Apion pulicare* Pascoe
Garden Island, Western Australia
Female, compared with holotype; 2.5 mm

3, 4 *Apion quadricolor* Lea
Darwin, Northern Territory
Paratype male; 2.6 mm

5, 6 *Apion rivulare* Lea
Roper River, Northern Territory
Paratype female; 2.7 mm

7, 8 *Apion solani* Lea
Sydney, New South Wales
Holotype male; 2.4 mm

1	2
3	4
5	6
7	8

Plate 88 Apionidae

1, 2 *Apion subopacum* Lea
Chillagoe, Queensland
Paratype male; 2.4 mm

3, 4 *Apion tasmanicum* Lea
Huon River, Tasmania
Paratype male; 2.1 mm

5, 6 *Apion* teretirostre Lea
Bunbury, Western Australia
Holotype female; 2.6 mm

7, 8 *Apion terraereginae* Blackburn
Magnetic Island, Queensland
Female; 4.4 mm; identified by Lea.

1	2
3	4
5	6
7	8

Plate 89 Apionidae

1, 2 *Apion turbidum* Lea
South Australia
Paratype female; 2.2 mm

3, 4 *Apion vertebrale* Lea
Ourimbah, New South Wales
Holotype male; 3.2 mm
Note the long hair on the profemur.

5, 6 *Apion* new species
Townsville, Queensland
Male; 2.9 mm
Reared from *Acacia* seeds.

7, 8 *Perapion violaceum* (Kirby)
Praia das Macas, Portugal
Female; 3.8 mm
Reared from *Emex spinosa*
This species is not at this writing known to be established in Australia, but it has been considered for importation to assist in the control of its host which is a spiny weed pest.

1	2
3	4
5	6
7	8

Plate 90 Apionidae

1, 2 *Kalainapion stilbum* (Lea)
Otford, New South Wales
Paratype male; 2.1 mm

3, 4 *Lissapion sculpticeps* (Lea)
Cairns, Queensland
Holotype male; 4.1 mm

5, 6 *Lissapion varistriatum* (Lea)
Tamborine, Queensland
Paratype female; 2.6 mm

7, 8 *Notapion (Notapion) mediopunctum* (Lea)
Cairns, Queensland
Syntype male; 2.9 mm

1	2
3	4
5	6
7	8

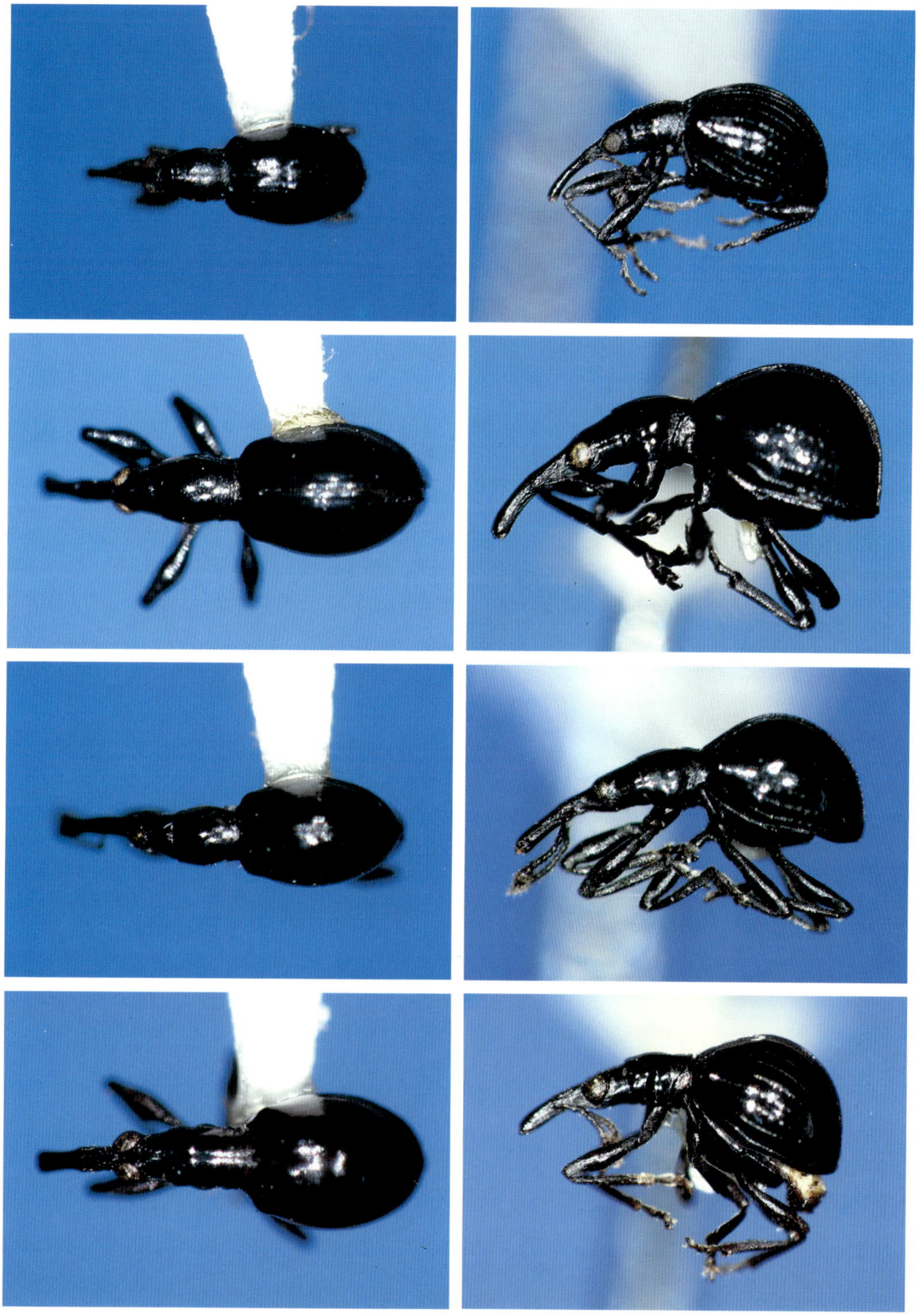

Plate 91 Apionidae

1, 2 *Notapion* (*Notapion*) *striatipenne* (Lea)
Stanwell Park, New South Wales
Holotype male; 2.4 mm

3, 4 *Notapion* (*Notapion*) *trilobicolle* (Lea)
Cairns, Queensland
Paratype male; 3.1 mm

5, 6 *Notapion* (*Notapion*) *varirostre* (Lea)
Tamborine, Queensland
Holotype male; 2.8 mm

7, 8 *Notapion* (*Ganoapion*) *mednostriatum* (Lea)
Cairns, Queensland
Holotype male; 3.3 mm

See Volume VI for more coloured illustrations of Apionidae.

1	2
3	4
5	6
7	8

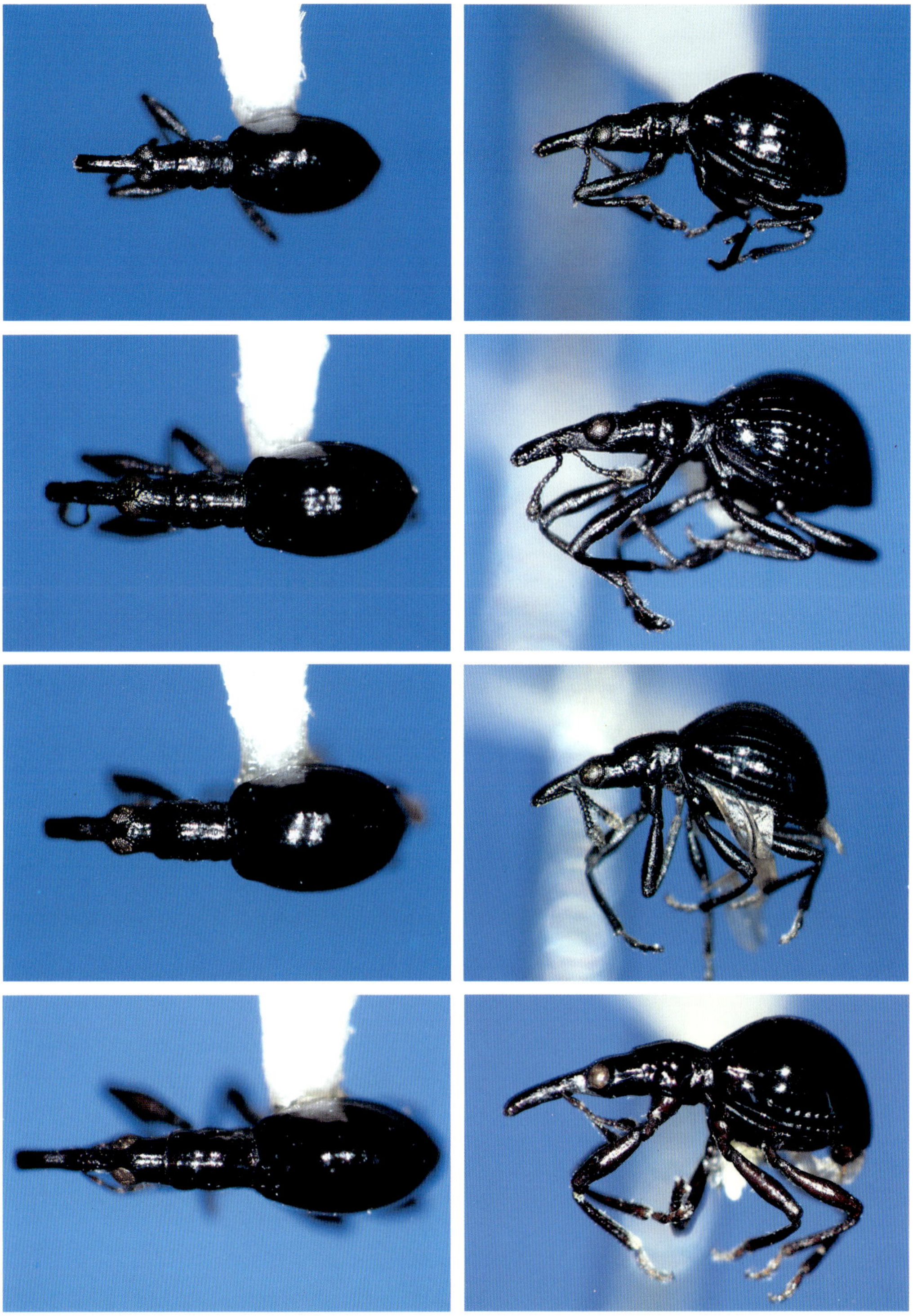

Plate 92 Apionidae

1, 2 *Nanophyes alleni* Lea
Cairns, Queensland
Holotype male; 2.5 mm

3, 4 *Nanophyes maurus* Pascoe
Cunnamulla, Queensland
Male, compared with holotype; 2.2 mm

5, 6 *Nanophyes nigrovarius* Lea
Melbourne River, Victoria
Male; 2.2 mm

7, 8 *Nanophyes pallidicornis* Lea
Forest Reefs, New South Wales
Paratype female; 1.5 mm

1	2
3	4
5	6
7	8

Plate 93 Apionidae – Rhynchophoridae

Apionidae

1, 2 *Nanophyes v-notatus* Lea
Nepean River, New South Wales
Holotype female; 2.5 mm

Rhynchophoridae

3, 4 *Diathetes signaticollis* Faust
Red Island Point, Cape York, Queensland
Female; 12.5 mm

5, 6 *Diathetes* new species
Katherine, Northern Territory
Male; 17.5 mm

7, 8 *Diathetes* new species
Stradbroke Island, Queensland
Male; 18 mm

1	2
3	4
5	6
7	8

Plate 94 Rhynchophoridae

1, 2 *Diocalandra frumenti* (Fabricius)
Moa Island, Torres Strait, Queensland
Female; 6.5 mm
This widely distributed species is occasionally common on coconut trees in many places, and it is often thought to be of more economic importance than is true.

3, 4 *Sitophilus linearis* (Herbst)
Townsville, Queensland
Male; 5 mm
This species infests the seeds of tamarinds.

5 *Sitophilus oryzae* (Linnaeus), the rice weevil
This notorious pest causes enormous loss to stored grain throughout the world. Untold millions of dollars are expended in attempting to control its ravages.

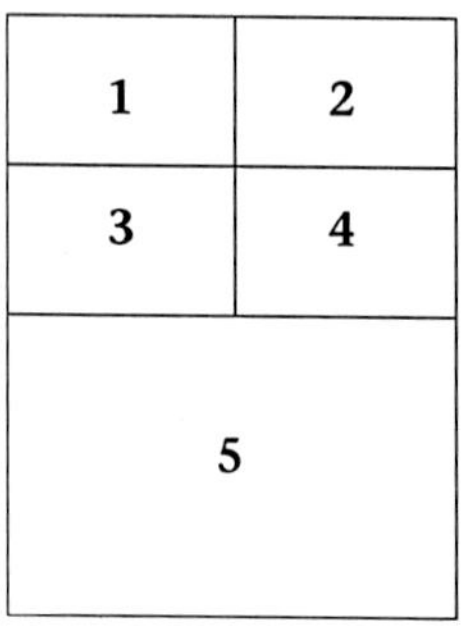

Plate 95 Erirhinidae

1, 2 *Aonychus argus* Lea
King Sound, Western Australia
Paratype female; 4.8 mm

3, 4 *Aonychus barbatus* Lea
Bathurst Island, Northern Territory
Holotype female; 6.3 mm

5, 6 *Aonychus hopei* Boheman
"Kareenda to Peebinga", South Australia
Male; 6.2 mm

7, 8 *Aonychus bicruciatus* Lea
Lake Austin, Western Australia
Male; 6 mm

The colour patterns of *Aonychus* are unusually variable, and care must be taken not to mistake colour forms for species.

1	2
3	4
5	6
7	8

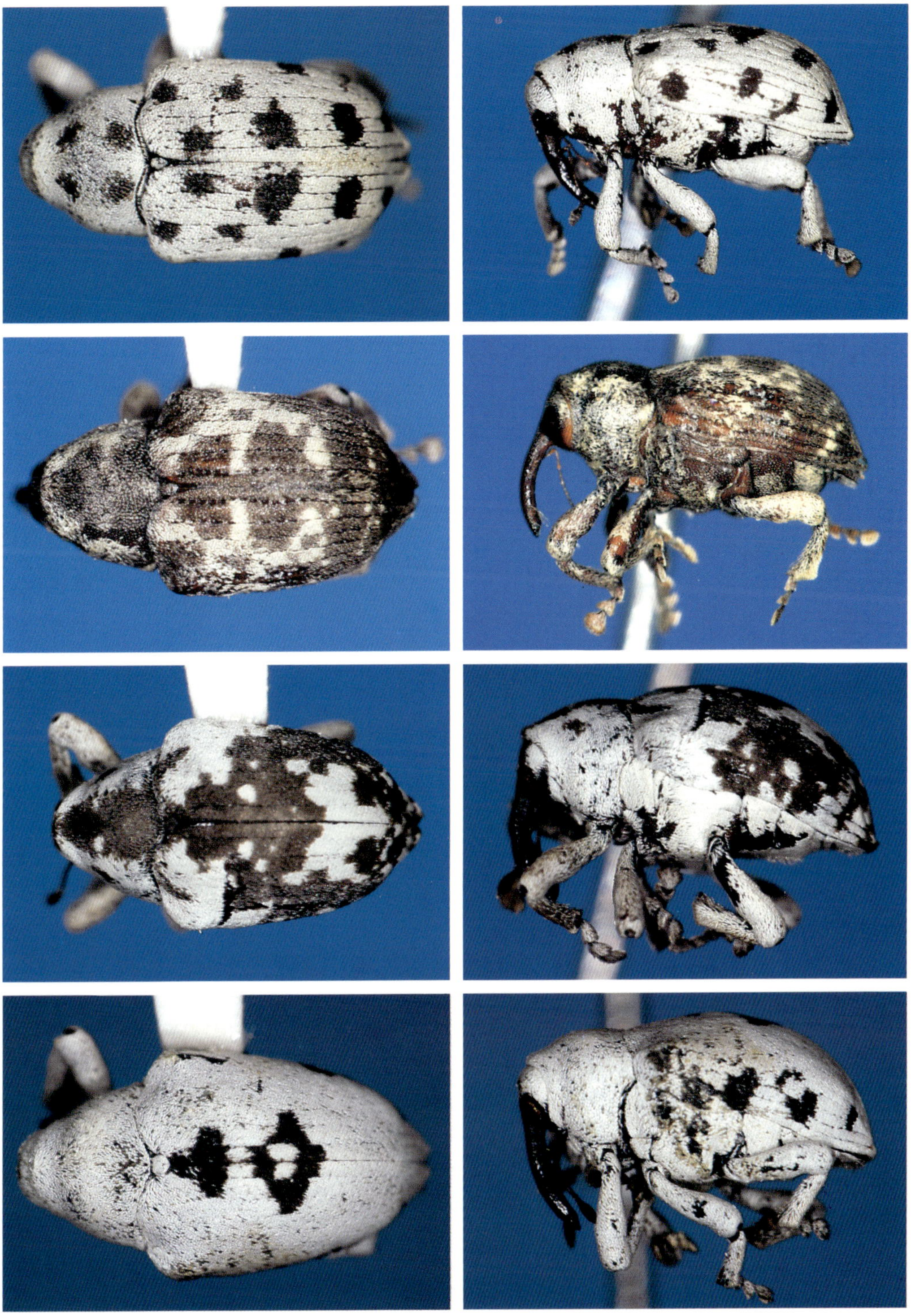

Plate 96 Erirhinidae

1, 2 *Aonychus lineatus* Pascoe
Cape Le Grand, Western Australia
Female; 8.8 mm

3, 4 *Aonychus pachypus* Lea
Swan River, Western Australia
Holotype female; 4.6 mm

5, 6 *Aonychus luctuosus* Pascoe
Cunderdin, Western Australia
From the holotype male of the synonym *picatus* Lea; 4.8 mm

7, 8 *Aonychus striatus* Lea
King Sound, Western Australia
Holotype male; 5.9 mm
See Volume VI for other species.

1	2
3	4
5	6
7	8

Plate 97 Amycterinae

1, 2 *Acantholophus adelaidae* Waterhouse
Adelaide, South Australia
Male; 13.5 mm

3, 4 *Acantholophus aureolus* Boheman
King George Sound, Western Australia
Female; 17.5 mm

5, 6 *Acantholophus blackburni* Ferguson
Palm Creek, Central Australia
Male; 16.5 mm

7, 8 *Acantholophus browni* Ferguson
Murchison, Western Australia
Male, compared with holotype; 21 mm

1	2
3	4
5	6
7	8

Plate 98 Amycterinae

1, 2 *Acantholophus crenaticollis* Macleay
South Australia
Male; 11.5 mm

3, 4 *Acantholophus cupreomicans* Ferguson
Parkerville, Western Australia
Holotype female; 20.5 mm

5, 6 *Acantholophus denticollis* Ferguson
New South Wales
Holotype male; 15 mm

7, 8 *Acantholophus doddi* Ferguson
Milmerran, Queensland
Female; 15.5 mm

1	2
3	4
5	6
7	8

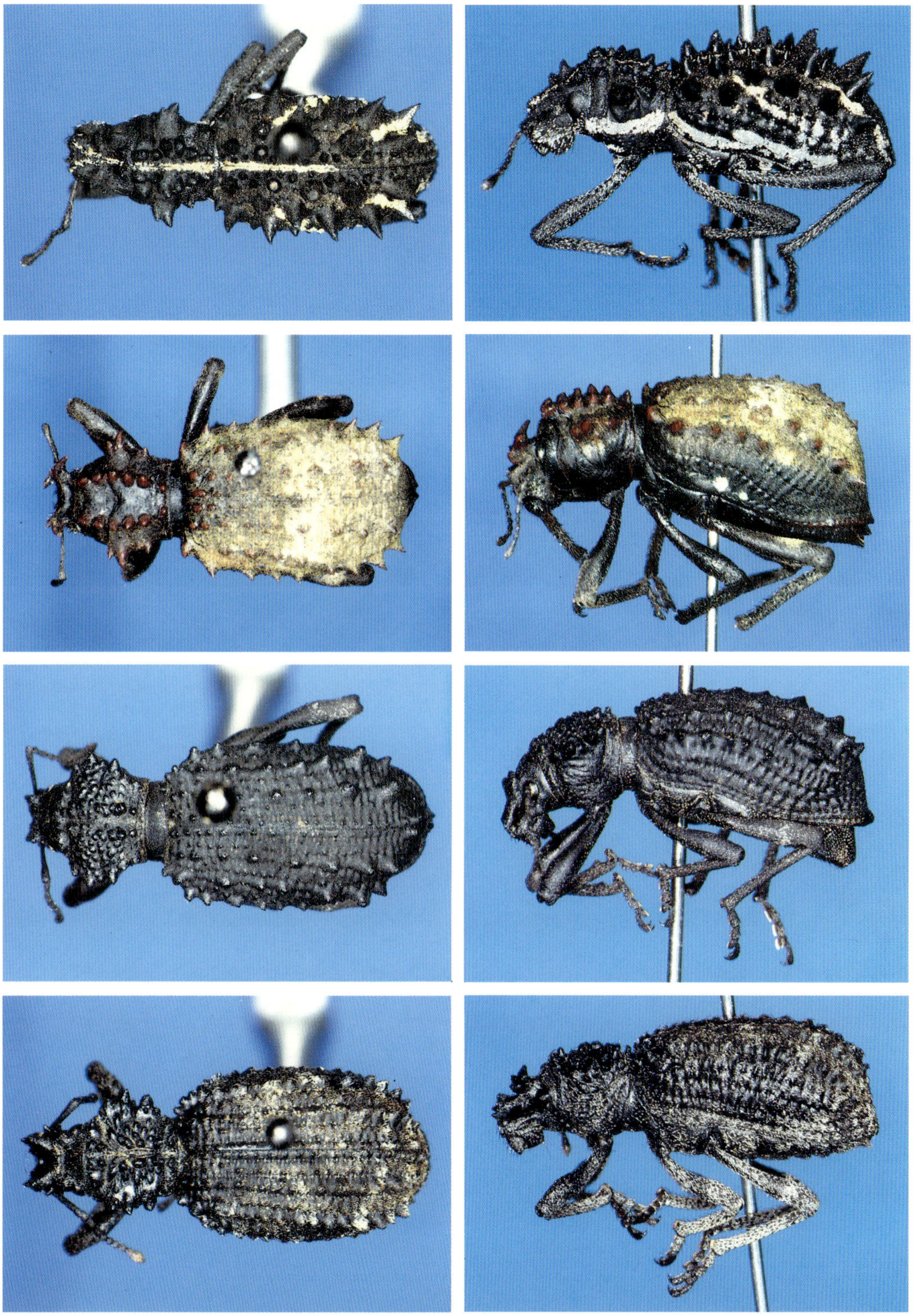

Plate 99 Amycterinae

1, 2 *Acantholophus dumosus* Boheman
Bunbury, Western Australia
Male; 15 mm

3, 4 *Acantholophus echinatus* Guérin-Méneville
New South Wales
Male; 17.5 mm
This is an extreme variety; no two specimens are alike.

5, 6 *Acantholophus foveirostris* Lea
South Australia
Holotype male; 14.5 mm

7, 8 *Acantholophus franklinensis* Blackburn
Yellanna, South Australia
Female; 17.5 mm

1	2
3	4
5	6
7	8

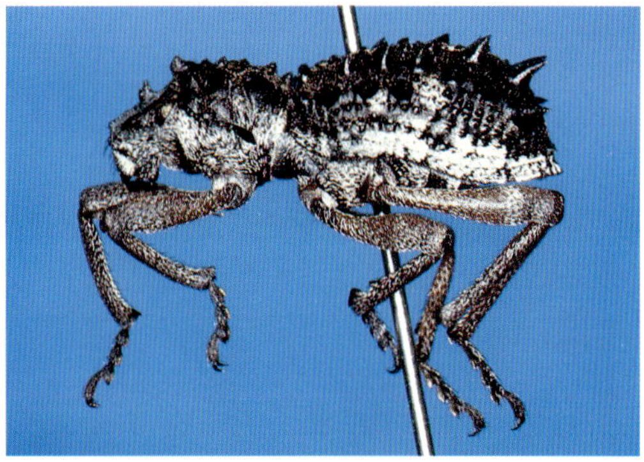

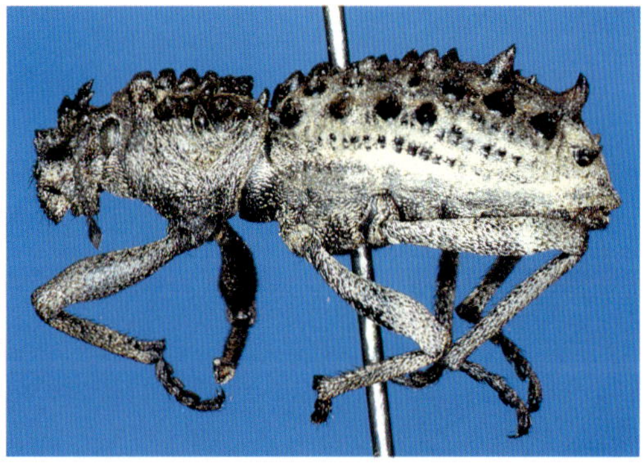

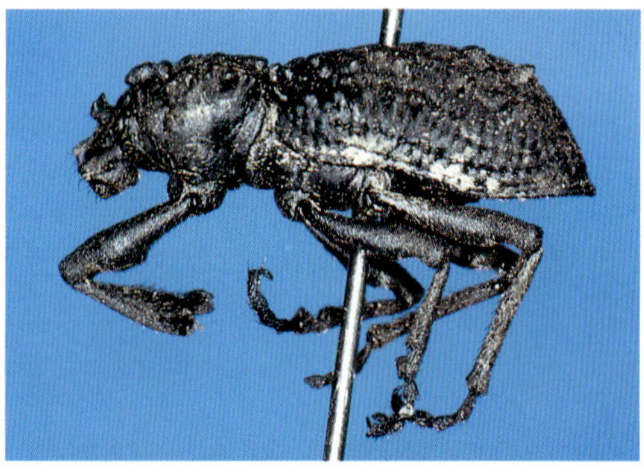
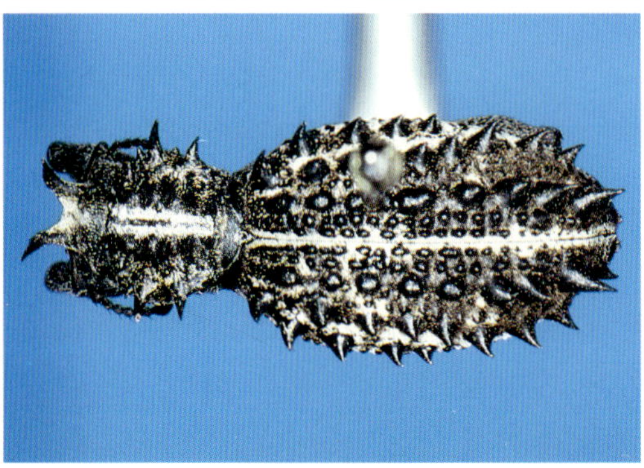
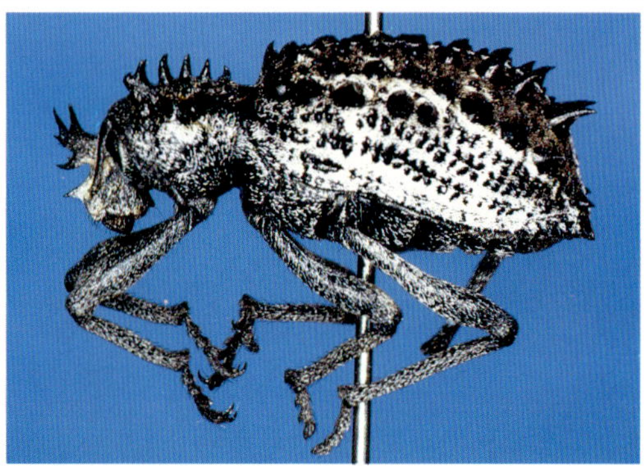

Plate 100 Amycterinae

1, 2 *Acantholophus granulatus* Sloane
Everard Range, South Australia
Holotype female; 20 mm

3, 4 *Acantholophus gravicollis* Macleay
South Australia
Holotype male; 15 mm

5, 6 *Acantholophus halmaturinus* Ferguson
Kangaroo Island, South Australia
Paratype male; 15 mm

7. 8 *Acantholophus humeralis* Macleay
Western Australia
Holotype male; 20 mm

1	2
3	4
5	6
7	8

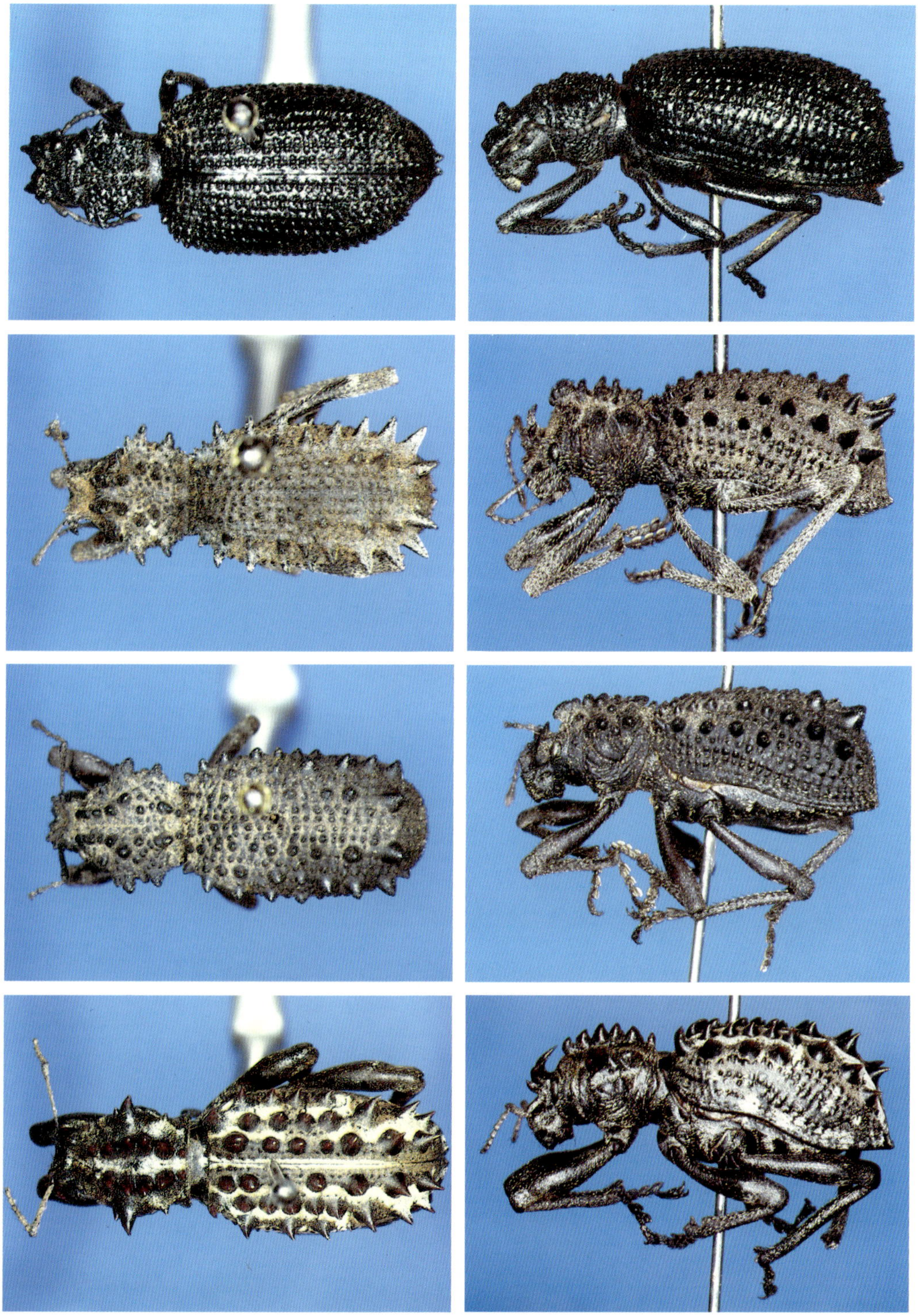

Plate 101 Amycterinae

1, 2 *Acantholophus hypoleucus* (Boheman)
Western Australia
Male; 15 mm

3, 4 *Acantholophus hystrix* (Boheman)
King George Sound, Western Australia
Female; 12.5 mm

5, 6 *Acantholophus mastersi* Macleay
Stirling Range, Western Australia
Paratype male; 15.5 mm

7, 8 *Acantholophus maximus* (Macleay)
Swan River, Western Australia
Female; 18 mm

1	2
3	4
5	6
7	8

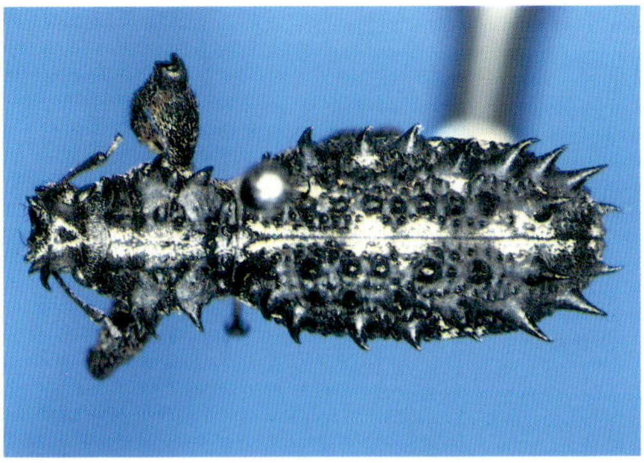
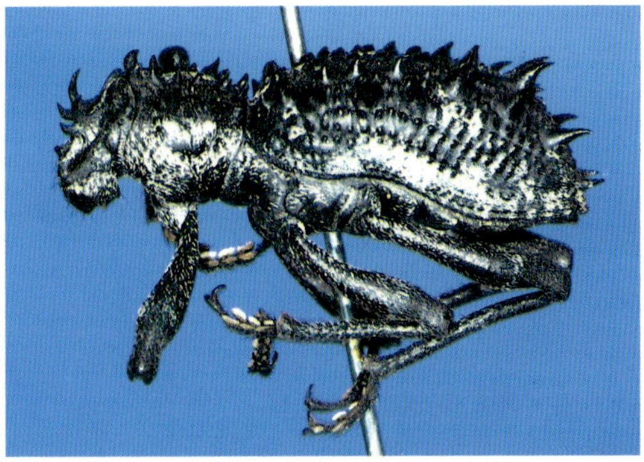

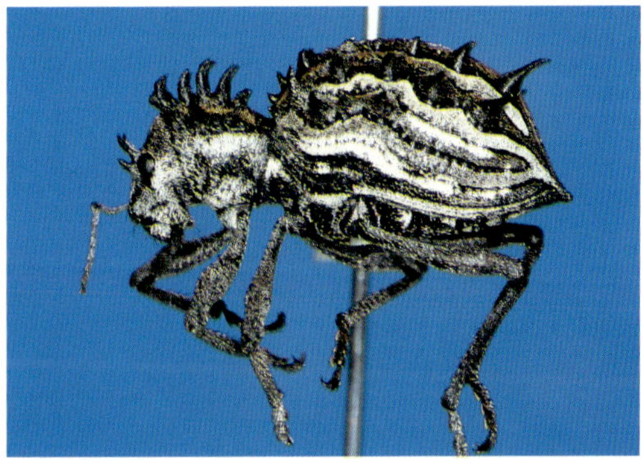

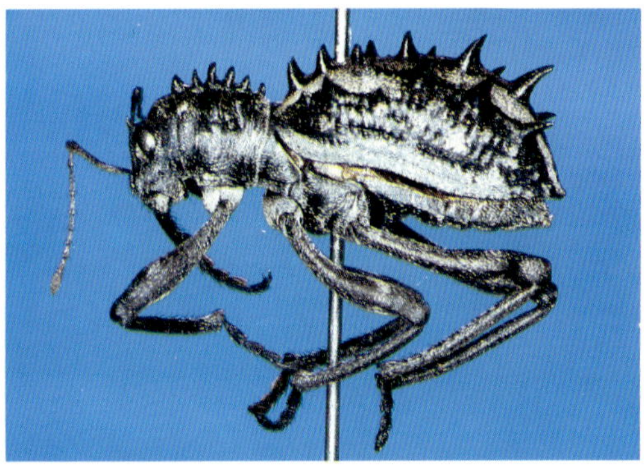

Plate 102 Amycterinae

1, 2 *Acantholophus nanus* Ferguson
Blue Mountains, New South Wales
Holotype male; 12 mm

3, 4 *Acantholophus niveovittatus* Blackburn
Conderdin, Western Australia
Female; 18.5 mm

5, 6 *Acantholophus ocelliger* Ferguson
Western Australia
Holotype male; 16.5 mm

7, 8 *Acantholophus parvulus* Ferguson
Mittagong, New South Wales
Holotype male; 12.5 mm

1	2
3	4
5	6
7	8

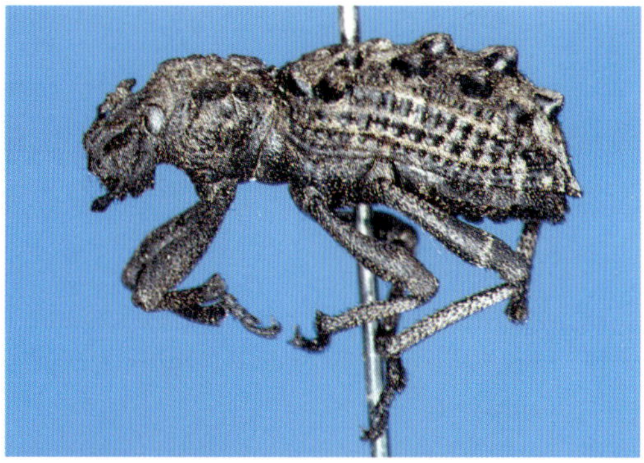
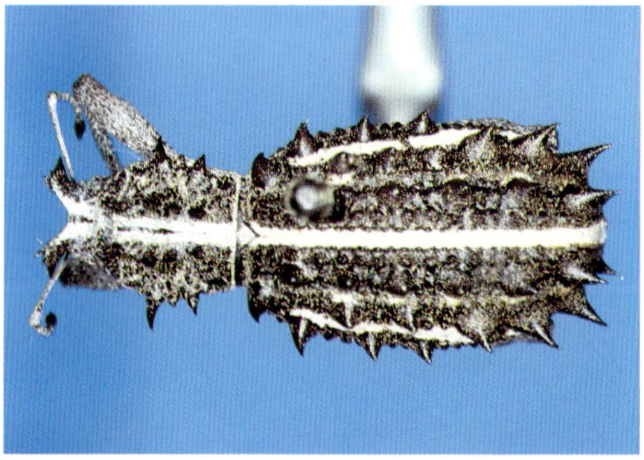
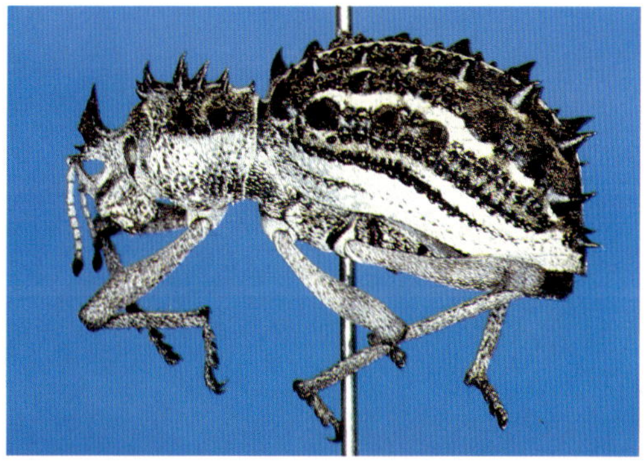

Plate 103 Amycterinae

1, 2 *Acantholophus planicollis* Waterhouse
Adelaide, South Australia
Male; 13 mm

3, 4 *Acantholophus scabrosus* Macleay
Portland, New South Wales
Male; 16 mm

5, 6 *Acantholophus scaphirostris* Ferguson
Karridale, Western Australia
Female; 17.5 mm

7, 8 *Acantholophus sellatus* Ferguson
Inverell, New South Wales
Holotype male; 17.5 mm

1	2
3	4
5	6
7	8

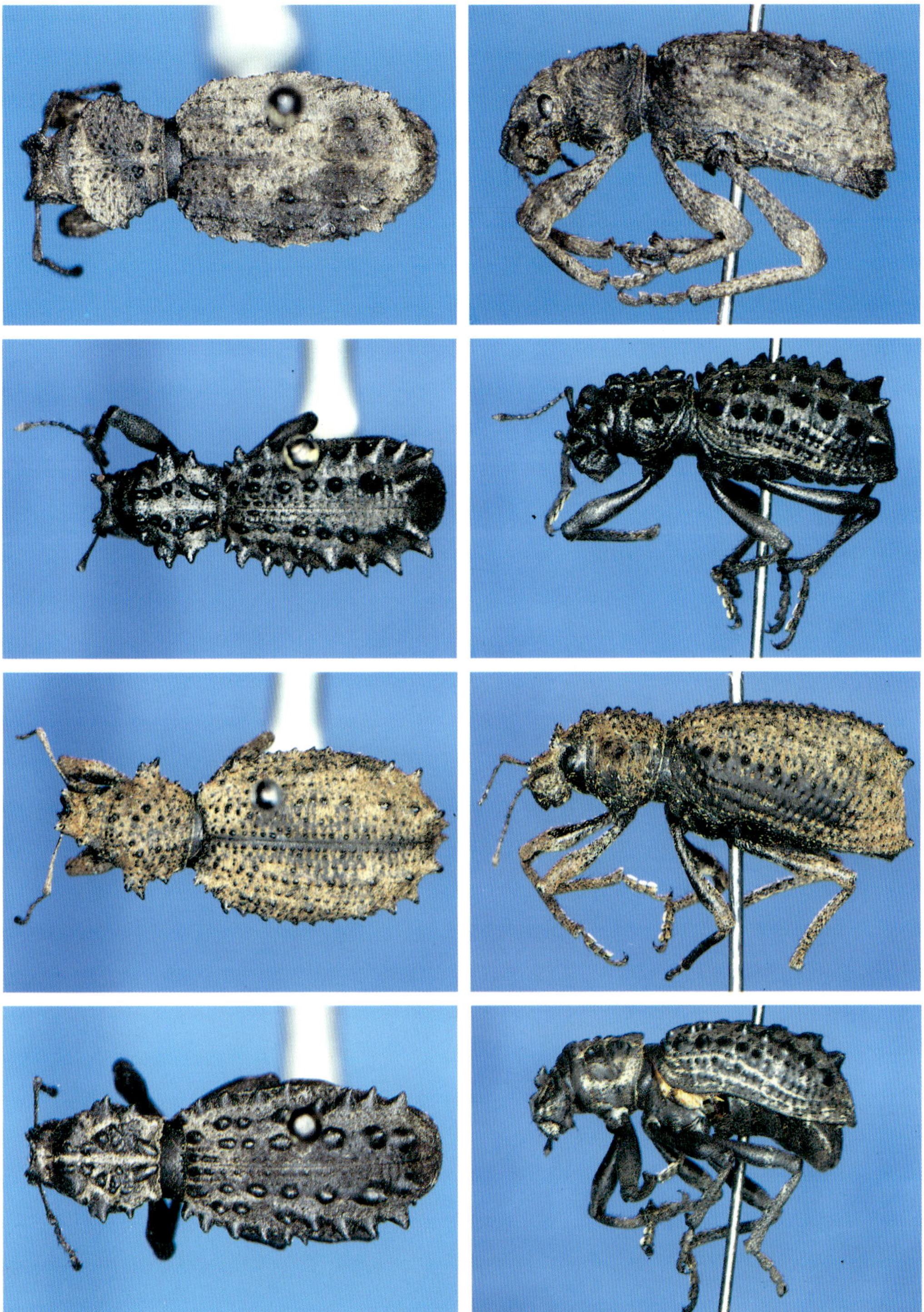

Plate 104 Amycterinae

1, 2 *Acantholophus simplex* Pascoe
Western Australia
Male; 14.5 mm

3, 4 *Acantholophus simulator* Ferguson
Kangaroo Island, South Australia
Paratype female; 16.5 mm

5, 6 *Acantholophus sordidus* Ferguson
Jamison, Victoria
Holotype male; 15 mm

7, 8 *Acantholophus spinifer* Macleay
Victoria
Holotype male; 16 mm

1	2
3	4
5	6
7	8

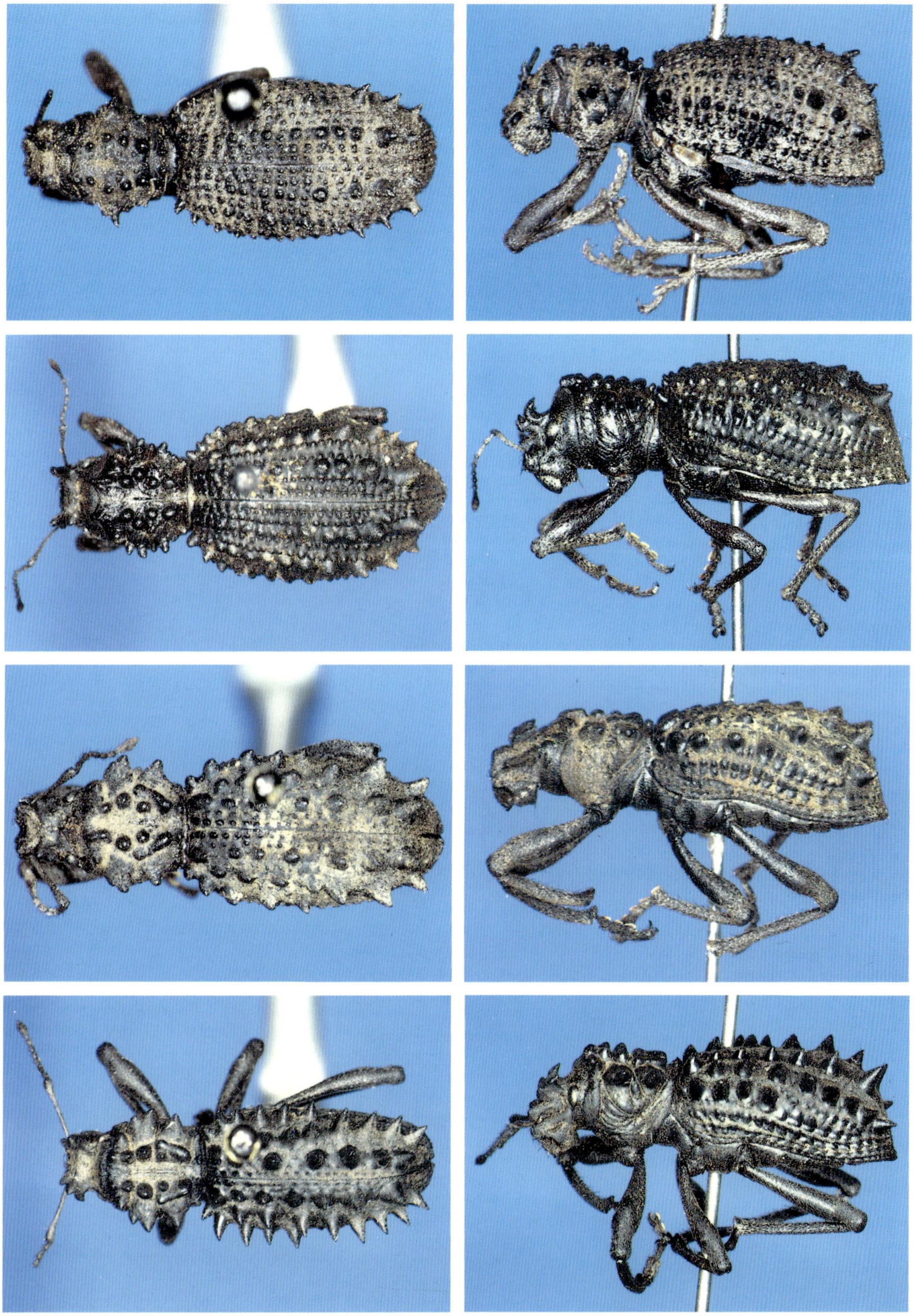

Plate 105 Amycterinae

1, 2 *Acantholophus squalidus* Macleay
Murrumbidgee, New South Wales
Holotype male; 14 mm

3, 4 *Acantholophus squamosus* Macleay
Cheltenham, Victoria
Male; 15.5 mm

5, 6 *Acantholophus subtridentatus* Ferguson
Walcha, New South Wales
Holotype male; 14 mm

7, 8 *Acantholophus suttoni* Carter
Fletcher, near Stanthorpe, Queensland
Male; 20.5 mm

1	2
3	4
5	6
7	8

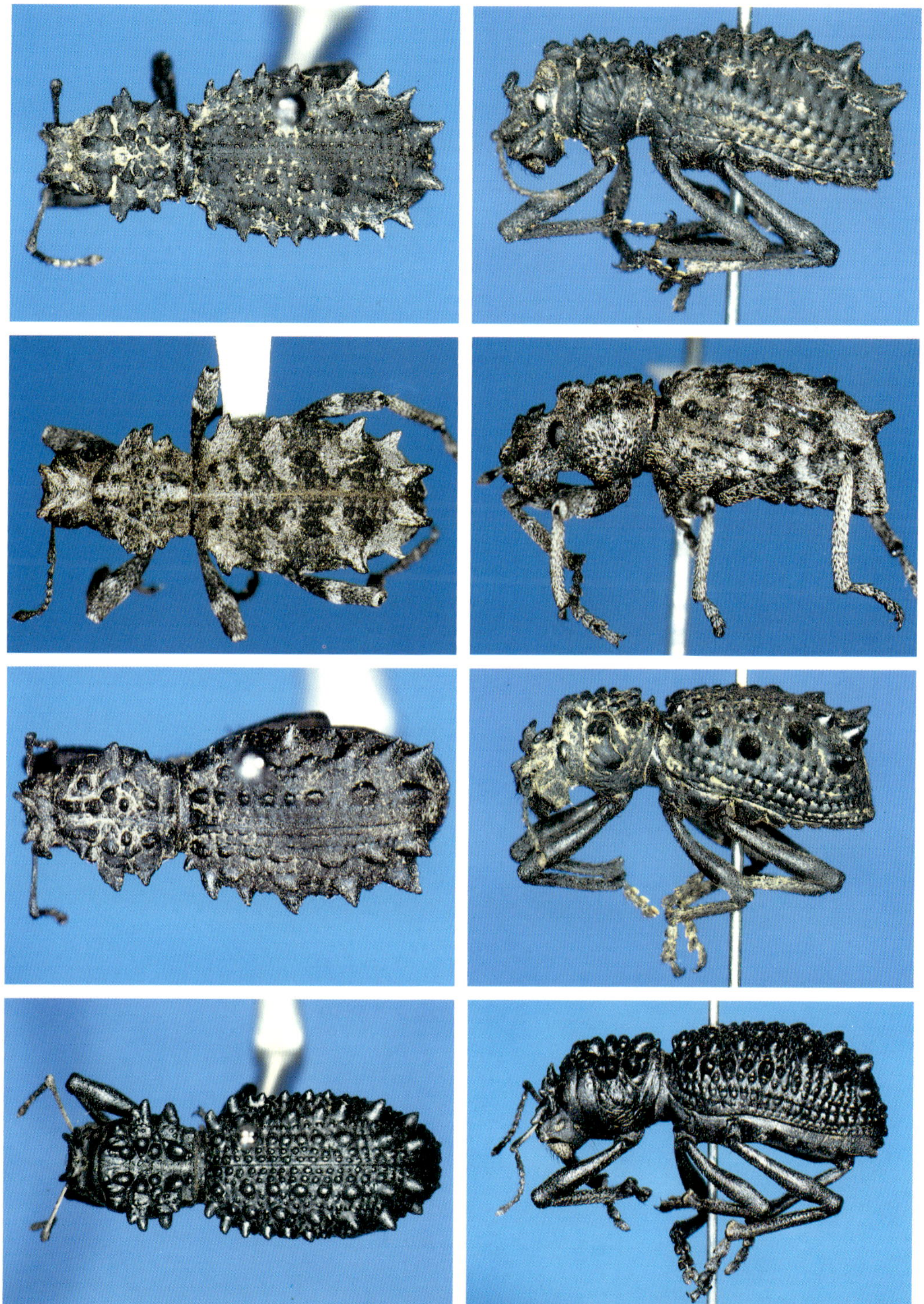

Plate 106 Amycterinae

1, 2 *Acantholophus suturalis* (Boheman)
Western Australia
Male, compared with holotype; 19 mm

3, 4 *Acantholophus tasmanicus* Lea
Tasmania
Male; 15.5 mm

5, 6 *Acantholophus tatei* "var." *murchisoni* Ferguson
Ankertel, Western Australia
Holotype male; 19 mm

7, 8 *Acantholophus terraereginae* Ferguson
Chinchilla, Queensland
Holotype male; 15 mm

1	2
3	4
5	6
7	8

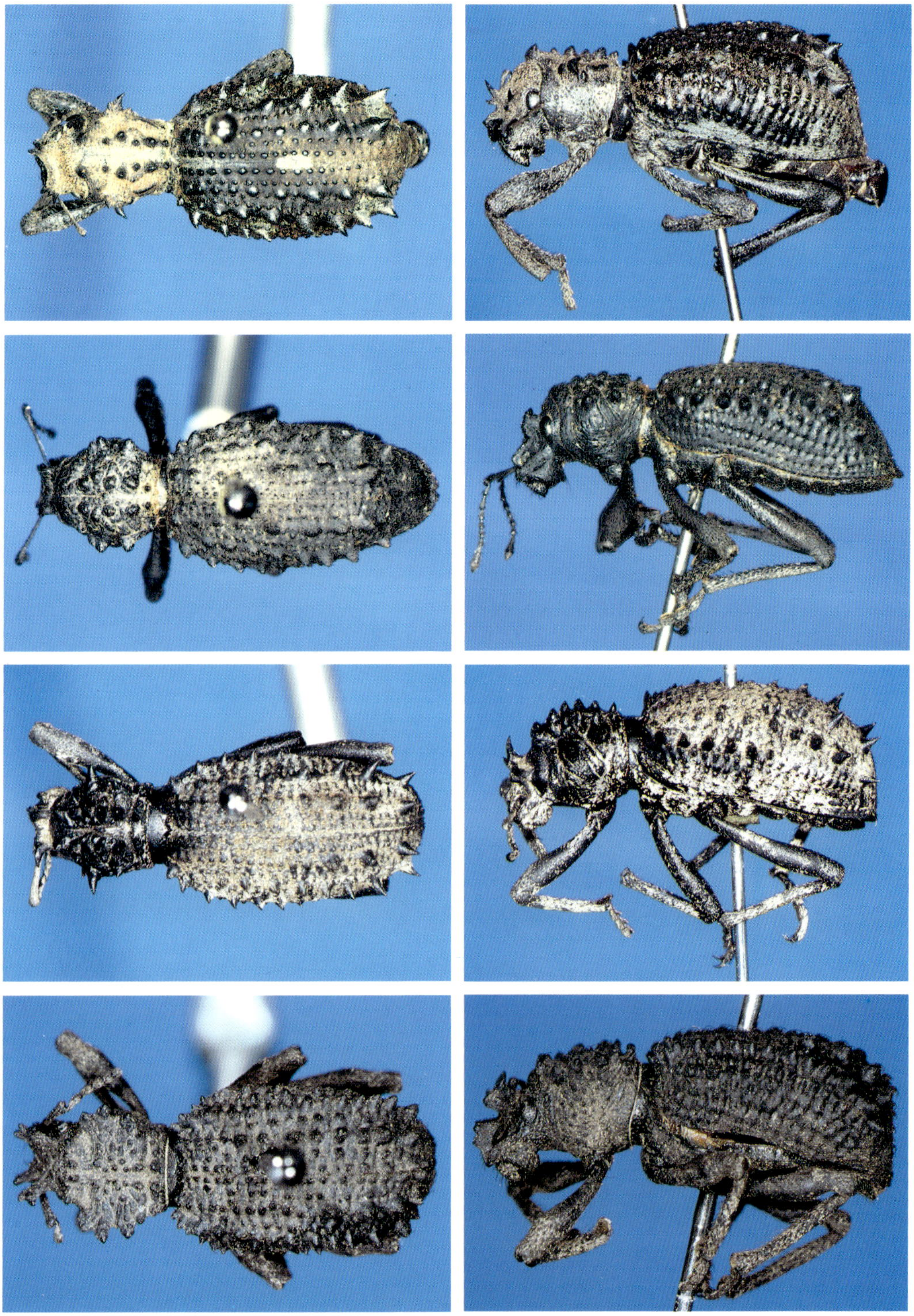

Plate 107 Amycterinae

1, 2 *Acantholophus tragocephalus* Ferguson
Onslow, Western Australia
Holotype male; 14.5 mm

3, 4 *Acantholophus transitus* Macleay
Western Australia
Syntype male; 21 mm

5, 6 *Acantholophus tribulus* Macleay
South Australia
Holotype female; 12.5 mm

7, 8 *Acantholophus tridentatus* Macleay
Victoria River,Queensland
Female; 16 mm

1	2
3	4
5	6
7	8

Plate 108 Amycterinae

1, 2 *Acherres mamillatus* Pascoe
Geraldton, Western Australia
Female; 8 mm

3, 4 *Achorostoma hystricosum* (Boheman)
King George Sound, Western Australia
Female; 12 mm
From a paralectotype of "*Talaurinus spinosus*" Macleay.

5, 6 *Aedriodes crawshawi* Ferguson
Jandakot, Western Australia
Paratype female; 10 mm

7, 8 *Aedriodes humeralis* Lea
Mount Barker, Western Australia
Male; 11 mm

1	2
3	4
5	6
7	8

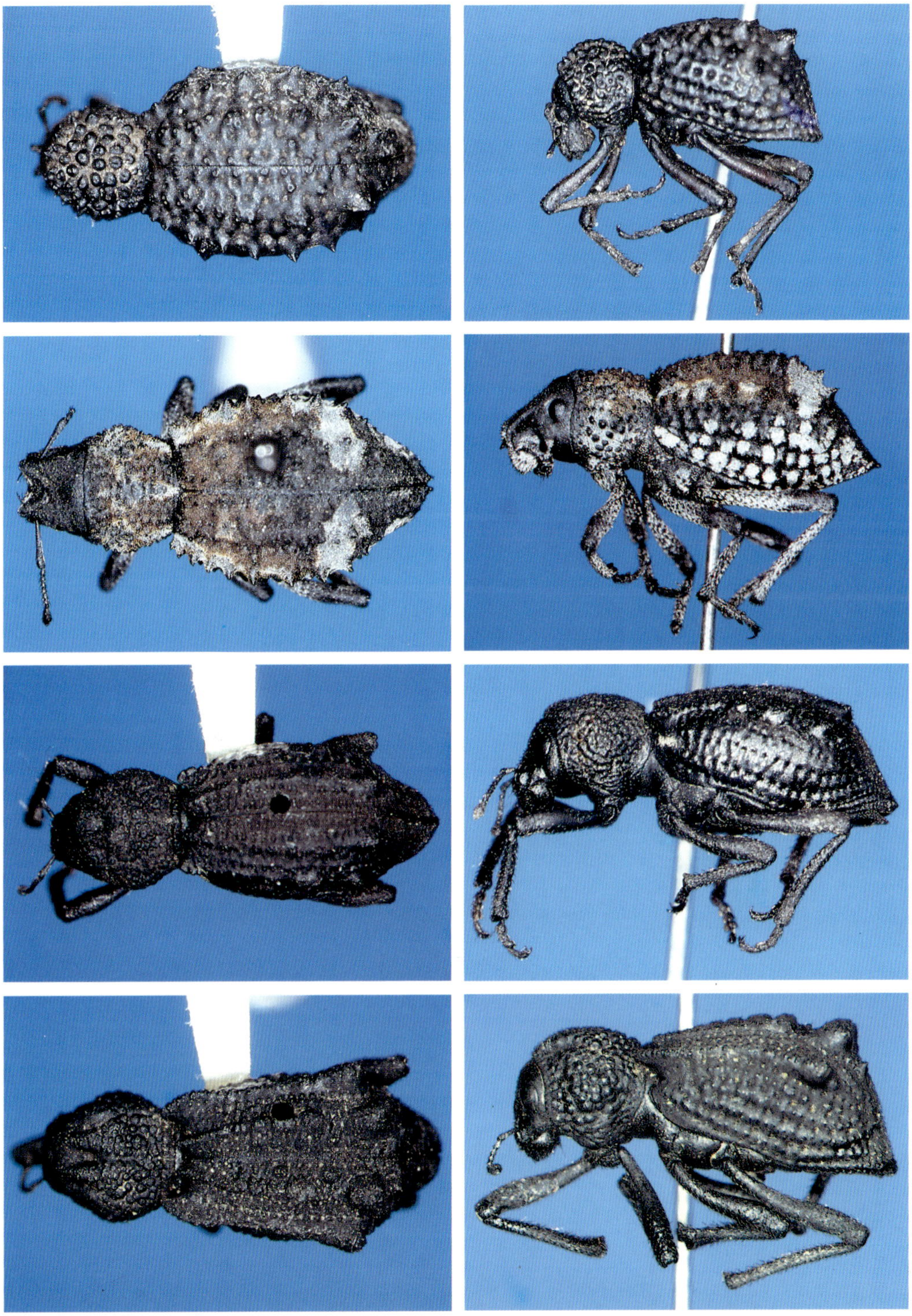

Plate 109 Amycterinae

1, 2 *Aedriodes inuus* Pascoe
Western Australia
Female, compared with holotype; 9.5 mm

3, 4 *Aedriodes mendosus* Pascoe
King George Sound, Western Australia
Female; 11.5 mm

5, 6 *Aedriodes mucronatus* Ferguson
Walpole road, Western Australia
Male; 10.5 mm

7, 8 *Aedriodes nodipennis* (Boheman)
Walpole Road, Western Australia
Female; 11 mm

1	2
3	4
5	6
7	8

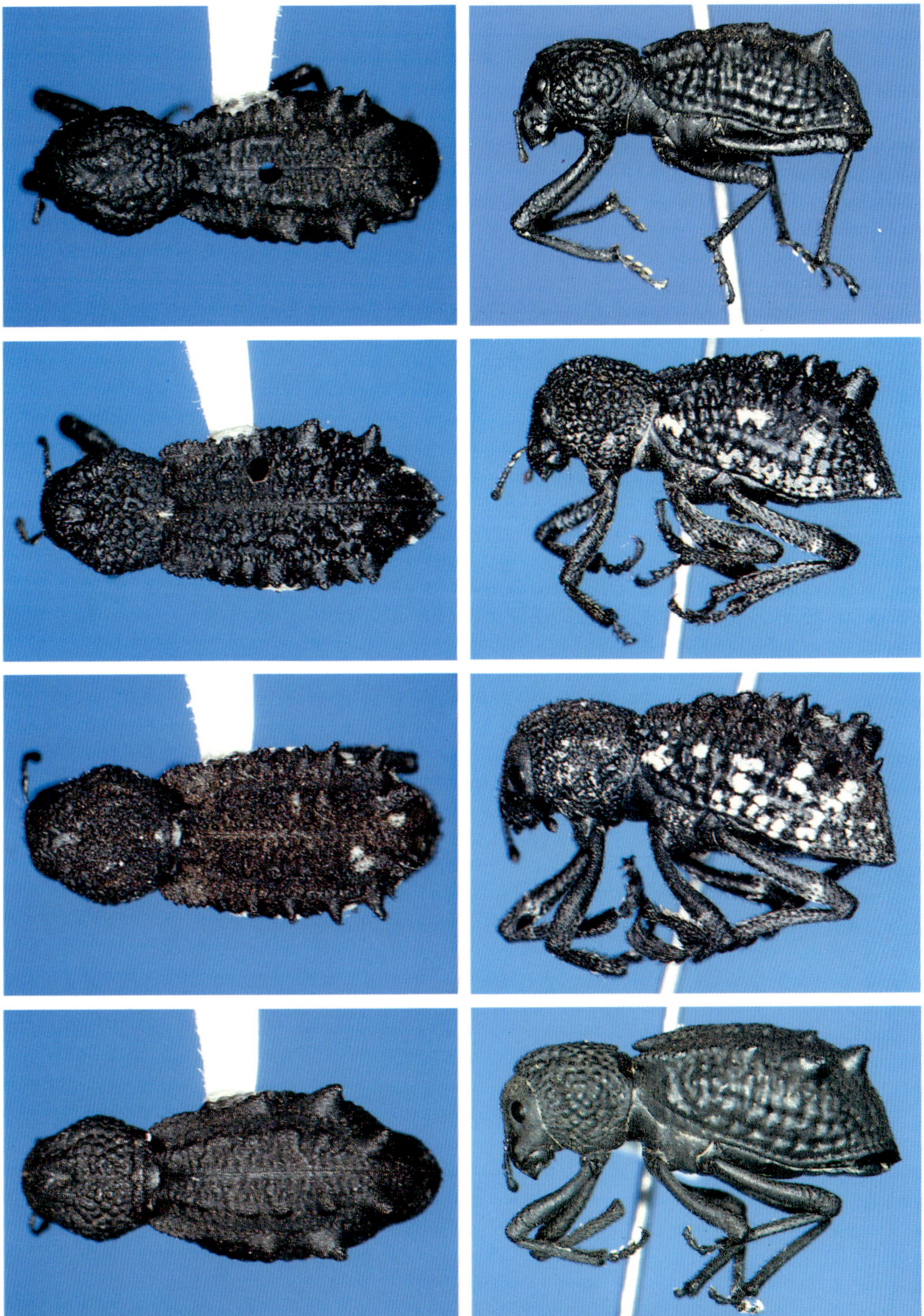

Plate 110 Amycterinae

1, 2 *Aedriodes tibialis* Ferguson
Conderdin, Western Australia
Holotype female; 11 mm

3, 4 *Alexirhea falsifica* Pascoe
Western Australia
Holotype female; 11 mm

5, 6 *Amorphorhinus arcanus* Pascoe
Western Australia
Holotype female; 11 mm

7, 8 *Amorphorhinus australis* (Germar)
Condobolin, New South Wales
Female; 10.5 mm

1	2
3	4
5	6
7	8

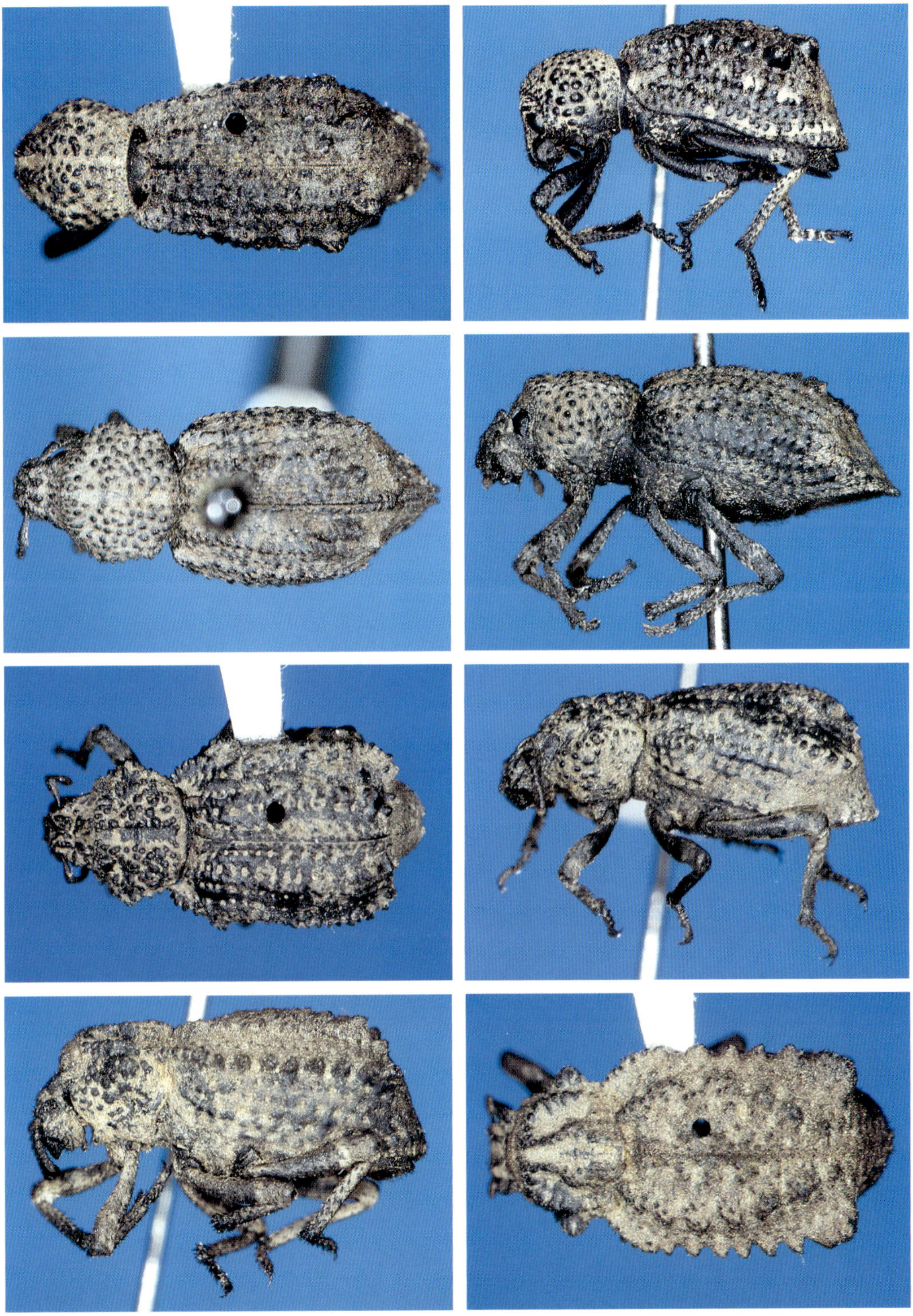

Plate 111 Amycterinae

1, 2 *Amorphorhinus rugicollis* (Lea)
Victoria
Male; 10.5 mm

3, 4 *Anascoptes muricatus* Pascoe
Mount Barker, Western Australia
Female?; 8.5 mm

5, 6 *Anascoptes obliteratus* Ferguson
Western Australia
Holotype male; 7 mm

7, 8 *Antalaurinus clavicornis* (Ferguson)
Garah, New South Wales
Male; 12.5 mm
From a paratype of *Talaurinus alaticornis* Ferguson.

1	2
3	4
5	6
7	8

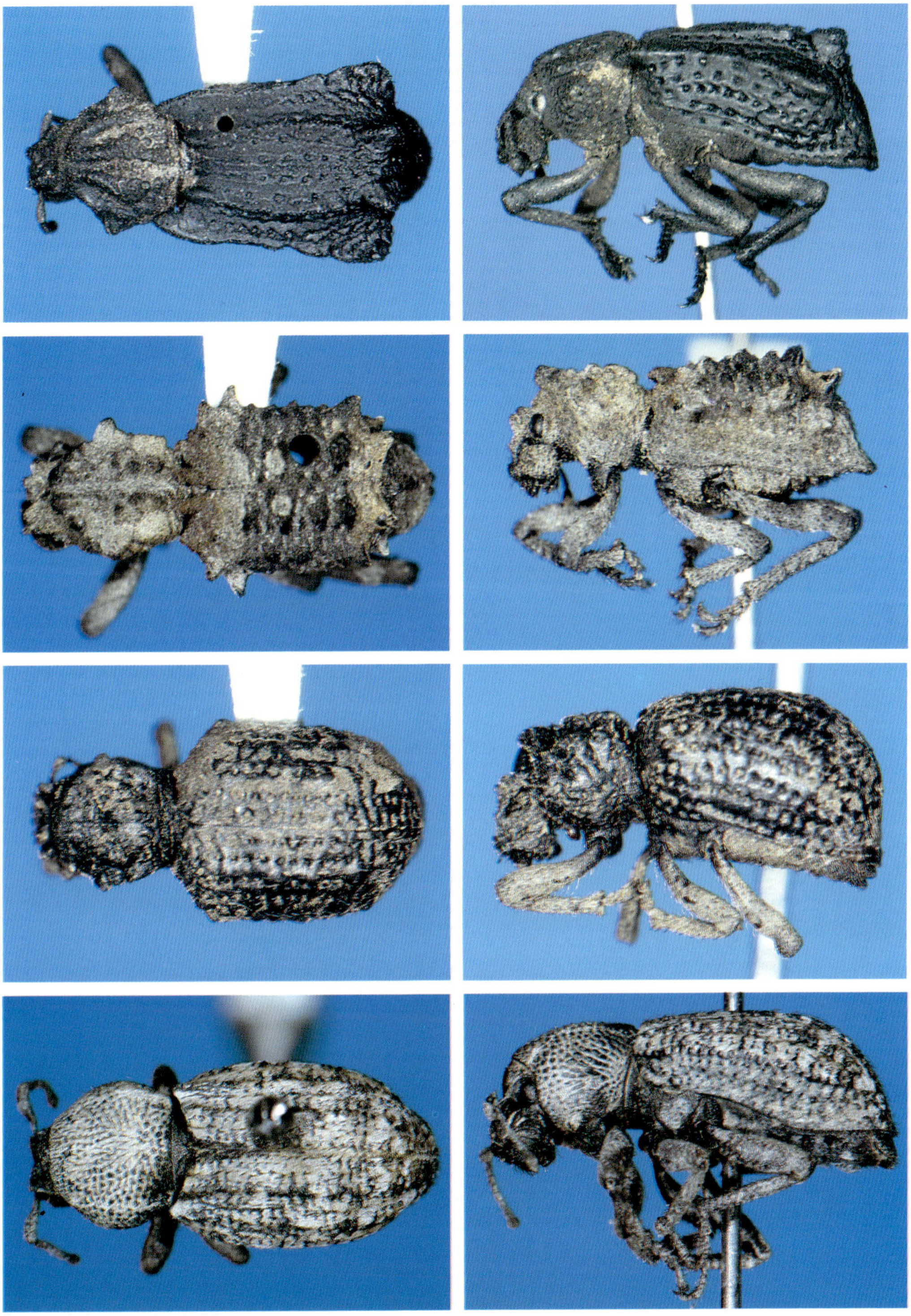

Plate 112 Amycterinae

1, 2 *Antalaurinus pulverulentus* (Macleay)
Cunnamulla, Queensland
Male; 13 mm

3, 4 *Atychoria funesta* Pascoe
Kangaroo Island, South Australia
Sex?; 10 mm
From a paratype of *Melanegis halmaturinus* Ferguson; abdomen missing.

5, 6 *Brachyrothus auritus* (Ferguson)
Cue, Western Australia
Paratype female; 7.5 mm

7, 8 *Chriotyphus tibialis* Ferguson
Onslow, Western Australia
Holotype female; 14 mm

1	2
3	4
5	6
7	8

Plate 113 Amycterinae

1, 2 *Cubicorhynchus aureomaculatus* Ferguson
Winning Station, Ashburton, Western Australia
Female; 14 mm

3, 4 *Cubicorhynchus auriculatus* Ferguson
Fletcher, south Queensland
Female; 11.5 mm

5, 6 *Cubicorhynchus aurigena* Blackburn
Kalgoorlie, Western Australia
Female?; 10.5 mm

7, 8 *Cubicorhynchus bohemani* (Boheman)
Mullewa, Western Australia
Female; 16.5 mm

1	2
3	4
5	6
7	8

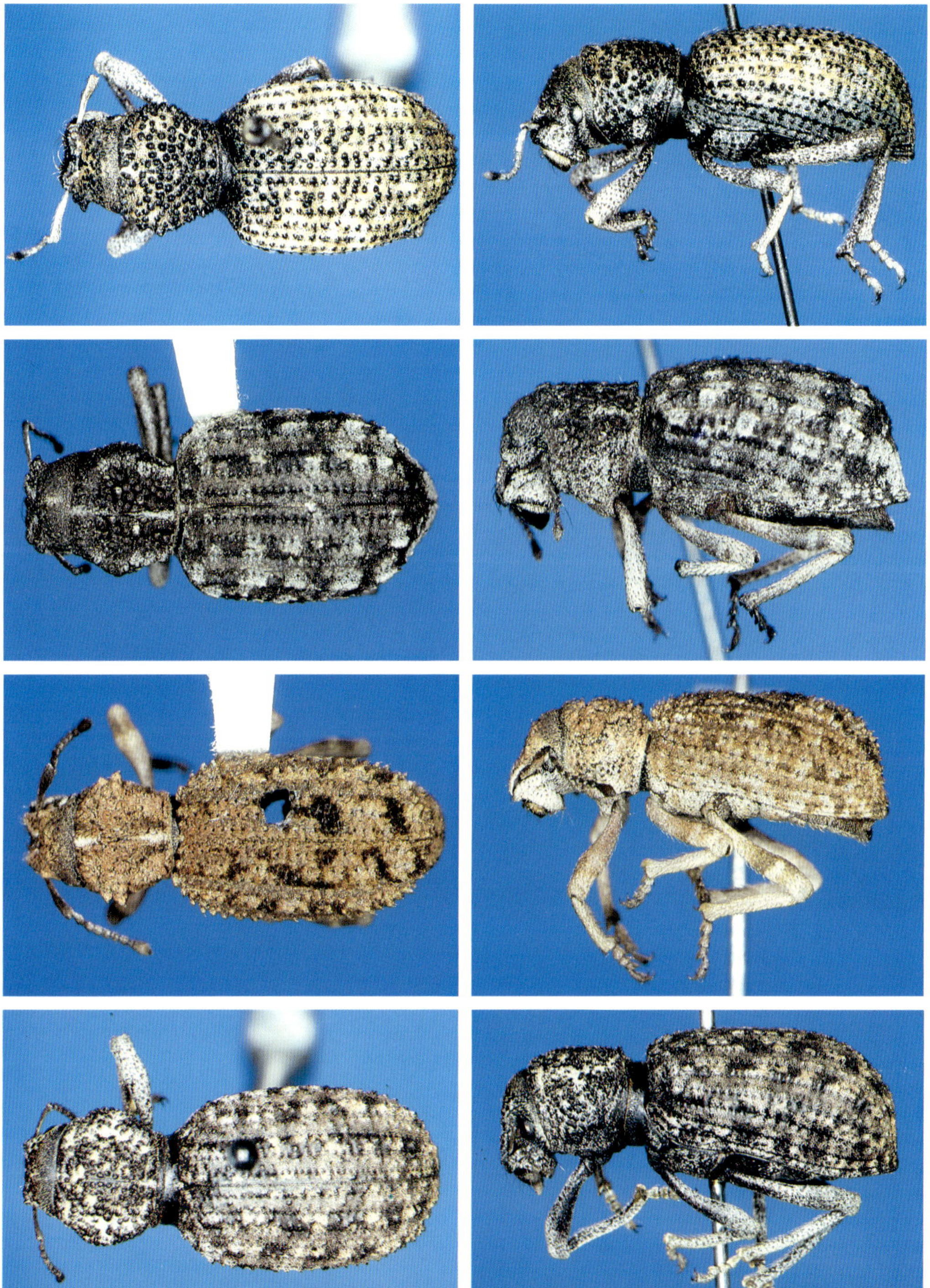

Plate 114 Amycterinae

1, 2 *Cubicorhynchus crenicollis* (Waterhouse)
Swan River, Western Australia
Female; 14 mm

3, 4 *Cubicorhynchus curvipes* Ferguson
Geraldton, Western Australia
Paratype male; 10 mm

5, 6 *Cubicorhynchus dohrni* (Waterhouse)
Perth, Western Australia
Male; 19.5 mm

7, 8 *Cubicorhynchus globicollis* Lea
New South Wales
Paratype female; 9 mm

1	2
3	4
5	6
7	8

Plate 115 Amycterinae

1, 2 *Cubicorhynchus illidgei* Ferguson
Tamborine, Queensland
Holotype male; 8.5 mm

3, 4 *Cubicorhynchus maculatus* Macleay
Murrumbidgee, New South Wales
Paratype female; 14 mm

5, 6 *Cubicorhynchus minor* Ferguson
Ankertell, Western Australia
Holotype male; 8 mm

7, 8 *Cubicorhynchus modestus* Sloane
Kellerberrin, Western Australia
Female; 12.5 mm

1	2
3	4
5	6
7	8

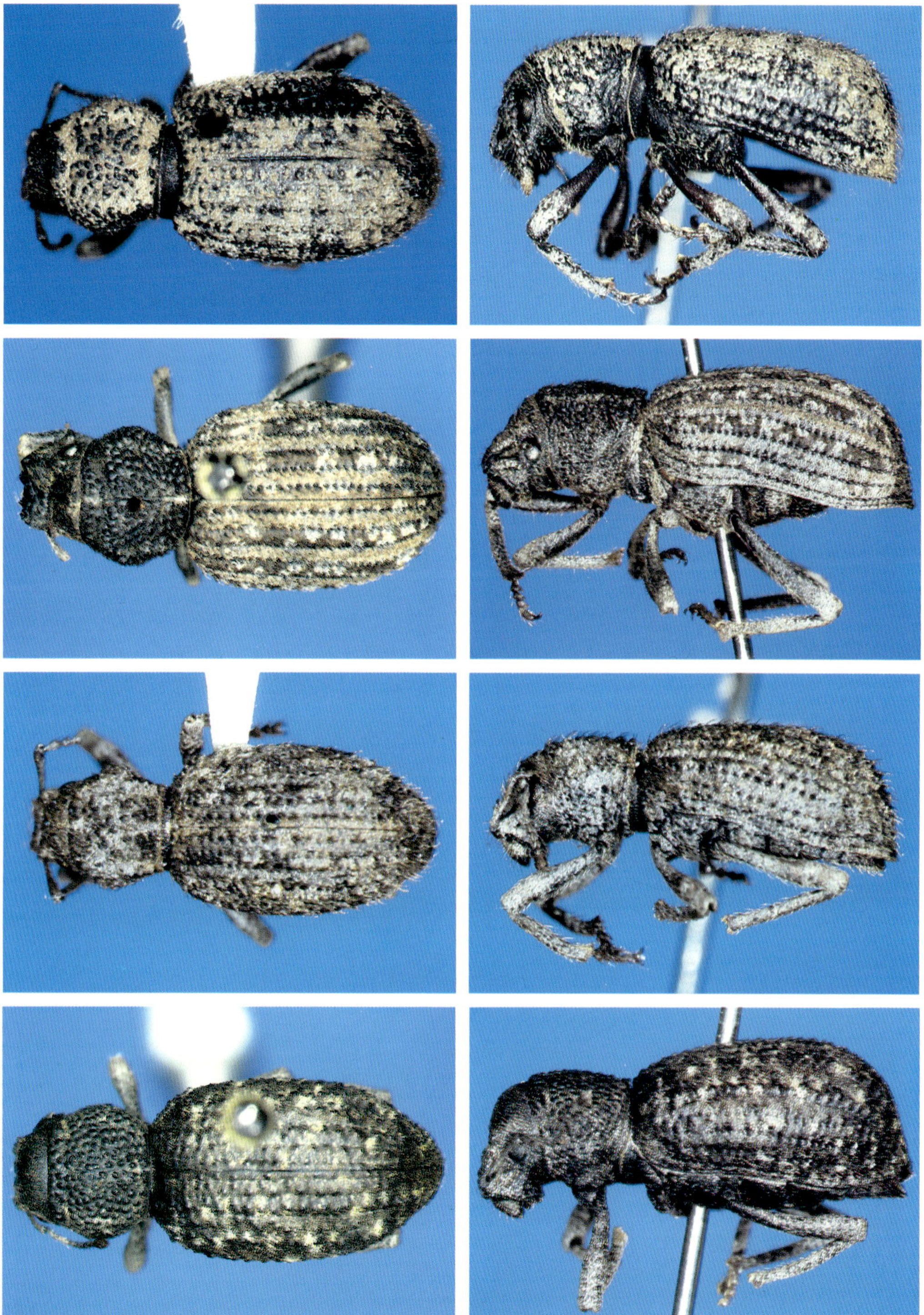

Plate 116 Amycterinae

1, 2 *Cubicorhynchus mussoni* Blackburn
Glen Innes, New South Wales
Female; 12 mm

3, 4 *Cubicorhynchus occultus* Sloane
Fraser Range, Western Australia
Female; 15.5 mm

5, 6 *Cubicorhynchus piceosetosus* Macleay
New South Wales
Syntype female; 13.5 mm

7,8 *Cubicorhynchus rectipes* Ferguson
Cue, Western Australia
Paratype female; 14.5 mm

1	2
3	4
5	6
7	8

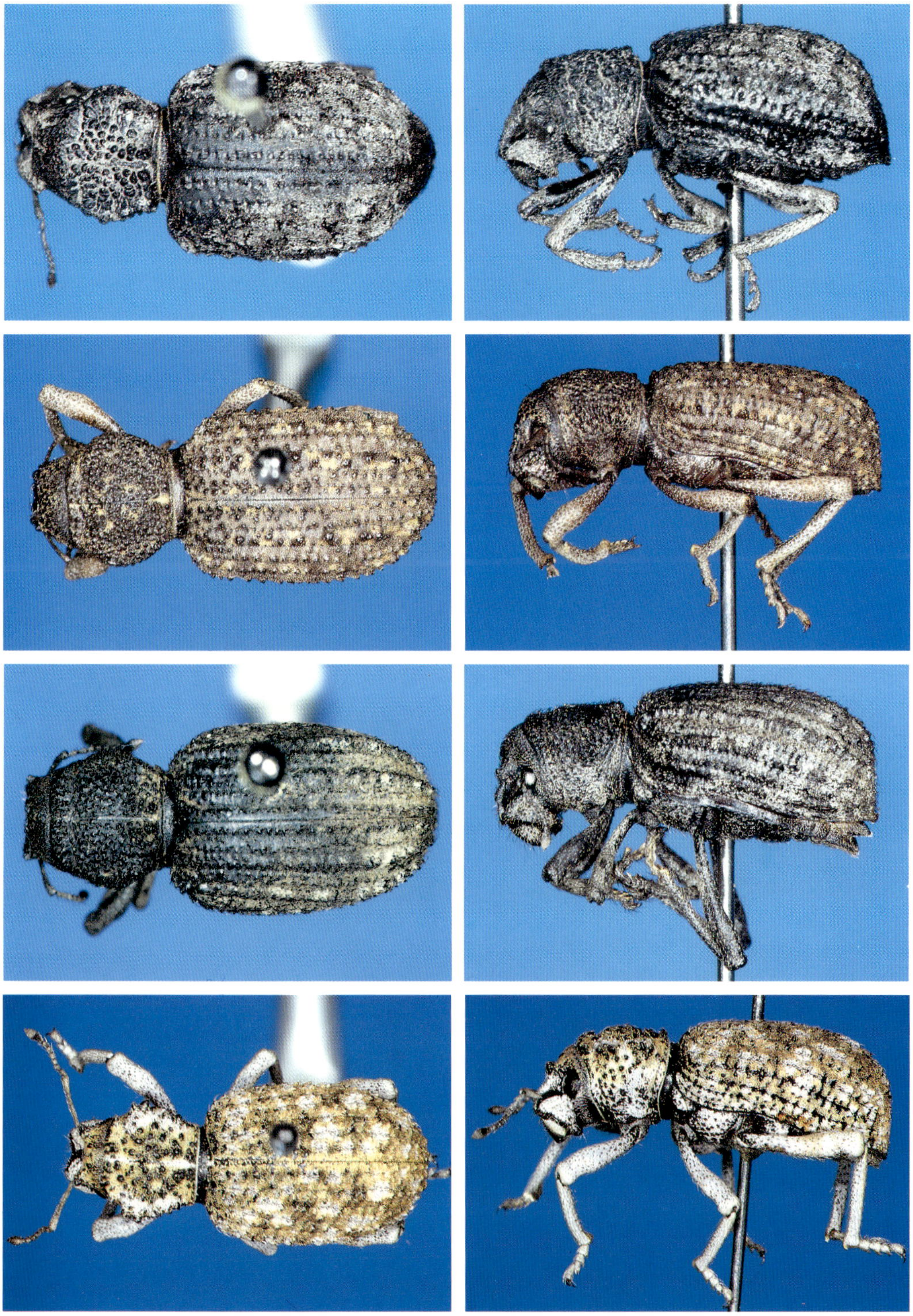

Plate 117 Amycterinae

1, 2 *Cubicorhynchus serratipes* Ferguson
Kookynie, Western Australia
Holotype male; 13 mm

3, 4 *Cubicorhynchus setosus* Ferguson
Northwest Western Australia
Holotype male; 11 mm

5, 6 *Cubicorhynchus sordidus* Ferguson
Jindabyne, New South Wales
Holotype male; 11.5 mm
Abraded and of misleading appearance.

7, 8 *Cubicorhynchus sparsus* Ferguson
Darling River, Western Australia
Allotype female; 12.5 mm

1	2
3	4
5	6
7	8

Plate 118 Amycterinae

1, 2 *Cubicorhynchus spinicollis* Macleay
Magadup, Western Australia
Syntype male; 10 mm

3, 4 *Cubicorhynchus sterilis* Pascoe
Mount Lofty, South Australia
Female; 11.5 mm

5, 6 *Cubicorhynchus strigicollis* Ferguson
Nathalin, Victoria
Holotype male; 12 mm

7, 8 *Cubicorhynchus substrigatus* Ferguson
Murray River, South Australia
Paratype female; 11 mm

1	2
3	4
5	6
7	8

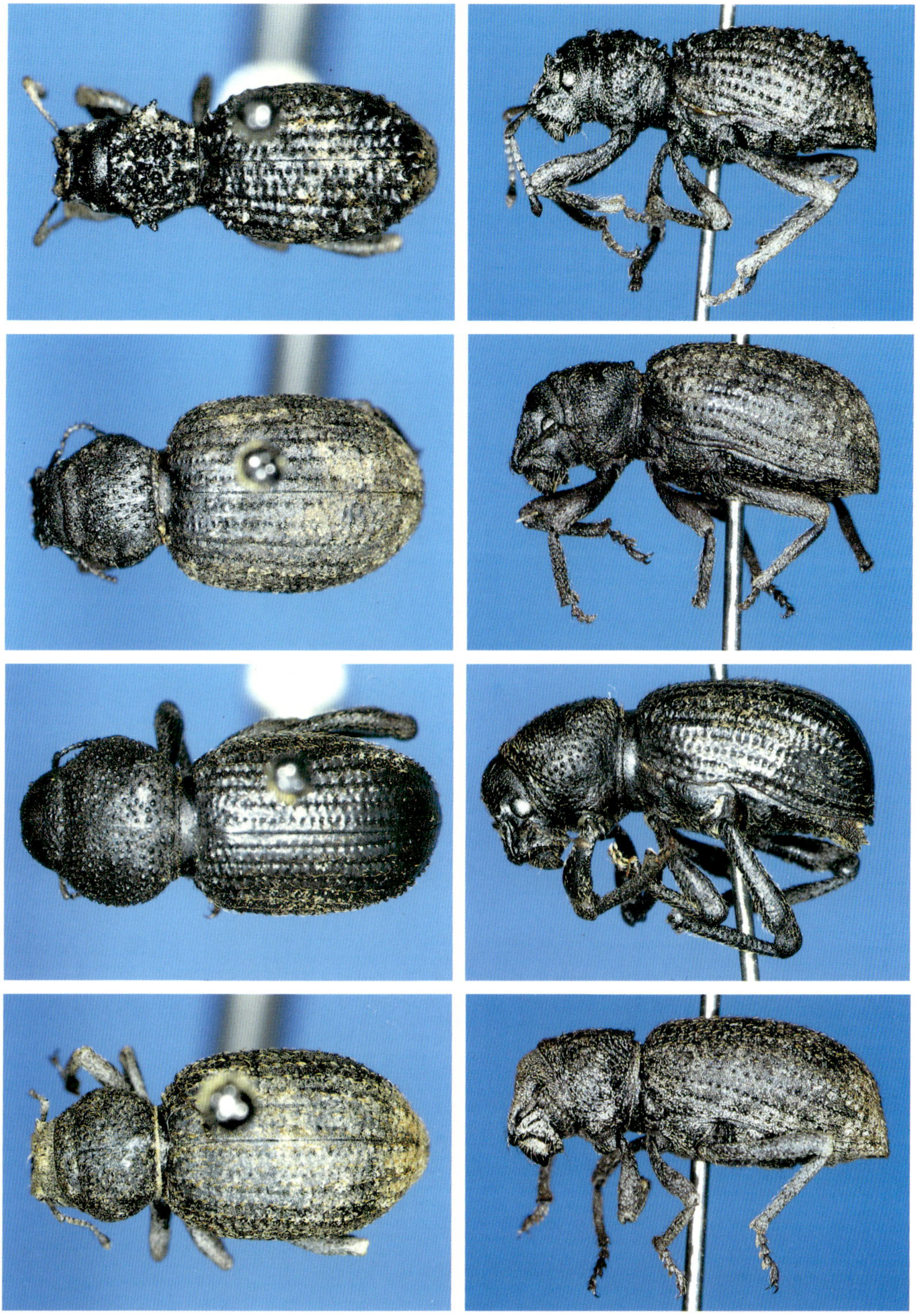

Plate 119 Amycterinae

1, 2 *Cubicorhynchus taurus* Blackburn
Lucindale, South Australia
Female; 11 mm

3, 4 *Cubicorhynchus valgus* Lea
Day Dawn, Western Australia
Holotype female; 1 mm

5, 6 *Cucullothorax horridus* Ferguson
Wilga, Western Australia
Female; 12 mm
The sulcus on the pronotum is an abnormality.

7, 8 *Dialeptopus ferreus* Pascoe
Port Lincoln, South Australia
Female; 14 mm

1	2
3	4
5	6
7	8

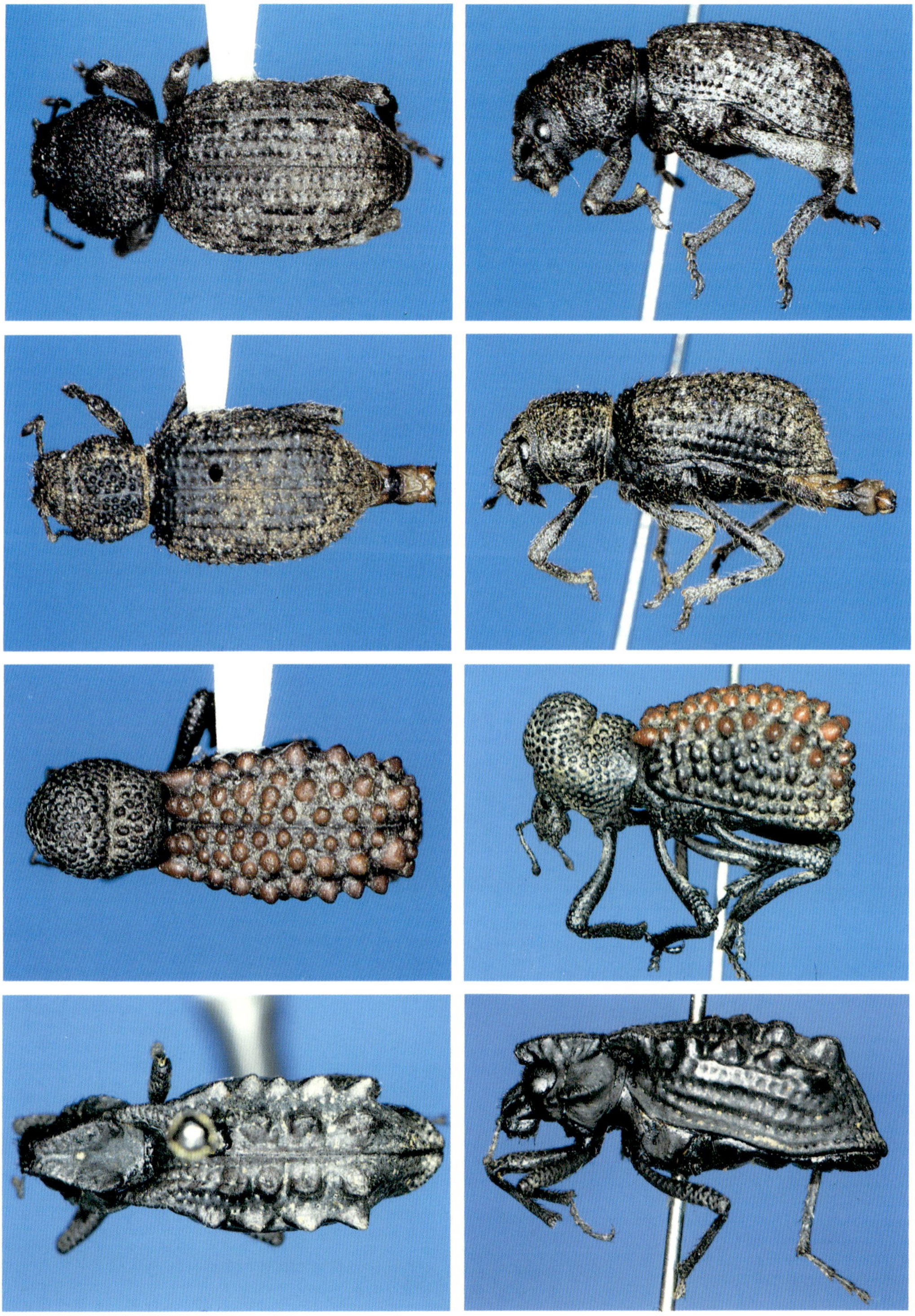

Plate 120 Amycterinae

1, 2 *Dialeptopus monachus* Pascoe
South Australia
Female; 17.5 mm

3, 4 *Dialeptopus planitarsis* Pascoe
Geraldton, Western Australia
Female; 17.5 mm

5, 6 *Dialeptopus serricollis* Pascoe
No locality data.
Holotype female; 13.5 mm

7, 8 *Dicherotropis cavirostris* (Lea)
Western Australia
Holotype male; 12.5 mm

1	2
3	4
5	6
7	8

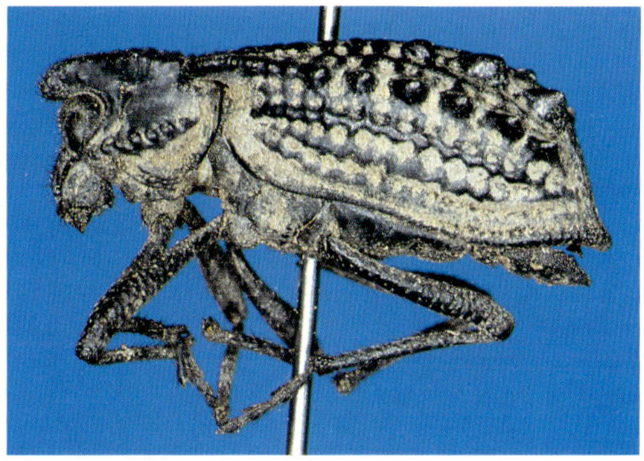

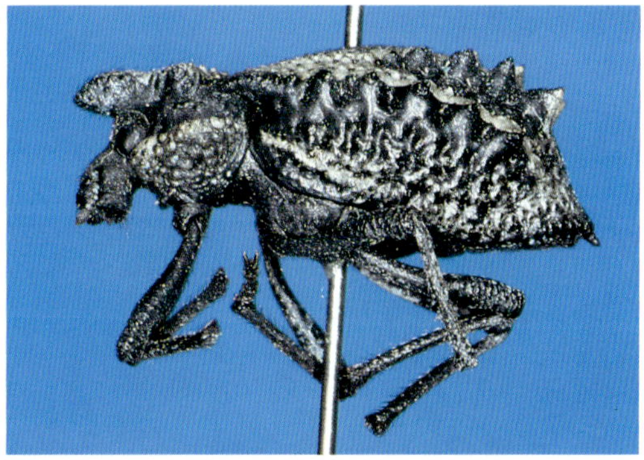

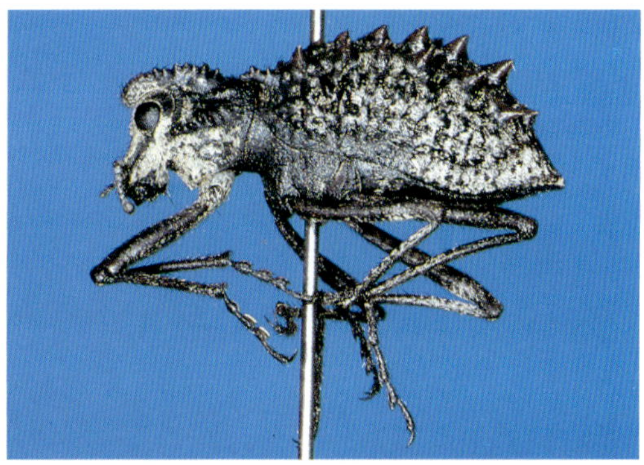

Plate 121 Amycterinae

1, 2 *Ennothus fallax* Pascoe
Western Australia
Female; 6 mm

3, 4 *Euomella retusa* Pascoe
Kellerberin, Western Australia
Male; 15.5 mm

5, 6 *Euomus scorpio* (Boisduval)
Dwellingup, Western Australia
Male; 15.5 mm

7, 8 *Euomus stephensii* (Gyllenhal)
Guangara, Western Australia
Female; 22 mm

1	2
3	4
5	6
7	8

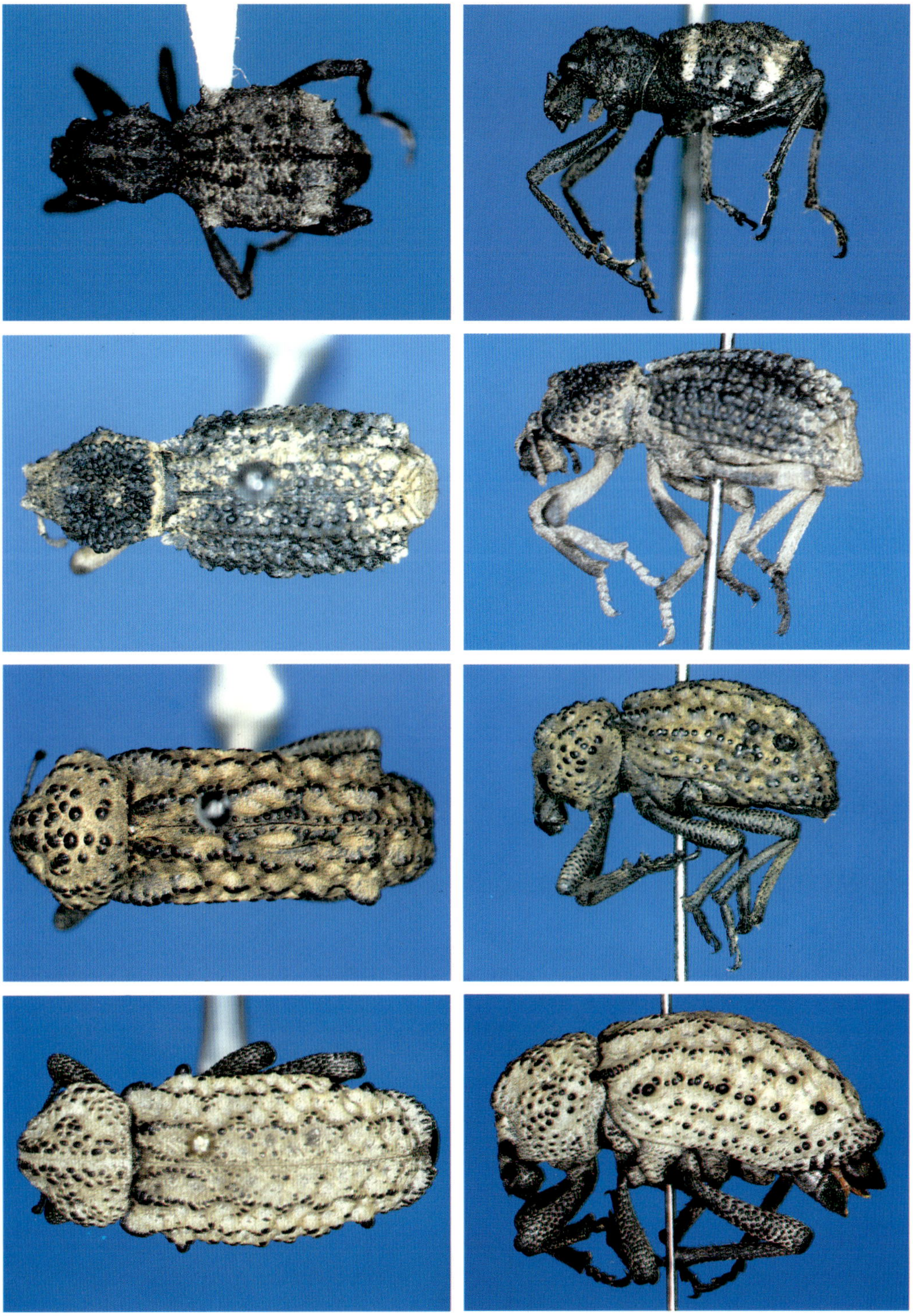

Plate 122 Amycterinae

1, 2 *Gagatophorus boisduvali* (Boisduval)
King George Sound, Western Australia
Male; 21.5 mm

3, 4 *Gagatophorus insignis* (Ferguson)
Western Australia
Holotype male; 27 mm

5, 6 *Gagatophorus leichardtii* (Macleay)
Western Australia
Male; 24 mm

7, 8 *Gagatophorus* obsoletus (Ferguson)
Yilgarn, Western Australia
Holotype male; 29 mm

1	2
3	4
5	6
7	8

Plate 123 Amycterinae

1, 2 *Gagatophorus schoenherri* (Hope)
Western Australia
Female; 30 mm

3, 4 *Gagatophorus tibialis* (Ferguson)
Shark Bay, Western Australia
Holotype male; 24 mm

5, 6 *Hyborrhinus coenosus* (Boheman)
King George Sound, Western Australia
Male; 12 mm

7, 8 *Hypotomops angustior* (Ferguson)
Northwest Western Australia
Holotype male; 13 mm

1	2
3	4
5	6
7	8

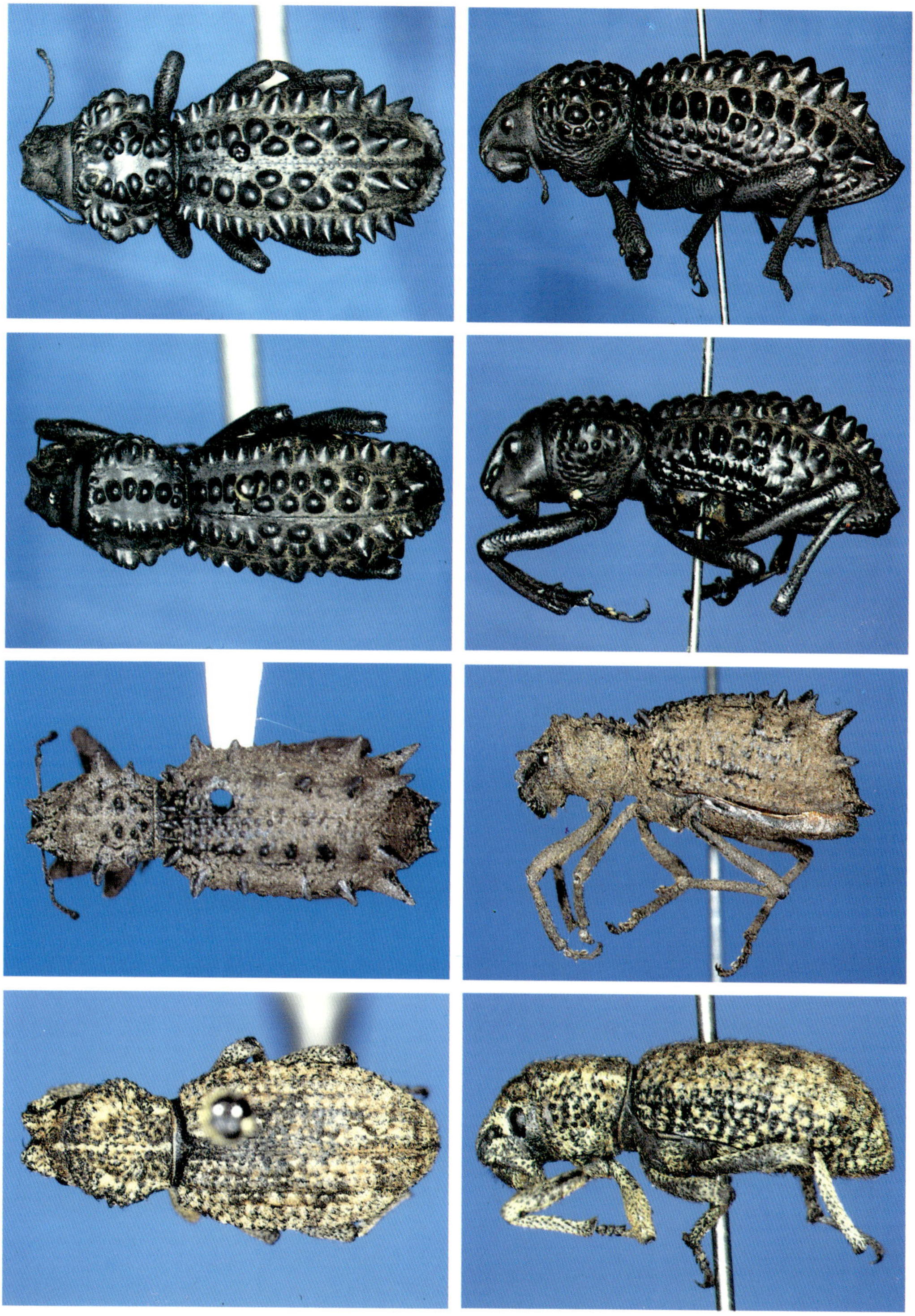

Plate 124 Amycterinae

1, 2 *Hypotomops leai* (Ferguson)
Daly River, Northern Territory
Male; 13 mm

3, 4 *Lataurinus macrocephalus* (Ferguson)
Western Australia
Male, compared with holotype; 22.5 mm

5, 6 *Melanegus stygius* Paascoe
King George Sound, Western Australia
Male; 10 mm

7, 8 *Molochtus hercules* Ferguson
Kookynie, Western Australia
Holotype male; 25.5 mm

1	2
3	4
5	6
7	8

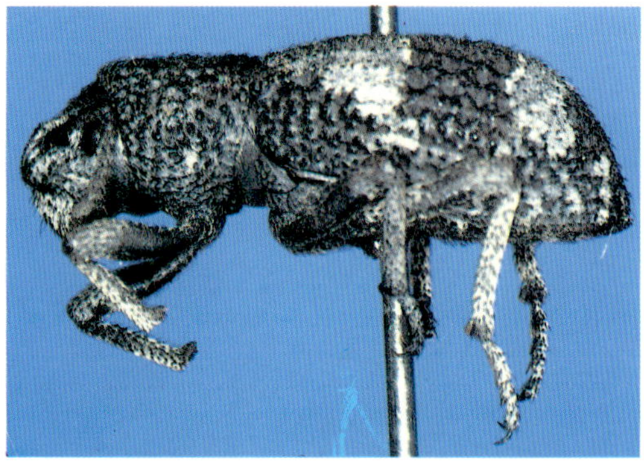

Plate 125 Amycterinae

1, 2 *Molochtus rotundicollis* Ferguson
Onslow, Western Australia
Holotype male; 16.5 mm

3, 4 *Molochtus tibialis* Sloane
Fraser Range, Western Australia
Holotype female; 25.5 mm

5, 6 *Myotrotus obtusus* Pascoe
Bilvela, Queensland
Male; 12 mm

7, 8 *Mythites arboricola* Ferguson
Tamborine Mountains, Queensland
Holotype female; 17.5 mm

1	2
3	4
5	6
7	8

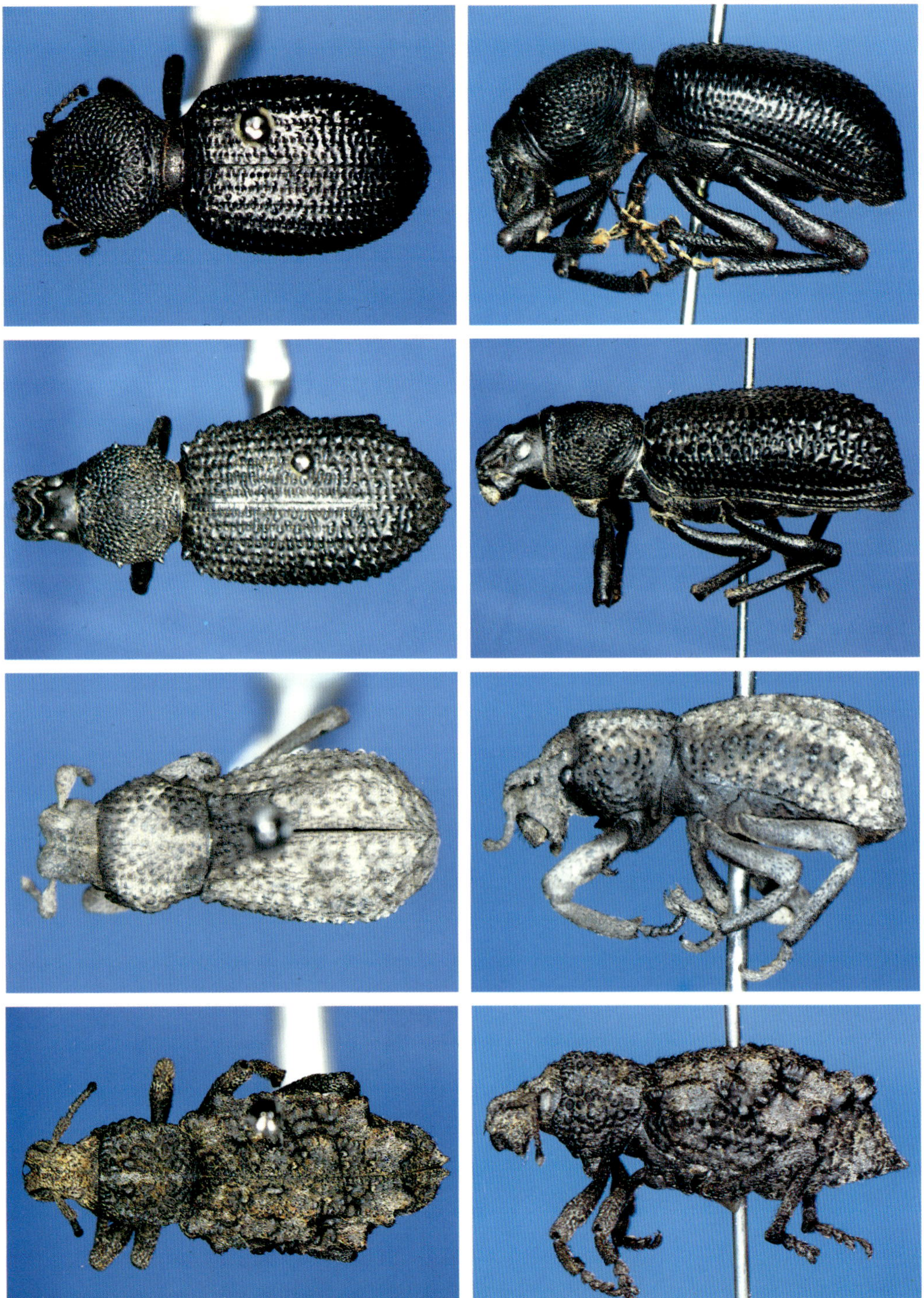

Plate 126 Amycterinae

1, 2 *Mythites asperatus* Pascoe
Randwick, Sydney, New South Wales
Male; 14 mm

3, 4 *Mythites nodosus* Ferguson
Yalgoo, Western Australia
Holotype male; 20 mm

5, 6 *Mythites coxi* Ferguson
Eccleston, New South Wales
Holotype female; 17 mm

7, 8 *Mythites foveipennis* Lea
Mittagong, New South Wales
Female; 20 mm

1	2
3	4
5	6
7	8

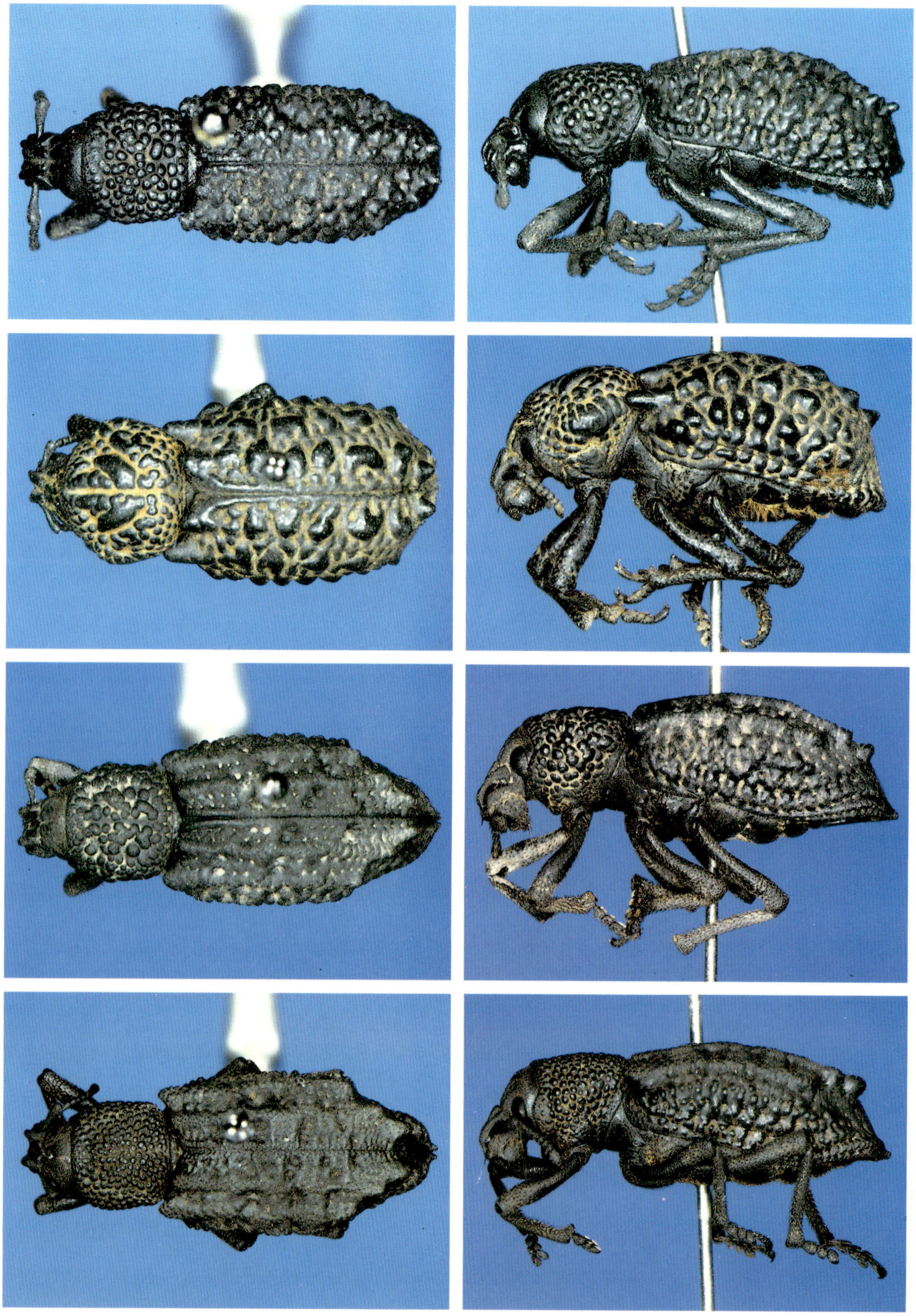

Plate 127 Amycterinae

1, 2 *Mythites frater* Lea
Coonabarabran, New South Wales
Paratype male; 15.5 mm

3, 4 *Mythites granulatus* Lea
Victoria
Holotype male; 15.5 mm

5, 6 *Mythites montanus* Ferguson
Blue Mountains, New South Wales
Holotype male; 14 mm

7, 8 *Mythites perfossus* (Germar)
South Australia
Male; 13 mm

1	2
3	4
5	6
7	8

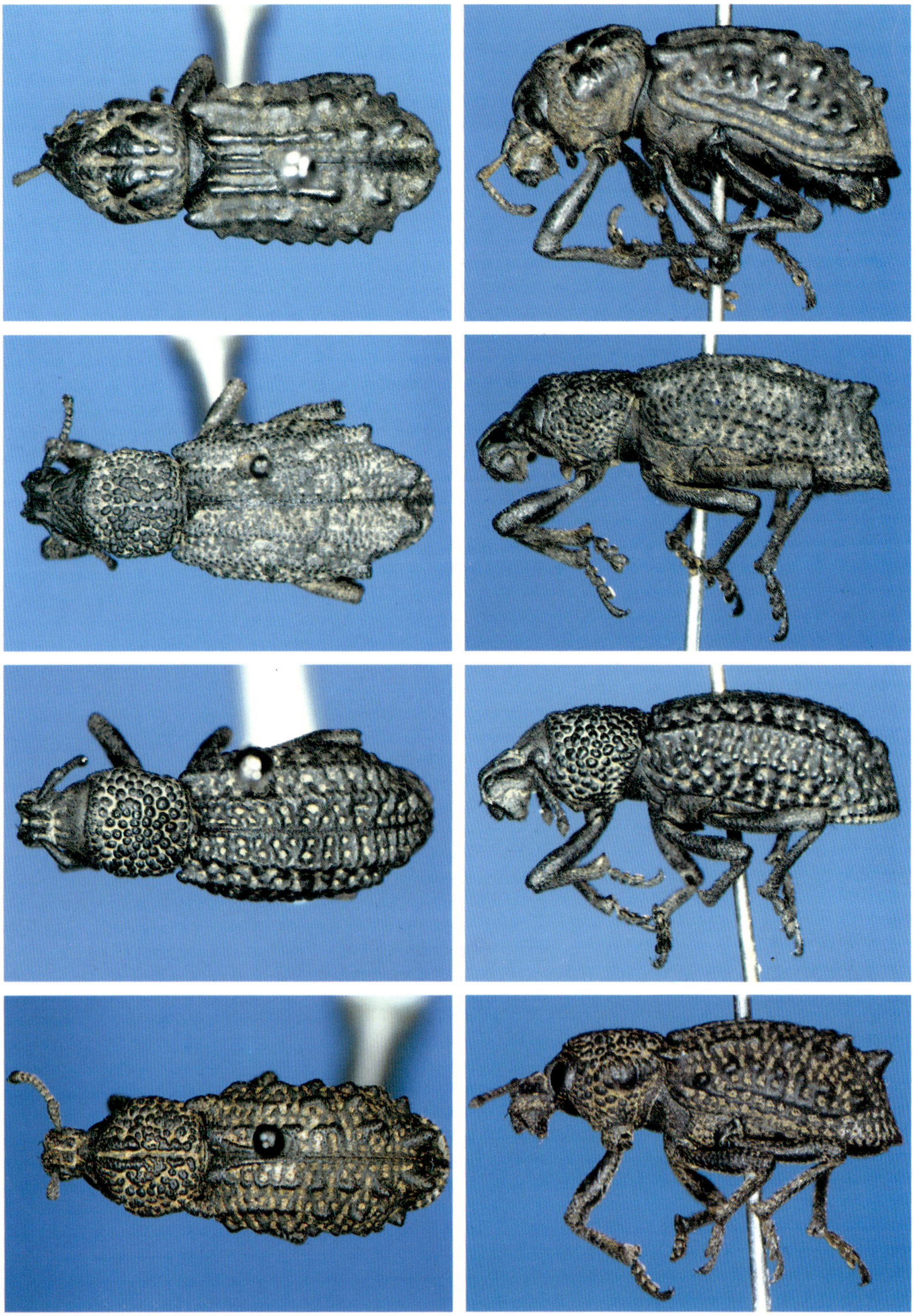

Plate 128 Amycterinae

1, 2 *Mythites poropteroides* Lea
Victoria
Holotype female; 19.5 mm

3, 4 *Mythites rugosus* Ferguson
Portland, New South Wales
Paratype male; 16 mm

5, 6 *Mythites sulcicollis* (Germar)
Yorke Peninsula, South Australia
Female; 20.5 mm

7, 8 *Neohyborrhynchus rugosus* (Macleay)
Walpole National Park, Western Australia
Male; 12.5 mm

1	2
3	4
5	6
7	8

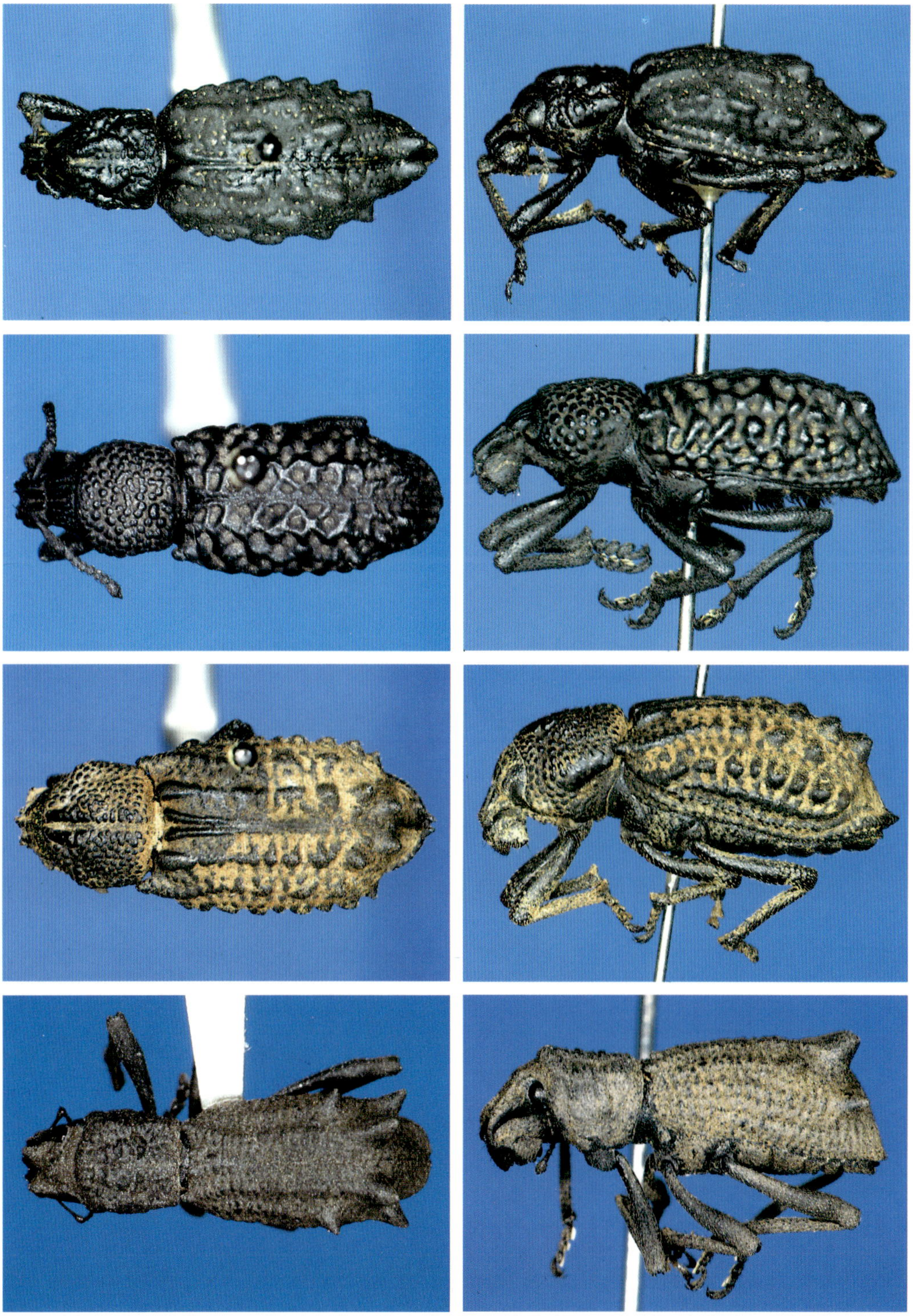

Plate 129 Amycterinae

1, 2 *Notonophes auriger* Ferguson
Northwest Western Australia
Paralectotype female; 14 mm

3, 4 *Notonophes dilaticeps* (Blackburn)
Central Australia
Holotype male; 13 mm

5, 6 *Notonophes gascoynensis* Baker
40 km north of Carnarvon, Western Australia
Paratype male; 11.5 mm

7, 8 *Notonophes hirsutinus* Baker
Winning Station, Western Australia
Paratype female; 13 mm

1	2
3	4
5	6
7	8

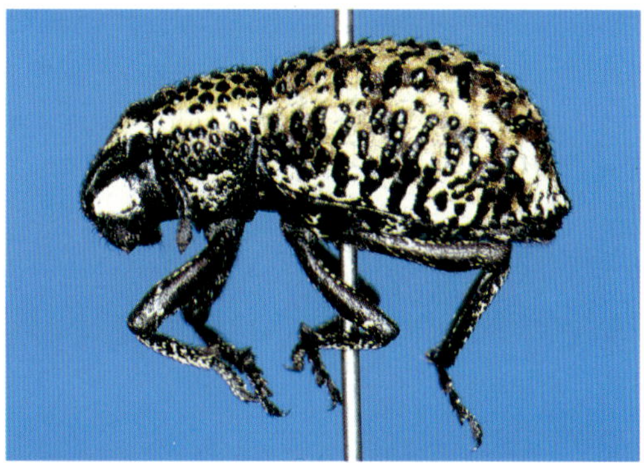

Plate 130 Amycterinae

1, 2 *Notonophes taurus* Ferguson
Northwest Western Australia
Paralectotype female; 15 mm

3, 4 *Oditesus buceros* Pascoe
King George Sound, Western Australia
Male; 10 mm

5, 6 *Ophthalamycterus laticeps* (Macleay)
King George Sound, Western Australia
Male; 19.5 mm

7, 8 *Parahyborrhynchus convexiusculus* (Macleay)
King George Sound, Western Australia
Female; 10 mm

1	2
3	4
5	6
7	8

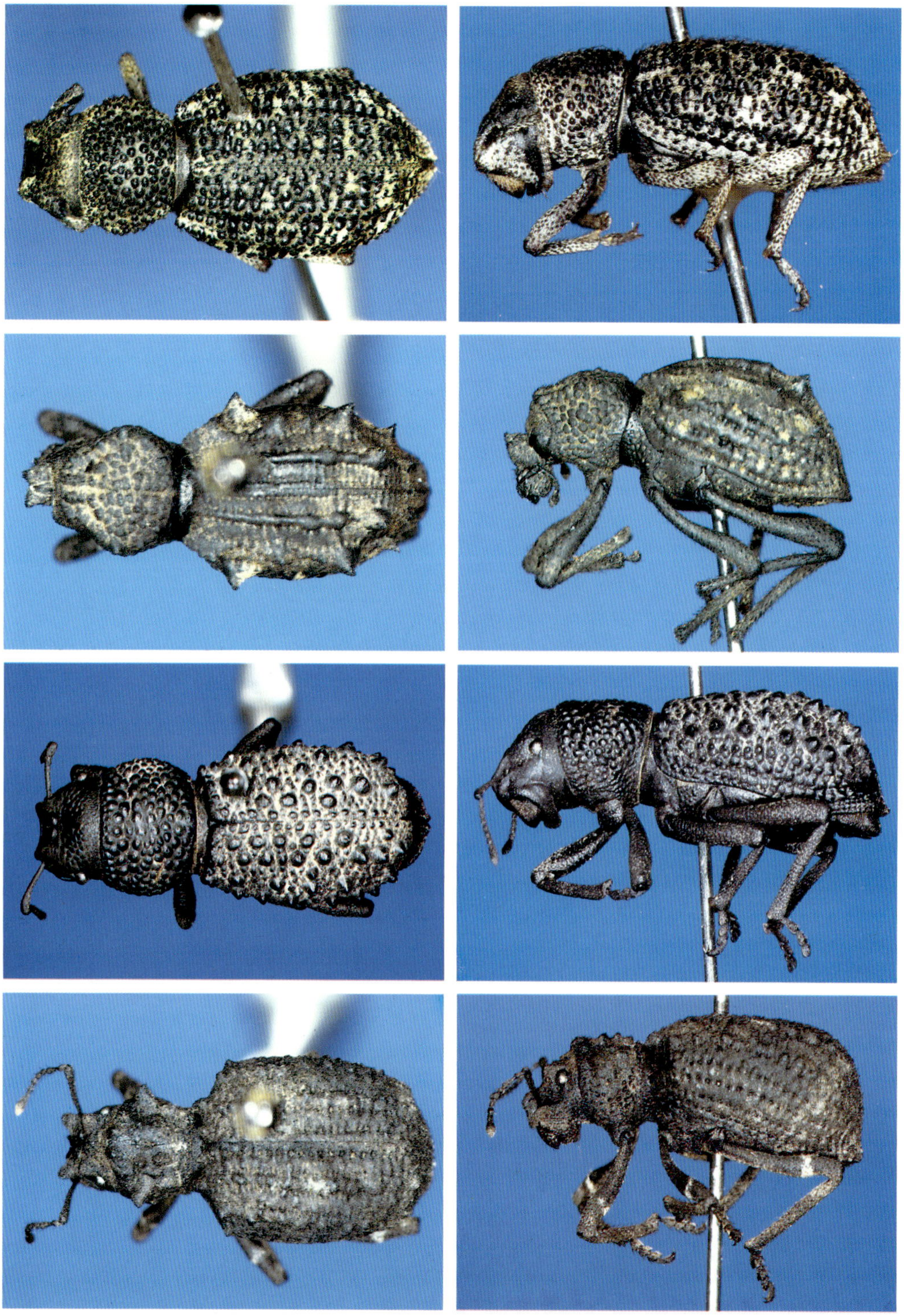

Plate 131 Amycterinae

1, 2 *Phalidura abnormis* (Macleay)
Manar Creek, New South Wales
Male; 20 mm

3, 4 *Phalidura amplicollis* (Ferguson)
Dalveen, Queensland
Holotype male; 23.5 mm

5, 6 *Phalidura amplipennis* (Ferguson)
Darling Downs, Queensland
Holotype male; 24.5 mm

For *Phalidura anthracoides* (Ferguson), see Plate 151:7, 8

7, 8 *Phalidura brevicauda* (Ferguson)
Dalby, Queensland
Holotype male; 20 mm

1	2
3	4
5	6
7	8

Plate 132 Amycterinae

1, 2 *Phalidura breviformis* (Ferguson)
Glen Innes, New South Wales
Holotype male; 19 mm

3, 4 *Phalidura carteri* (Ferguson)
Mount Kosciusko, New South Wales
Holotype male; 25.5 mm

5, 6 *Phalidura caudata* (Macleay)
Townsville, Queensland
Male; 28 mm

7, 8 *Phalidura cultrata* (Ferguson)
Victoria
Holotype male; 23.5 mm

1	2
3	4
5	6
7	8

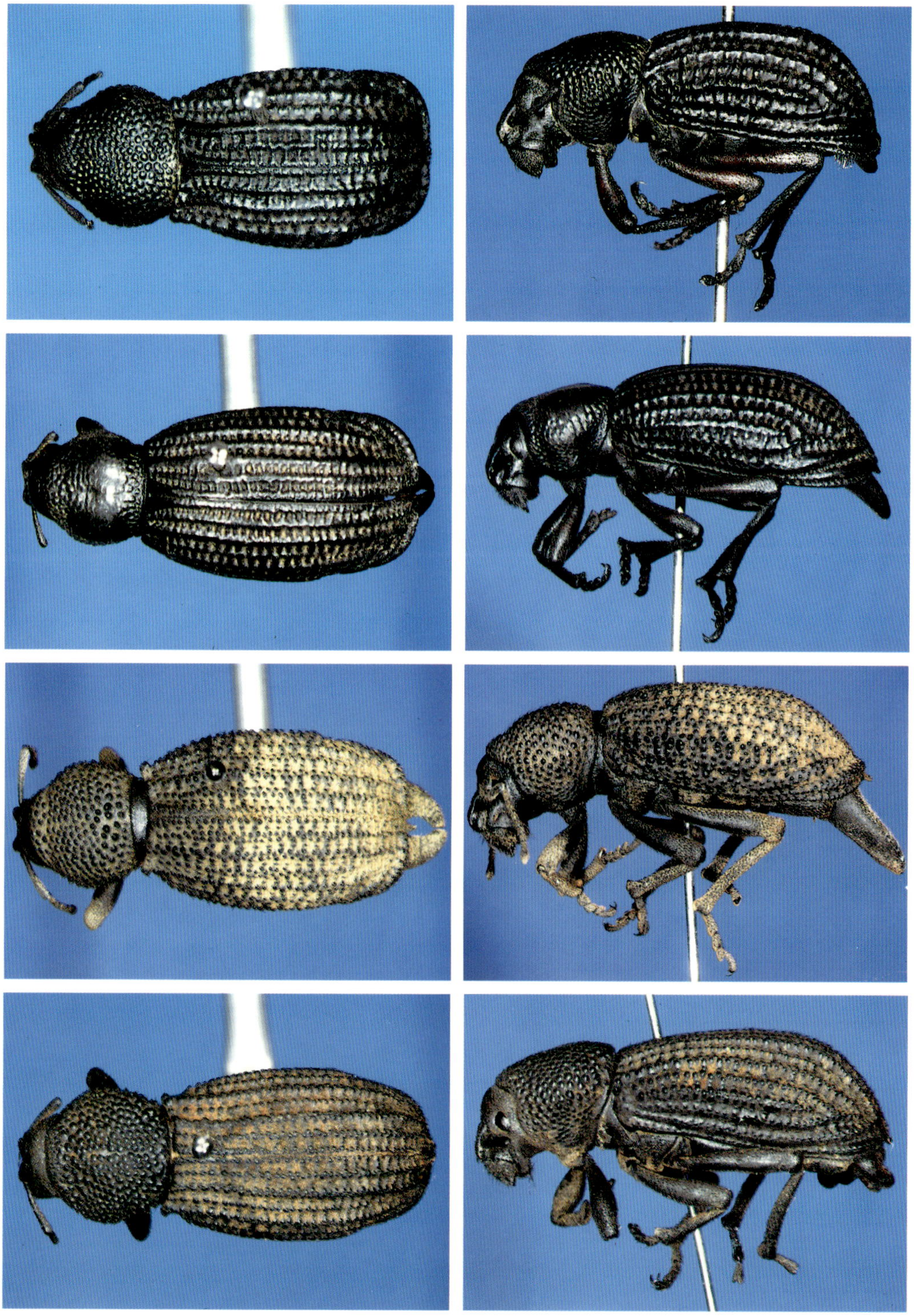

Plate 133 Amycterinae

1, 2 *Phalidura cuneicauda* (Ferguson)
Charters Towers, Queensland
Allotype female; 22.5 mm

3, 4 *Phalidura elongata* (Macleay)
No locality data.
Syntype male; 23.5 mm

5, 6 *Phalidura exasperata* (Erichson)
Tasmania
Paralectotype male; 12 mm

7, 8 *Phalidura falciformis* (Macleay)
Tamboroora, New South Wales
Male; 24 mm

1	2
3	4
5	6
7	8

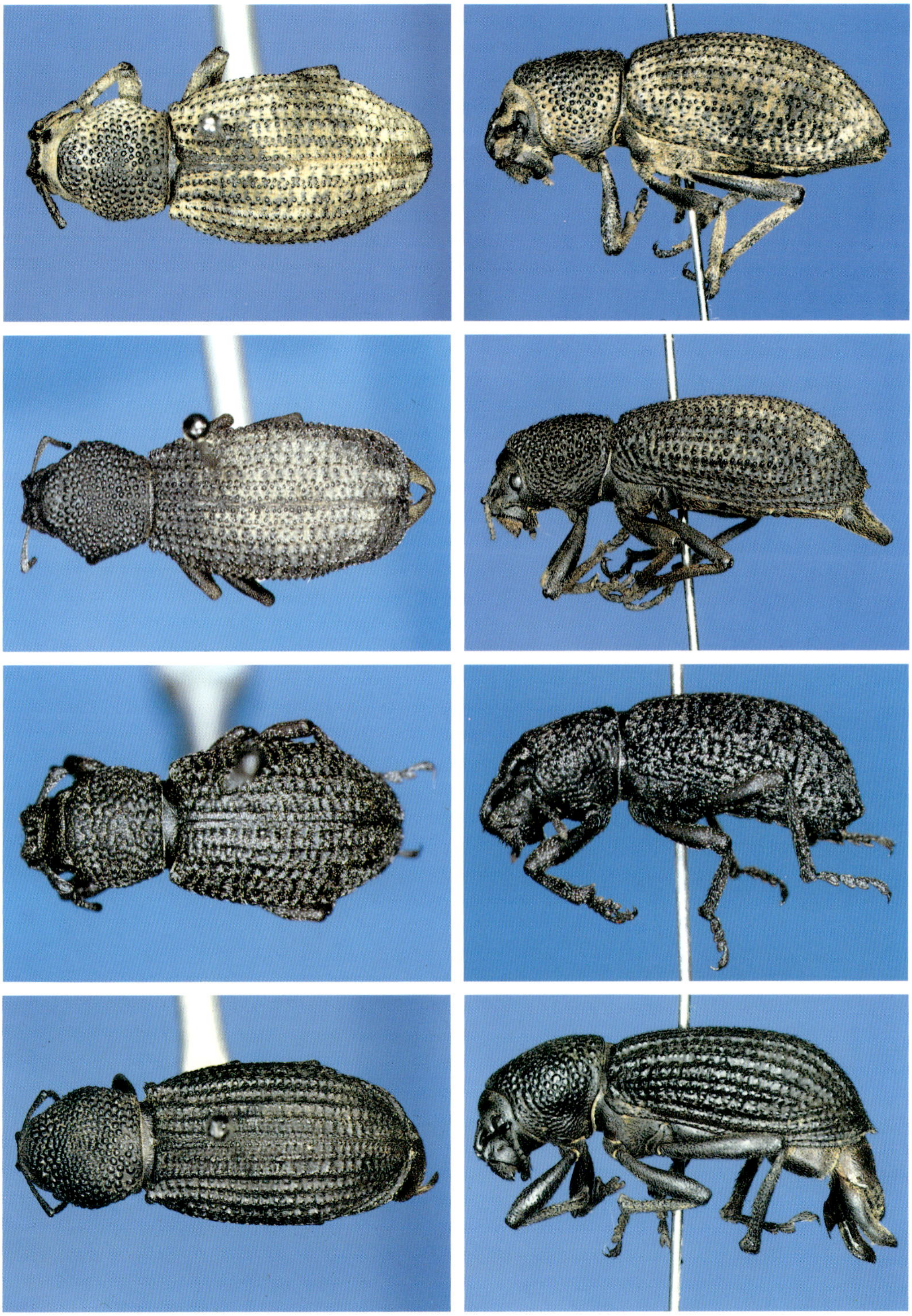

Plate 134 Amycterinae

1, 2 *Phalidura flavovaria* (Ferguson)
South Australia
Holotype male; 19 mm

3, 4 *Phalidura forficulata* (Macleay)
Rockhampton, Queensland
Syntype male; 22 mm

5, 6 *Phalidura foveata* (Macleay)
Northeast Coast of Australia
Syntype male; 21.5 mm

7, 8 *Phalidura frenchi* (Ferguson)
Rockhampton, Queensland
Holotype male; 25.5 mm

1	2
3	4
5	6
7	8

Plate 135 Amycterinae

1, 2 *Phalidura helmsi* (Ferguson)
Benalla, Victoria
Holotype male; 22 mm

3, 4 *Phalidura hopsoni* (Ferguson)
Barrington Tops, New South Wales
Holotype male; 17 mm

5, 6 *Phalidura kosciuskoana* (Ferguson)
Mount Kosciusko, New South Wales
Holotype male; 23 mm

7, 8 *Phalidura leai* (Ferguson)
Victoria
Holotype male; 21 mm

1	2
3	4
5	6
7	8

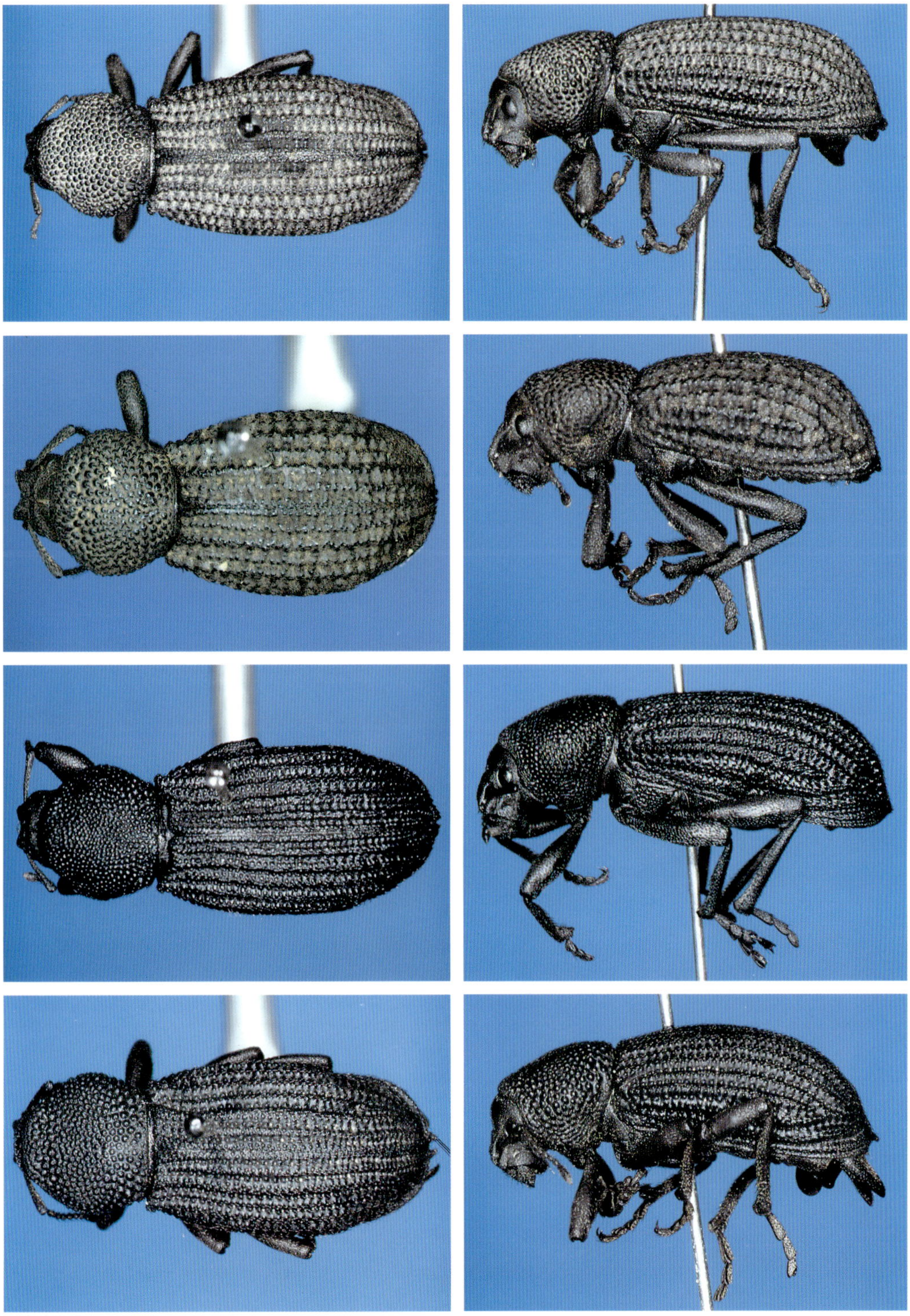

Plate 136 Amycterinae

1, 2 *Phalidura mastersii* (Macleay)
New South Wales
Male; 26 mm

3, 4 *Phalidura mira* (Boheman)
"Swan River" (labeled in error; it is a New South Wales species)
Male; 22.5 mm

5, 6 *Phalidura mirabilis* Kirby
Ferntree Gully, Victoria
Male; 22 mm

7, 8 *Phalidura miracula* (Macleay)
Blue Mountains, New South Wales
Male; 25 mm

1	2
3	4
5	6
7	8

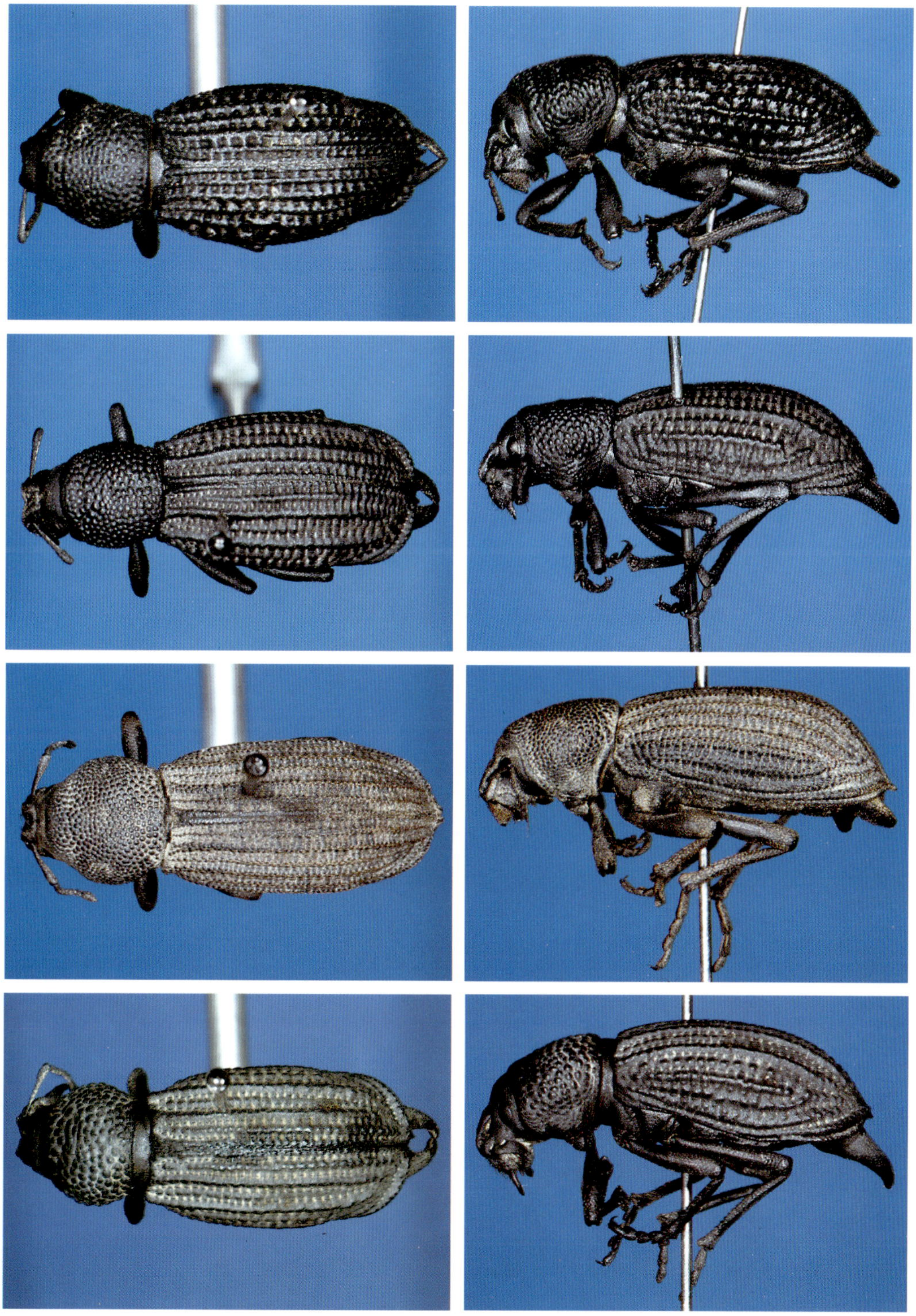

Plate 137 Amycterinae

1, 2 *Phalidura monticola* (Ferguson)
Buffalo Mountains, Victoria
Holotype male; 20.5 mm

3, 4 *Phalidura morbillosa* (Boisduval)
Moruya, New South Wales
Male, 15.5 mm

5, 6 *Phalidura orthodoxa* (Lea)
Australia
Holotype male; 16.5 mm

7, 8 *Phalidura panduriformis* (Ferguson)
North Queensland
Holotype male; 13 mm

1	2
3	4
5	6
7	8

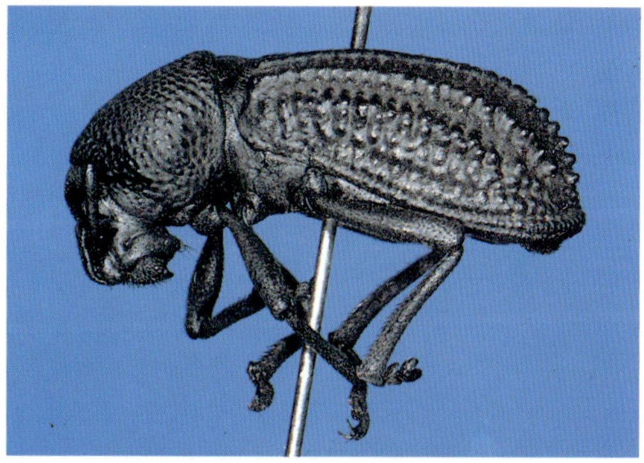

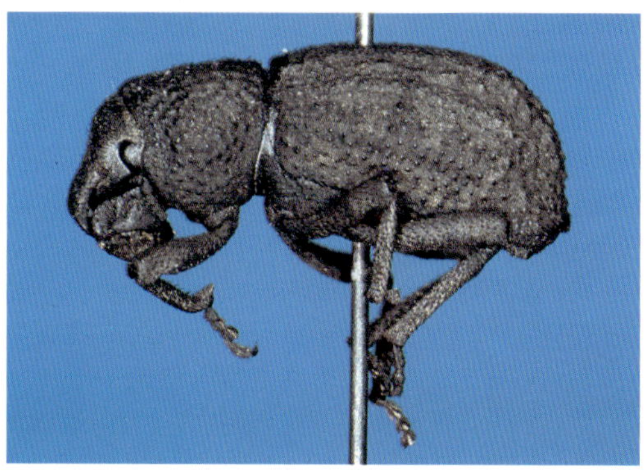

Plate 138 Amycterinae

1, 2 *Phalidura perlata* (Ferguson)
Eucla, Western Australia
Holotype male; 19 mm
Abraded; compare next figure.

3, 4 *Phalidura perlata* (Ferguson)
Northwest Victoria
Female; 26 mm
From a fresh specimen of the synonym *flavescens* Ferguson.

5, 6 *Phalidura rayneri* (Macleay)
New Holland
Holotype female; 22 mm

7, 8 *Phalidura reticulata* (Boisduval)
Queensland
Male; 24 mm
Note the so-called "gular horn". See also Plate 179:5, 6, 7, 8.

1	2
3	4
5	6
7	8

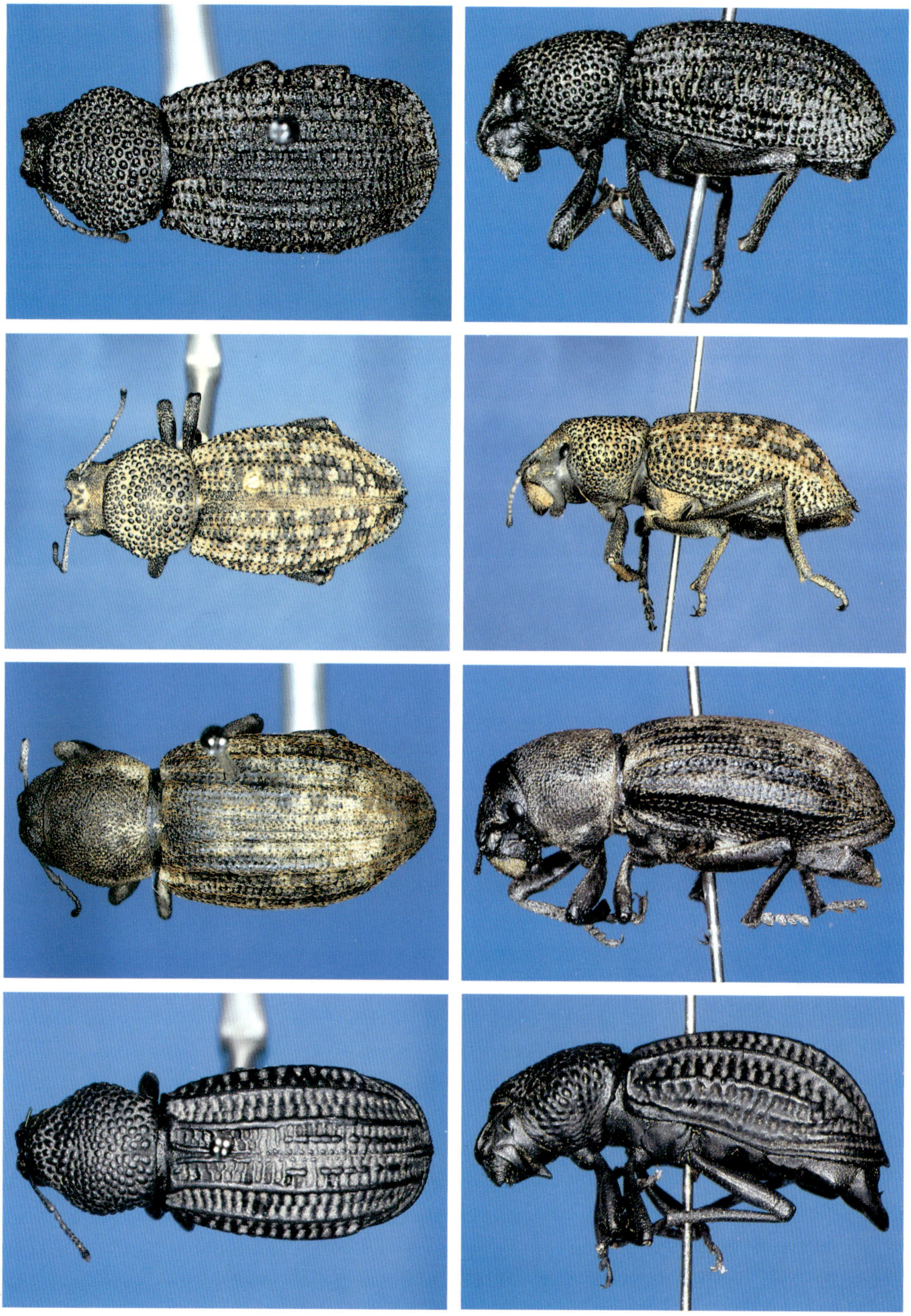

Plate 139 Amycterinae

1, 2 *Phalidura rufipes* (Blackburn)
Simpsons Gap, near Alice Springs, Northern Territory
Male; 19.5 mm

3, 4 *Phalidura sloanei* (Ferguson)
Wagga Wagga, New South Wales
Holotype male; 29.5 mm

5, 6 *Phalidura subcostata* (Macleay)
Parramatta, New South Wales
Holotype male; 20 mm

7, 8 *Phalidura sublaevigata* (Ferguson)
Mittagong, New South Wales
Holotype male; 23.5 mm

1	2
3	4
5	6
7	8

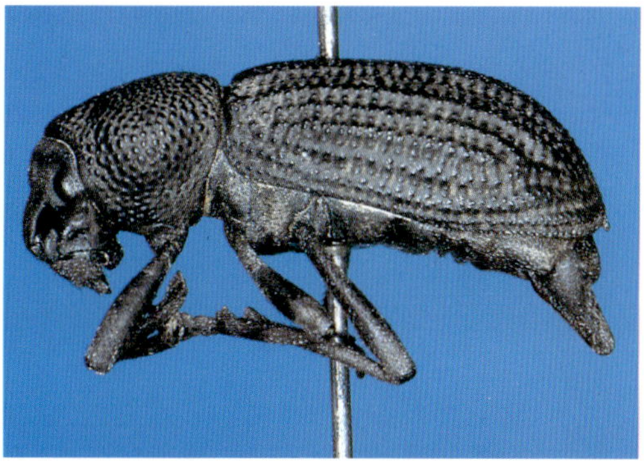

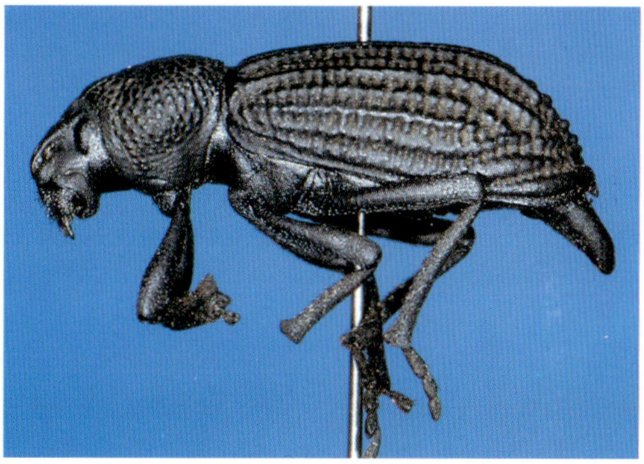

Plate 140 Amycterinae

1, 2 *Phalidura sulcipennis* (Ferguson)
Blue Mountains, New South Wales
Holotype male; 17.5 mm

3, 4 *Phalidura taylori* (Ferguson)
Guy Fawkes, New South Wales
Holotype male; 16 mm

5, 6 *Phalidura tenebricosa* (Ferguson)
Benalla, Victoria
Holotype male; 14.5 mm

7, 8 *Phalidura tessellata* (Pascoe)
Champion Bay, Western Australia
Syntype male; 14.5 mm

1	2
3	4
5	6
7	8

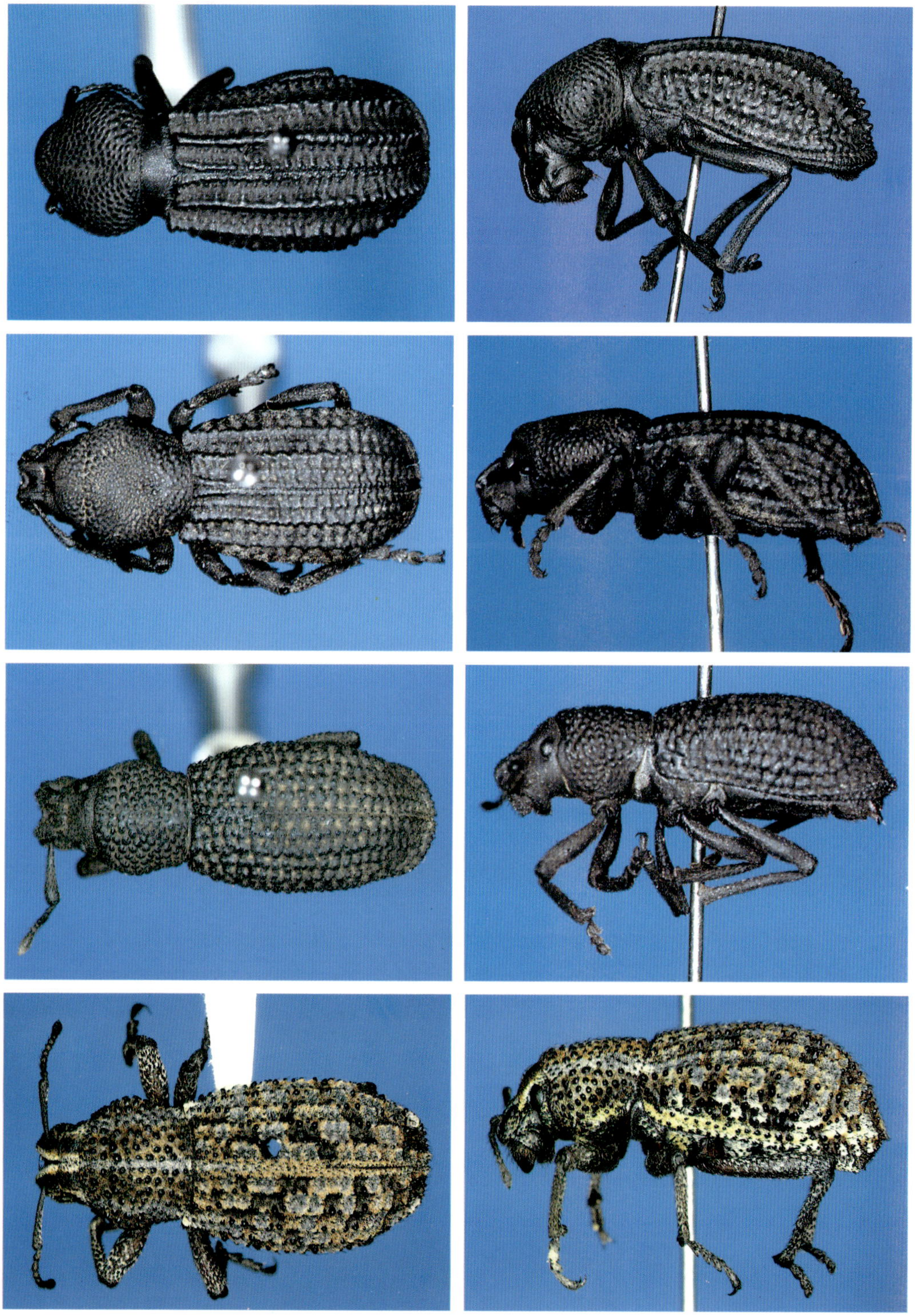

Plate 141 Amycterinae

1, 2 *Phalidura tomentosa* (Boisduval)
Cape Otway, Victoria
Female, compared with holotype; 18 mm

3, 4 *Phalidura variegata* (Macleay)
Dalby, Queensland
Male; 15 mm

5, 6 *Phalidura variolosa* (Ferguson)
Darling Downs, Queensland
Holotype male; 18.5 mm

7, 8 *Phalidura wilcoxi* (Macleay)
Clarence River, New South Wales
Holotype male; 16.5 mm
See Volume VI for additional photographs of *Phalidura* and male genitalia.

1	2
3	4
5	6
7	8

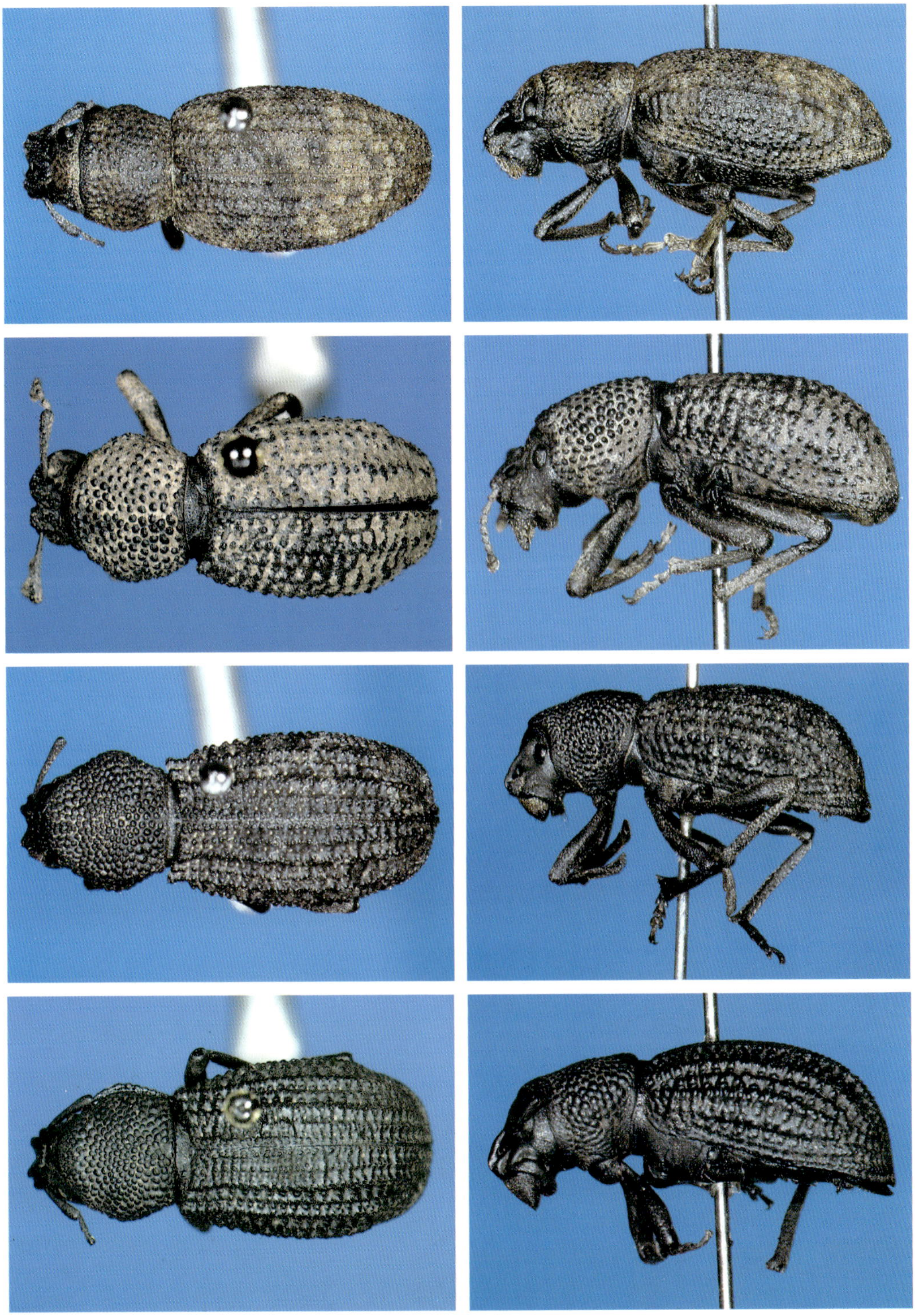

Plate 142 Amycterinae

1, 2 *Pseudonotonophes lemnus* (Pascoe)
Rockingham Park, Western Australia
Male; 6.5 mm

3, 4 *Sclerorinus albovittatus* Ferguson
Near Cocklebiddy, Western Australia
Male; 18 mm

5, 6 *Sclerorinus amycteroides* Ferguson
Portland, Victoria
Holotype male; 17 mm

7, 8 *Sclerorinus angustior* Ferguson
Whitlock, Western Australia
Holotype male; 22 mm

1	2
3	4
5	6
7	8

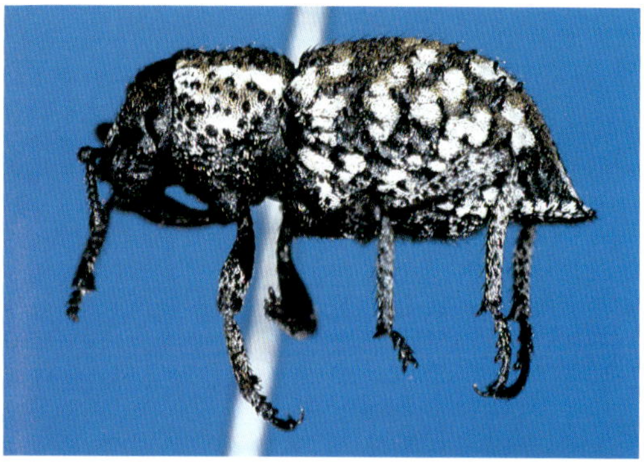

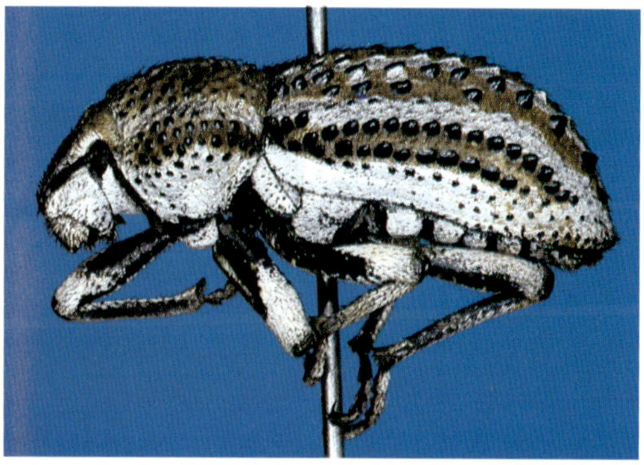

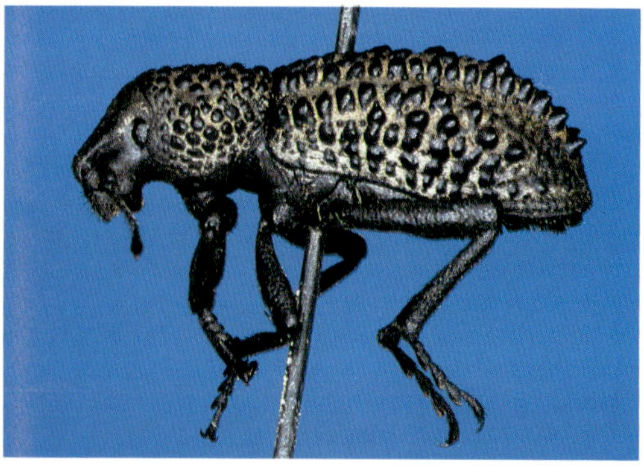

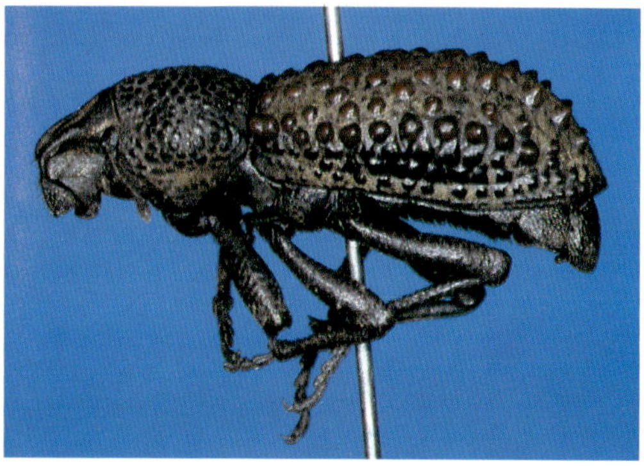

Plate 143 Amycterinae

1, 2 *Sclerorinus angustipennis* Sloane
Ropes Creek, New South Wales
Male; 17 mm

3, 4 *Sclerorinus apicalis* Macleay
New South Wales
Holotype female; 17 mm

5, 6 *Sclerorinus aterrimus* Ferguson
Mount Kosciusko, New South Wales
Holotype male; 18 mm

7, 8 *Sclerorinus besti* Ferguson
Portland, Victoria
Holotype male; 12.5 mm

1	2
3	4
5	6
7	8

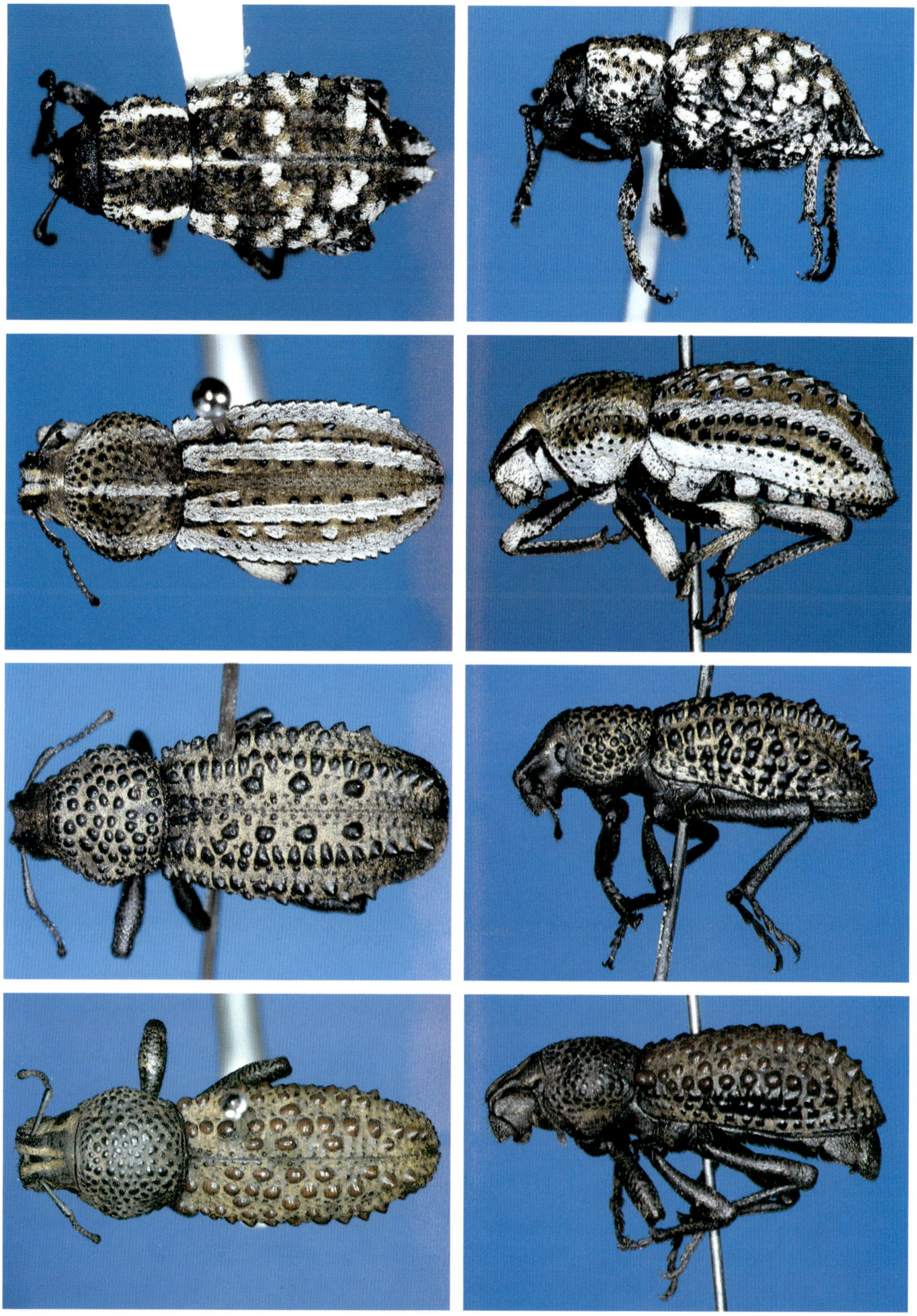

Plate 143 Amycterinae

1, 2 *Sclerorinus angustipennis* Sloane
Ropes Creek, New South Wales
Male; 17 mm

3, 4 *Sclerorinus apicalis* Macleay
New South Wales
Holotype female; 17 mm

5, 6 *Sclerorinus aterrimus* Ferguson
Mount Kosciusko, New South Wales
Holotype male; 18 mm

7, 8 *Sclerorinus besti* Ferguson
Portland, Victoria
Holotype male; 12.5 mm

1	2
3	4
5	6
7	8

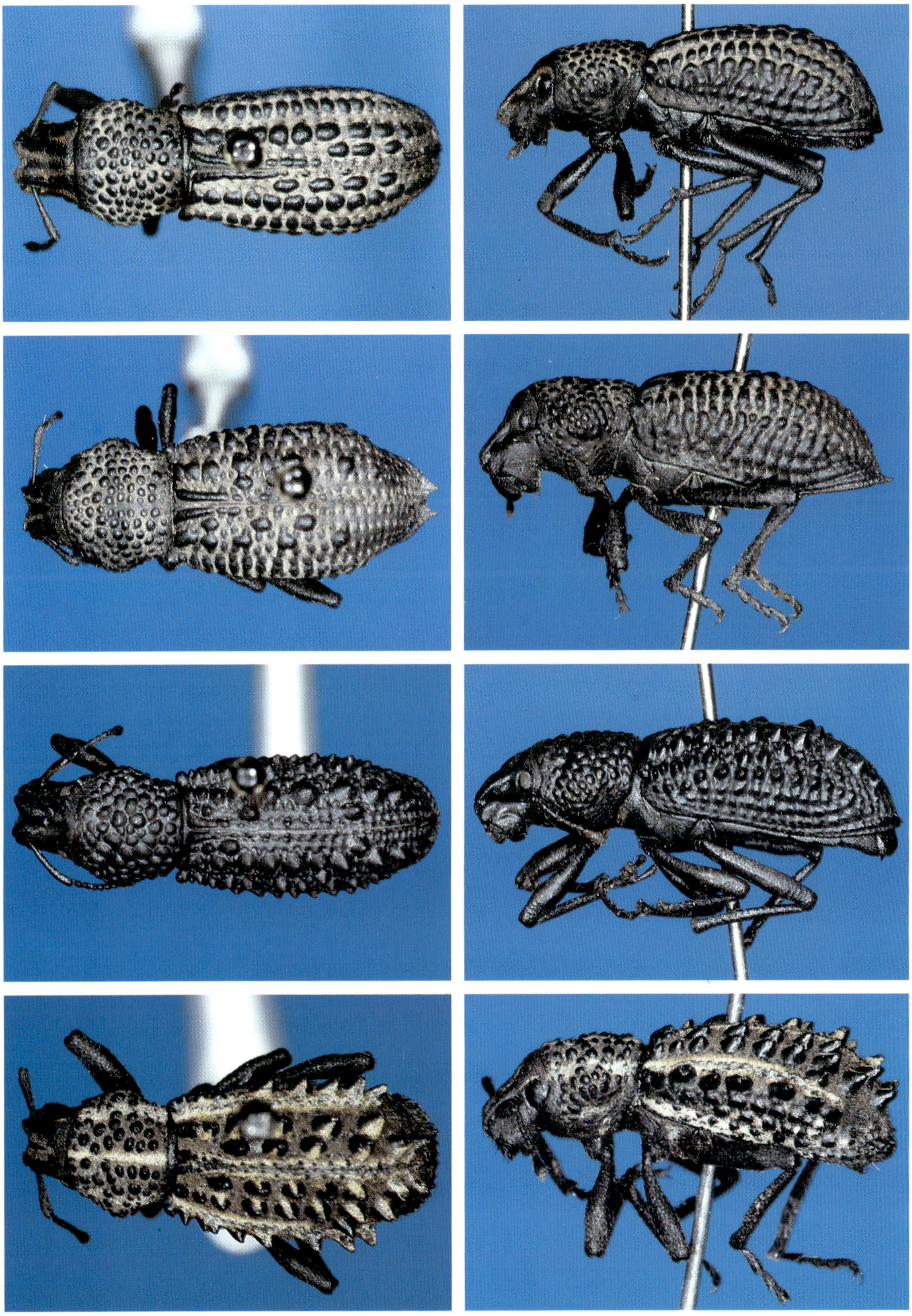

Plate 144 Amycterinae

1, 2 *Sclerorinus biordinatus* Macleay
Yorke Peninsula, South Australia
Holotype male; 22 mm

3, 4 *Sclerorinus browni* Ferguson
Yalgon, Western Australia
Holotype male; 22.5 mm

5, 6 *Sclerorinus bubalus* (Olivier)
Hobart, Tasmania
Male; 14 mm

7, 8 *Sclerorinus carteri* Ferguson
Bridgetown, Western Australia
Holotype male; 16.5 mm

1	2
3	4
5	6
7	8

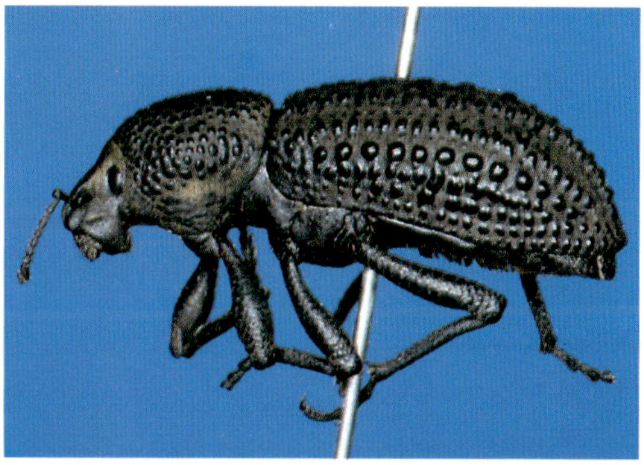

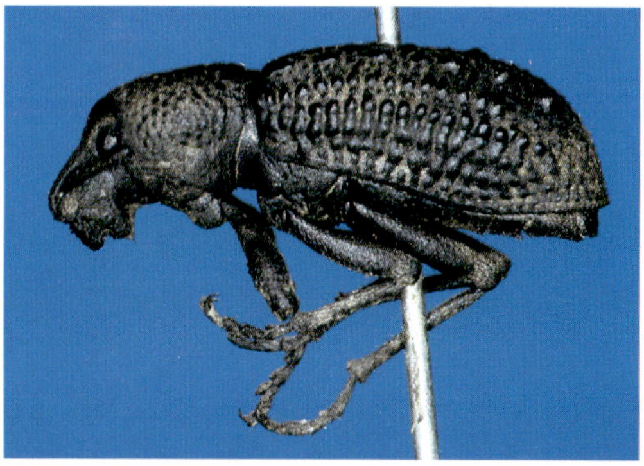

Plate 145 Amycterinae

1, 2 *Sclerorinus convexus* Sloane
MacDonnell Ranges, Northern Territory
Holotype; 22.5 mm

3, 4 *Sclerorinus daveyi* Ferguson
Portland, Victoria
Holotype male; 16.5 mm

5, 6 *Sclerorinus dilaticollis* Macleay
Victoria
Syntype female; 17 mm

7, 8 *Sclerorinus dimidiatus* Macleay
Flinders Range, South Australia
Male; 19 mm

1	2
3	4
5	6
7	8

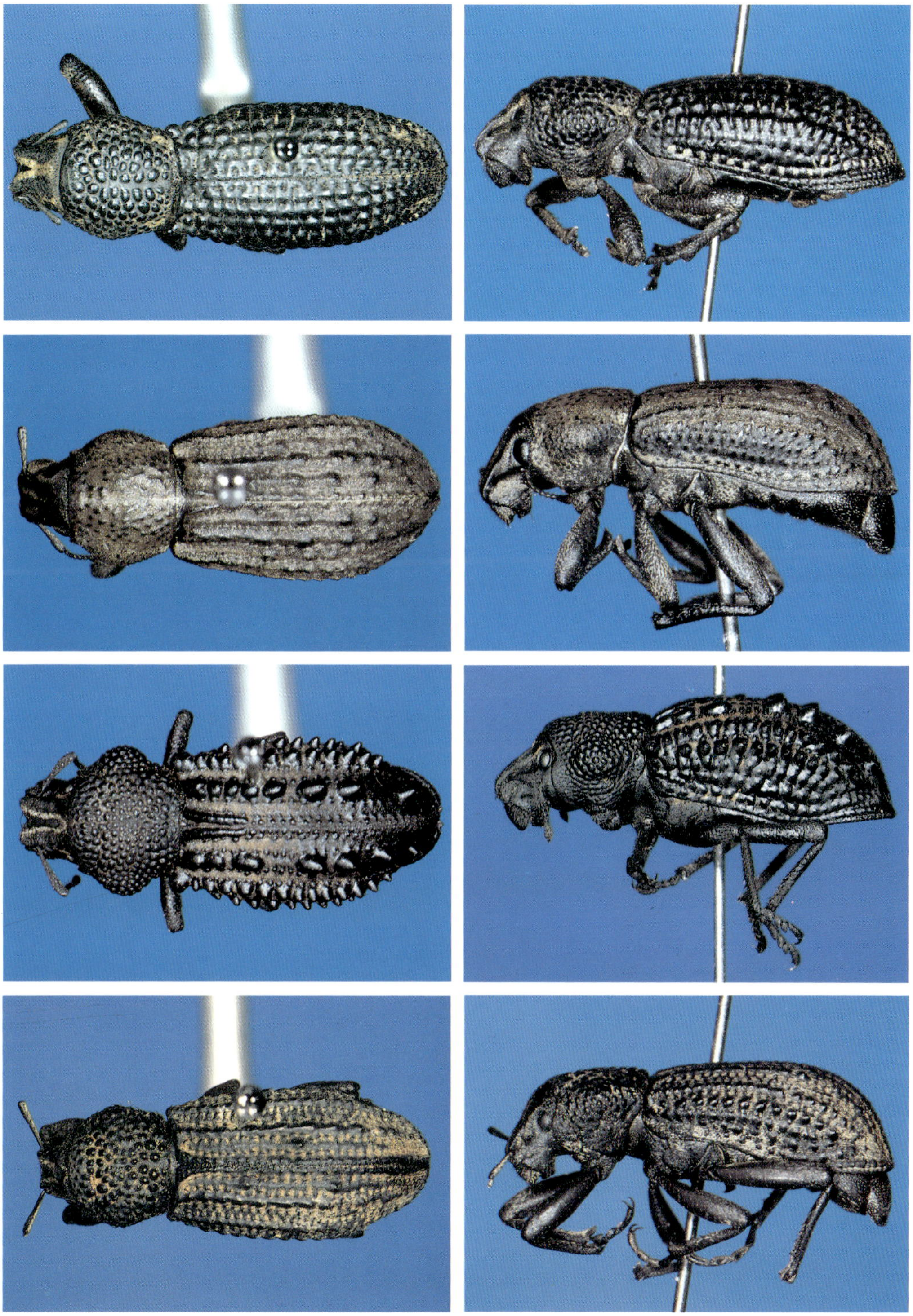

Plate 146 Amycterinae

1, 2 *Sclerorinus dixoni* Ferguson
Mallee District, Victoria
Holotype male; 25.5 mm

3, 4 *Sclerorinus elderi* Sloane
Western Australia
Paratype male; 23 mm

5, 6 *Sclerorinus exilis* Macleay
Murrumbidgee, New South Wales
Male; 19 mm

7, 8 *Sclerorinus germari* Macleay
Port Lincoln, South Australia
Syntype male; 13 mm

1	2
3	4
5	6
7	8

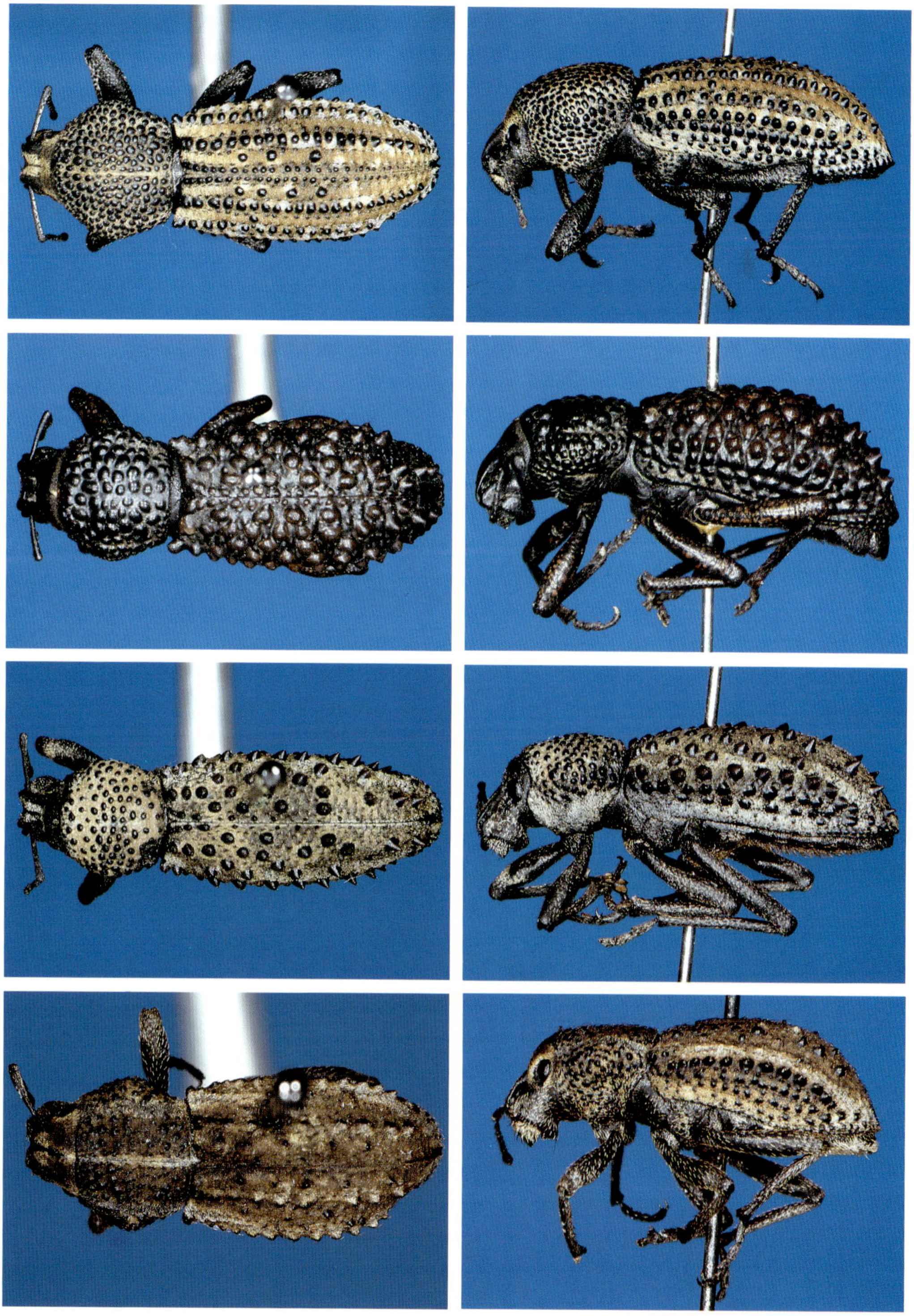

Plate 147 Amycterinae

1, 2 *Sclerorinus goudiei* Ferguson
Birchip, Victoria
Holotype male; 25 mm

3, 4 *Sclerorinus horridus* Macleay
South Australia
Syntype female; 14.5 mm

5, 6 *Sclerorinus inconstans* Lea
Mount Kosciusko, New South Wales
Holotype male; 19.5 mm

7, 8 *Sclerorinus inornatus* Ferguson
Ballarat, Victoria
Male; 14 mm

1	2
3	4
5	6
7	8

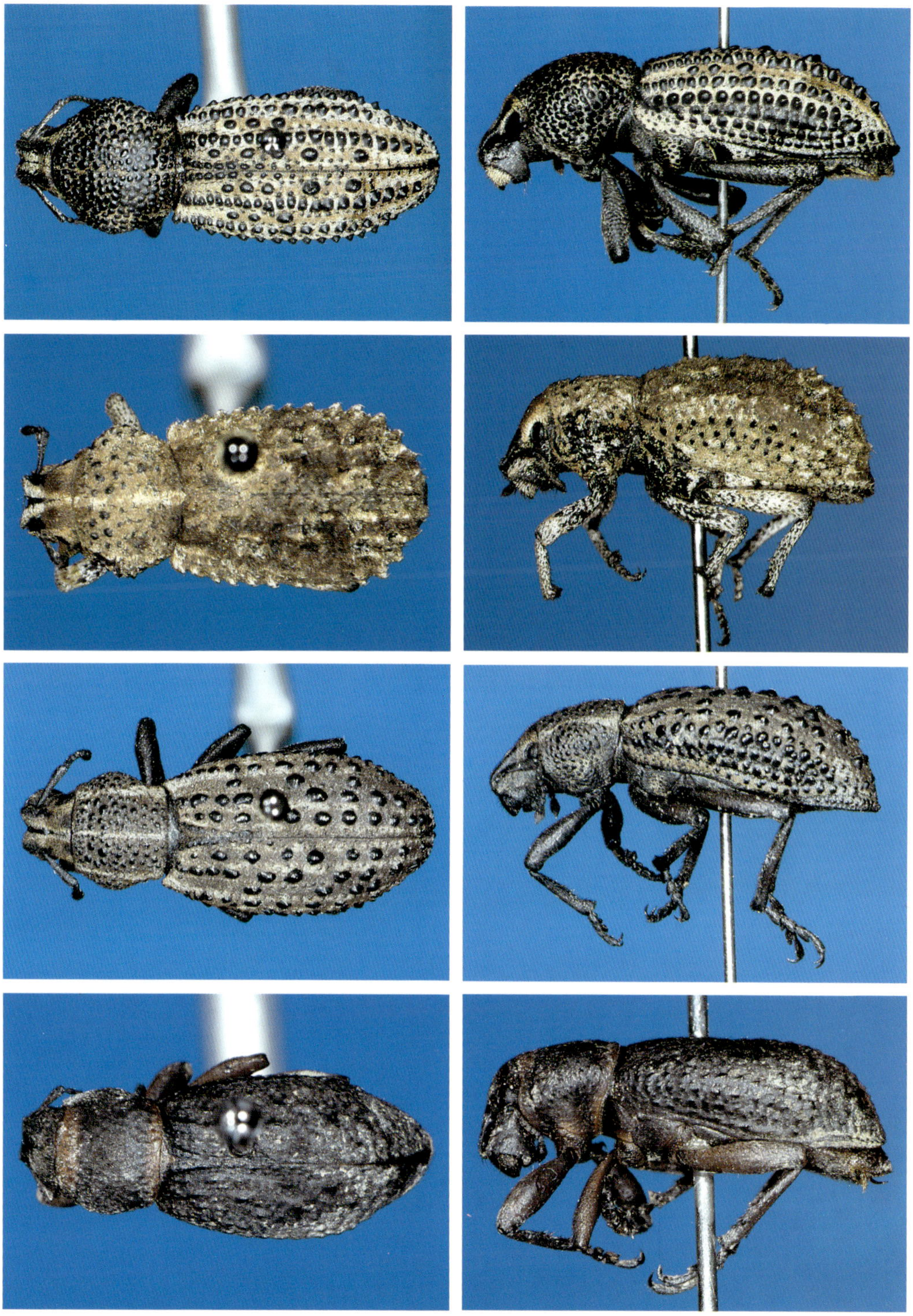

Plate 148 Amycterinae

1, 2 *Sclerorinus insignis* Sloane
18 km west of Gosses Bluff, Western Australia
Male; 22 mm

3, 4 *Sclerorinus kirbyi* (Boisduval)
New South Wales
Female; 18 mm
From a syntype of *Sclerorinus subcostatus* Macleay.

5, 6 *Sclerorinus laticollis* Macleay
King George Sound, Western Australia
Male, compared with lectotype; 20 mm

7, 8 *Sclerorinus longus* Macleay
South Australia
Syntype male; 21.5 mm

1	2
3	4
5	6
7	8

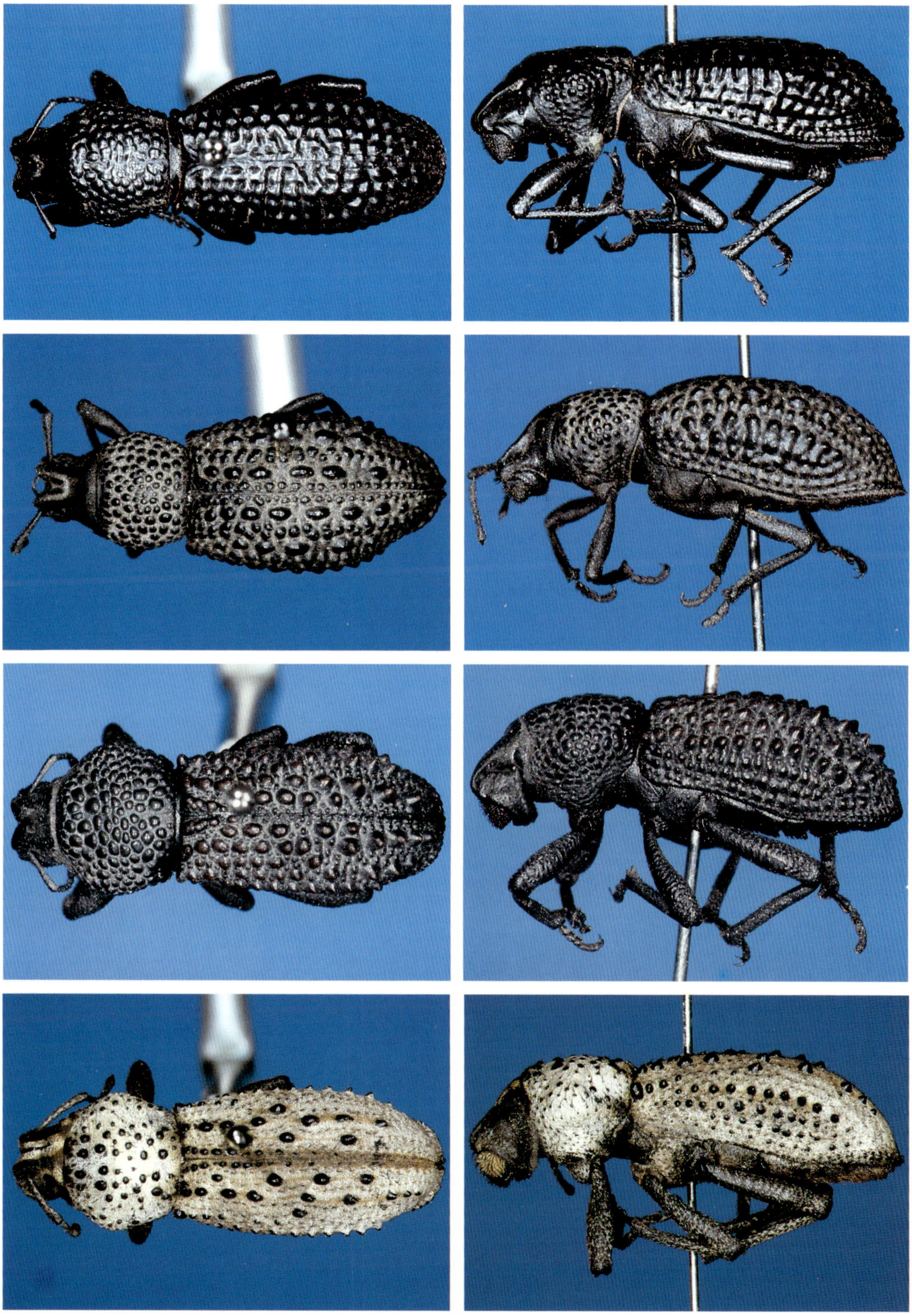

Plate 149 Amycterinae

1, 2 *Sclerorinus mastersii* Macleay
South Australia
Syntype male; 20 mm

3, 4 *Sclerorinus meliceps* Pascoe
Rockhampton, Queensland
Holotype male (teneral); 15.5 mm

5, 6 *Sclerorinus molestus* Pascoe
Near Cocklebiddy, Western Australia
Male; 22.5 mm

7, 8 *Sclerorinus mucronipennis* Ferguson
Nelson, Victoria
Holotype female; 14 mm

1	2
3	4
5	6
7	8

Plate 150 Amycterinae

1, 2 *Sclerorinus neglectus* Ferguson
South Australia
Holotype male; 17 mm

3, 4 *Sclerorinus noctis* Sloane
Barrow Range, Western Australia
Holotype female; 22.5 mm

5, 6 *Sclerorinus obliteratus* Macleay
Victoria
Holotype female; 19 mm

7, 8 *Sclerorinus oblongatus* Ferguson
Nelson, Victoria
Allotype female; 15 mm

1	2
3	4
5	6
7	8

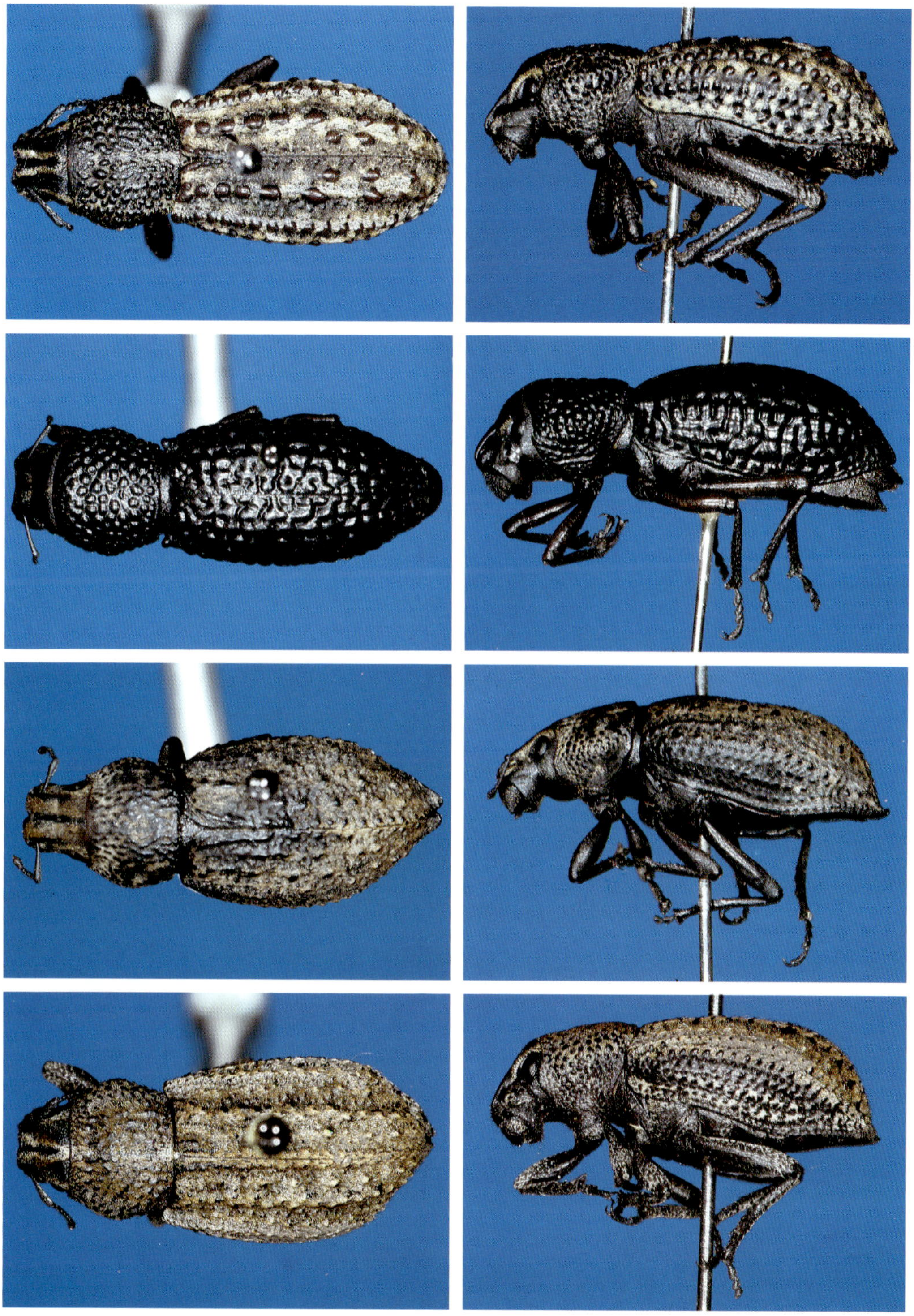

Plate 151 Amycterinae

1, 2 *Sclerorinus occidentalis* Sloane
Western Australia
Male; 23 mm

3, 4 *Sclerorinus parvulus* Macleay
South Australia
Syntype male; 13 mm

5, 6 *Sclerorinus pilularius* Macleay
South Australia
Syntype male; 19 mm

7, 8 *Phalidura anthracoides* (Ferguson)
Trawool, Victoria
Holotype female; 18 mm

1	2
3	4
5	6
7	8

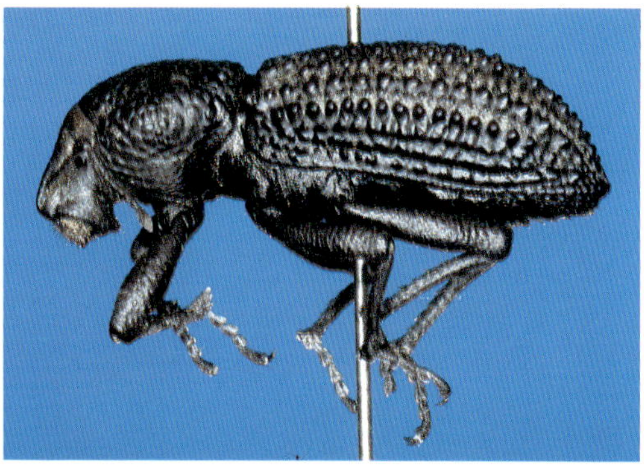

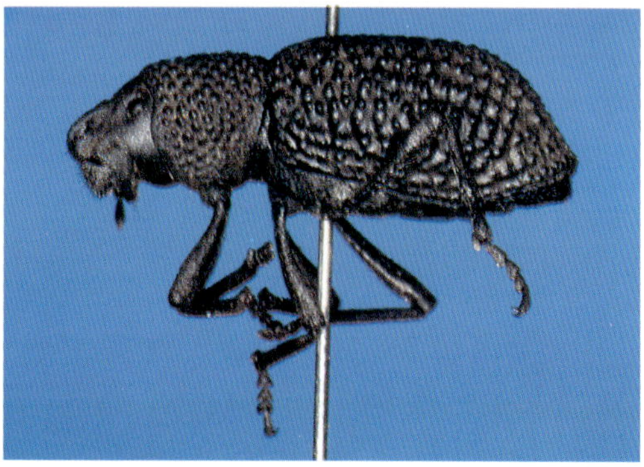

Plate 152 Amycterinae

1, 2 *Sclerorinus queenslandicus* Ferguson
Upper Logan River, Queensland
Holotype male; 14.5 mm

3, 4 *Sclerorinus regularis* Ferguson
Gawler Ranges, South Australia
Holotype male; 19 mm

5, 6 *Sclerorinus riverinae* Macleay
Murrumbidgee, New South Wales
Syntype female; 14 mm

7, 8 *Sclerorinus sabulosus* Macleay
Flinders Range, South Australia
Male; 23.5 mm
From the holotype of *arenosus* Macleay

1	2
3	4
5	6
7	8

Plate 153 Amycterinae

1, 2 *Sclerorinus sloanei* Ferguson
Capertee, New South Wales
Holotype male; 18 mm

3, 4 *Sclerorinus sordidus* Macleay
South Australia
Syntype male; 17 mm

5, 6 *Sclerorinus squalidus* Macleay
Lambing Flat, New South Wales
Syntype male; 17.5 mm

7, 8 *Sclerorinus stewartii* Macleay
South Australia
Female; 23 mm

1	2
3	4
5	6
7	8

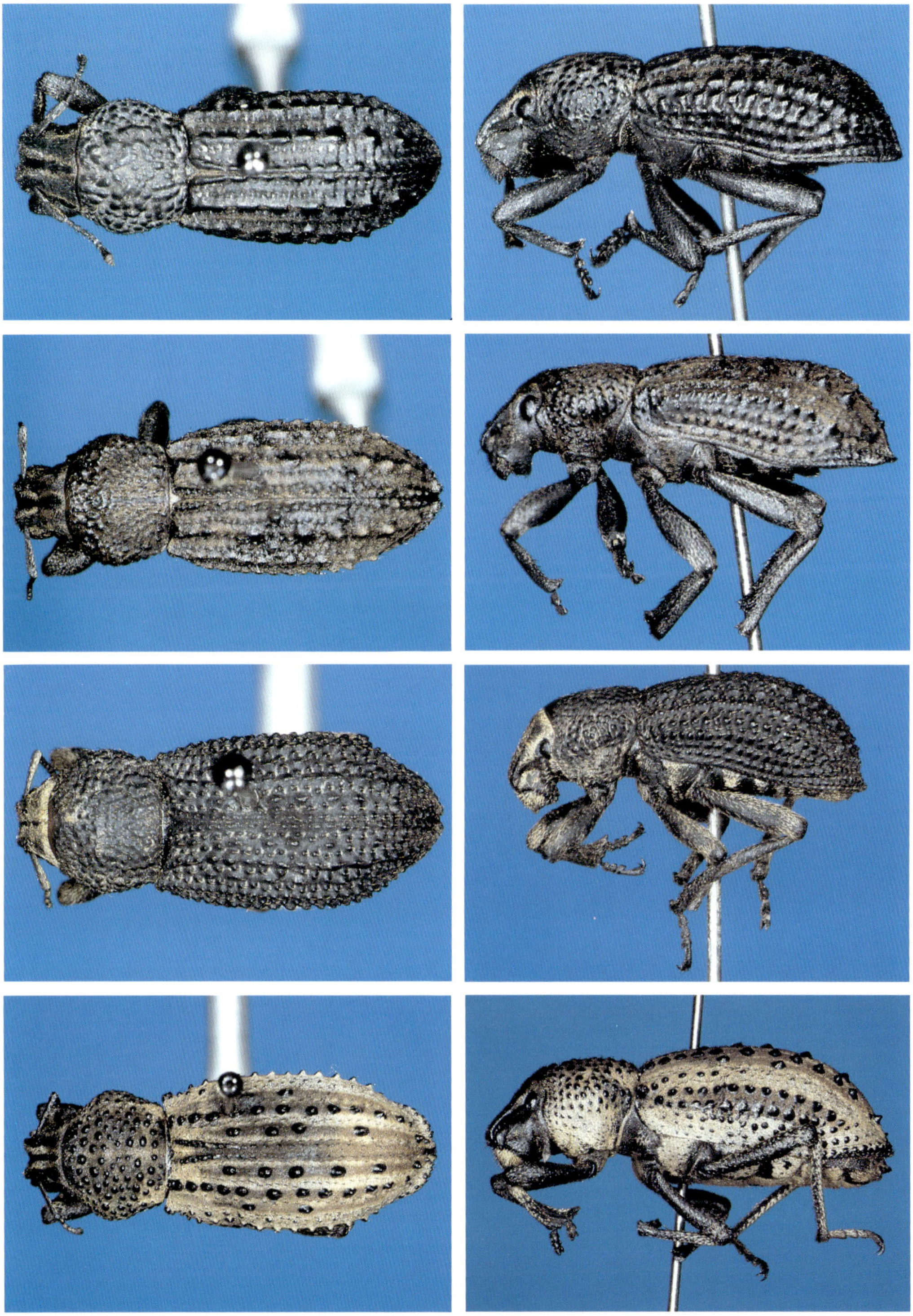

Plate 154 Amycterinae

1, 2 *Sclerorinus sublineatus* (Germar)
South Australia
Male; 19 mm

3, 4 *Sclerorinus tristis* (Boisduval)
South Australia
Male; 15.5 mm
From the holotype of *asper* Macleay.

5, 6 *Sclerorinus verrucosus* Macleay
New Holland
Holotype female; 17 mm

7, 8 *Sclerorinus vestitus* Macleay
Flinders Range, South Australia
Syntype male; 20.5 mm

1	2
3	4
5	6
7	8

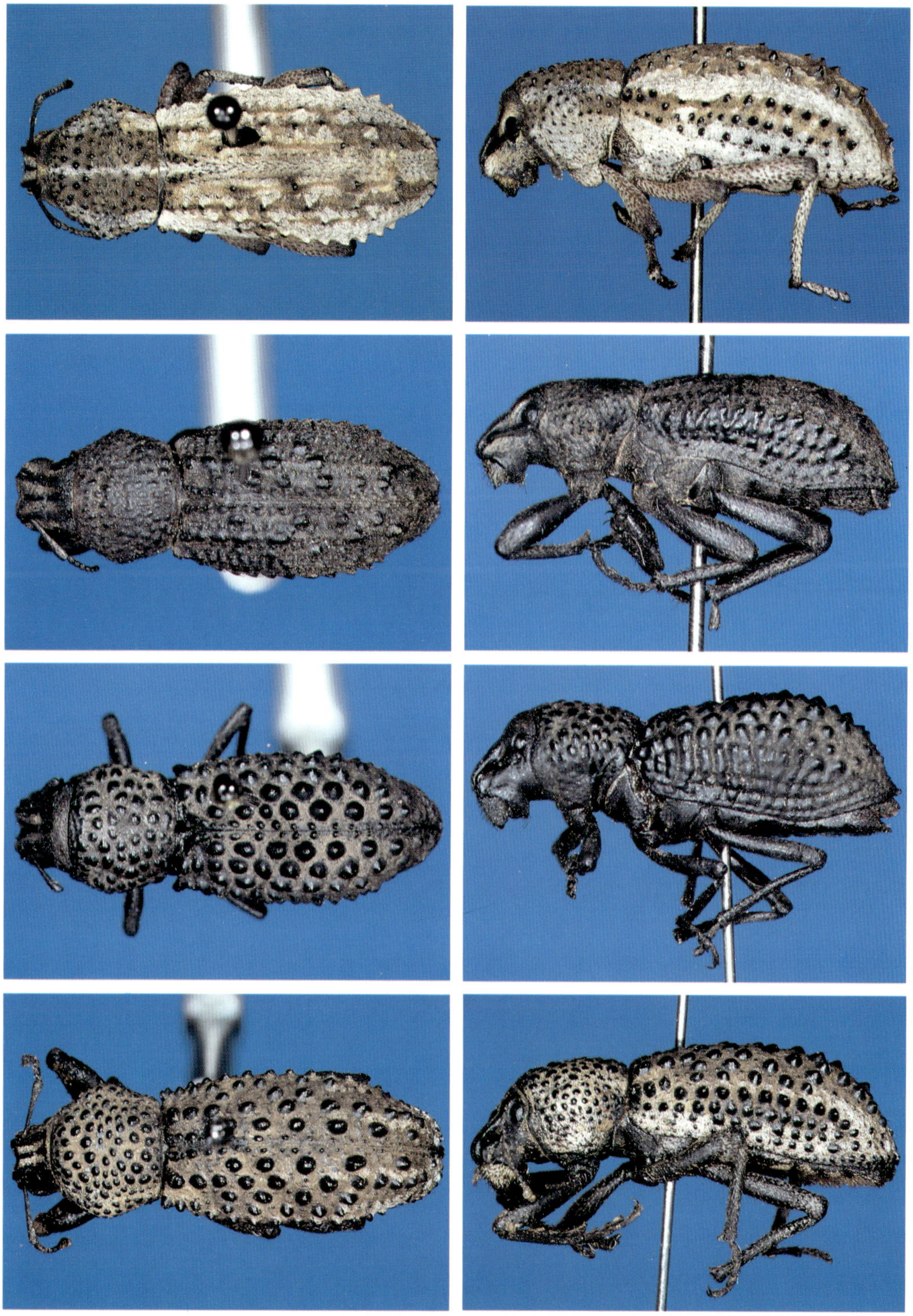

Plate 155 Amycterinae

1, 2 *Sclerorinella crawshawi* Ferguson
Jandikot, Western Australia
Holotype female; 17.5 mm

3, 4 *Sclerorinella geniculata* (Pascoe)
Western Australia
Male, compared with holotype; 14.5 mm

5, 6 *Sclerorinella granuliceps* Ferguson
Harvey Agricultural Area, Western Australia
Female, compared with holotype; 15 mm

7, 8 *Sclerorinella melanopis* (Pascoe)
Western Australia
Holotype male; 14 mm

1	2
3	4
5	6
7	8

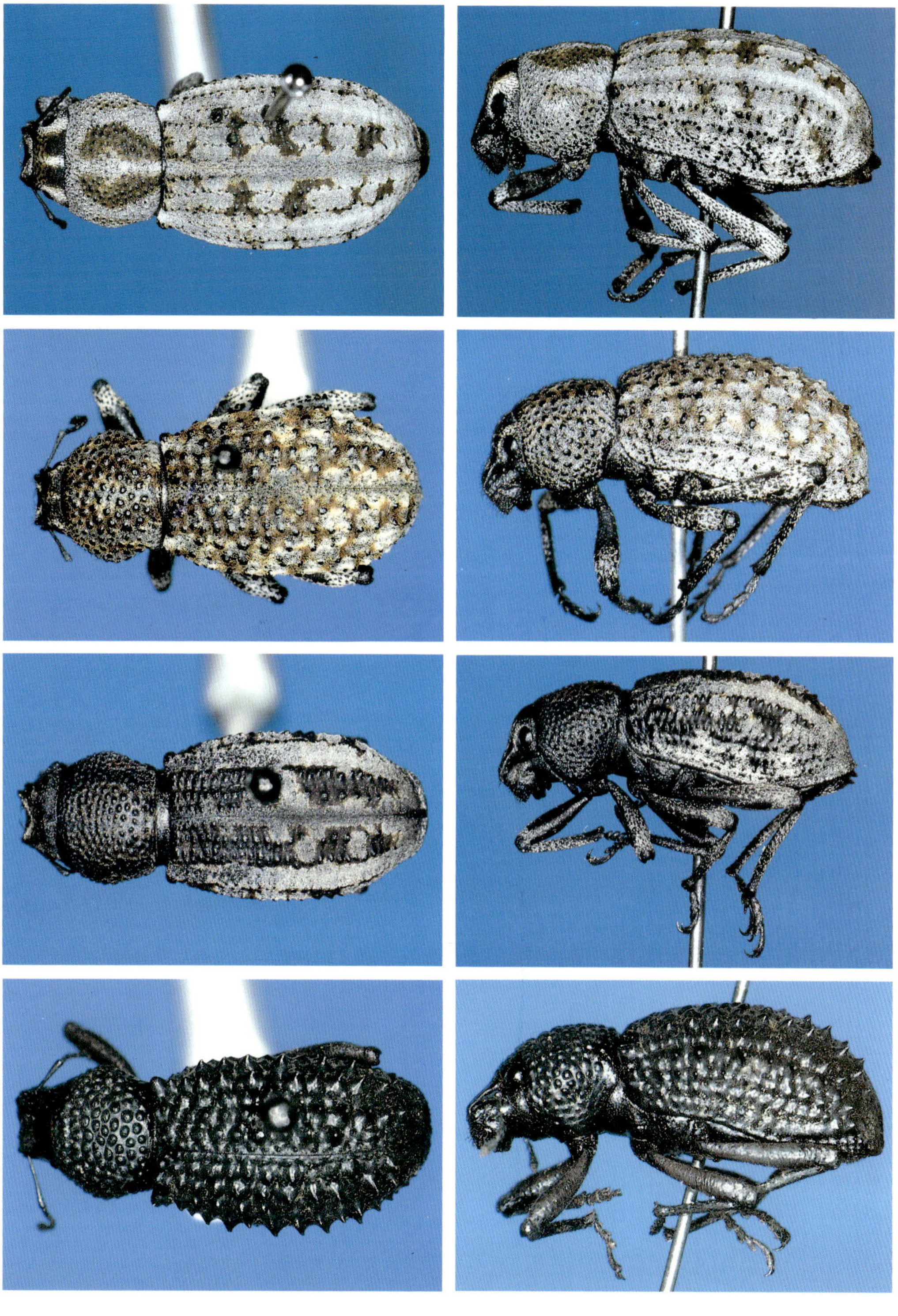

Plate 156 Amycterinae

1, 2 *Sosytelus hardcastlei* (Lea)
Cunnamulla, Queensland
Holotype female; 10 mm

3, 4 *Sosytelus pithecius* (Pascoe)
Forest Reefs, New South Wales
Female; 12.5 mm

5, 6 *Sosytelus pubescens* (Lea)
Bourke, New South Wales
Holotype male; 10 mm

7, 8 *Sosytelus rudis* (Blackburn)
Northern Australia ("Central Australia")
Syntype female; 9 mm

1	2
3	4
5	6
7	8

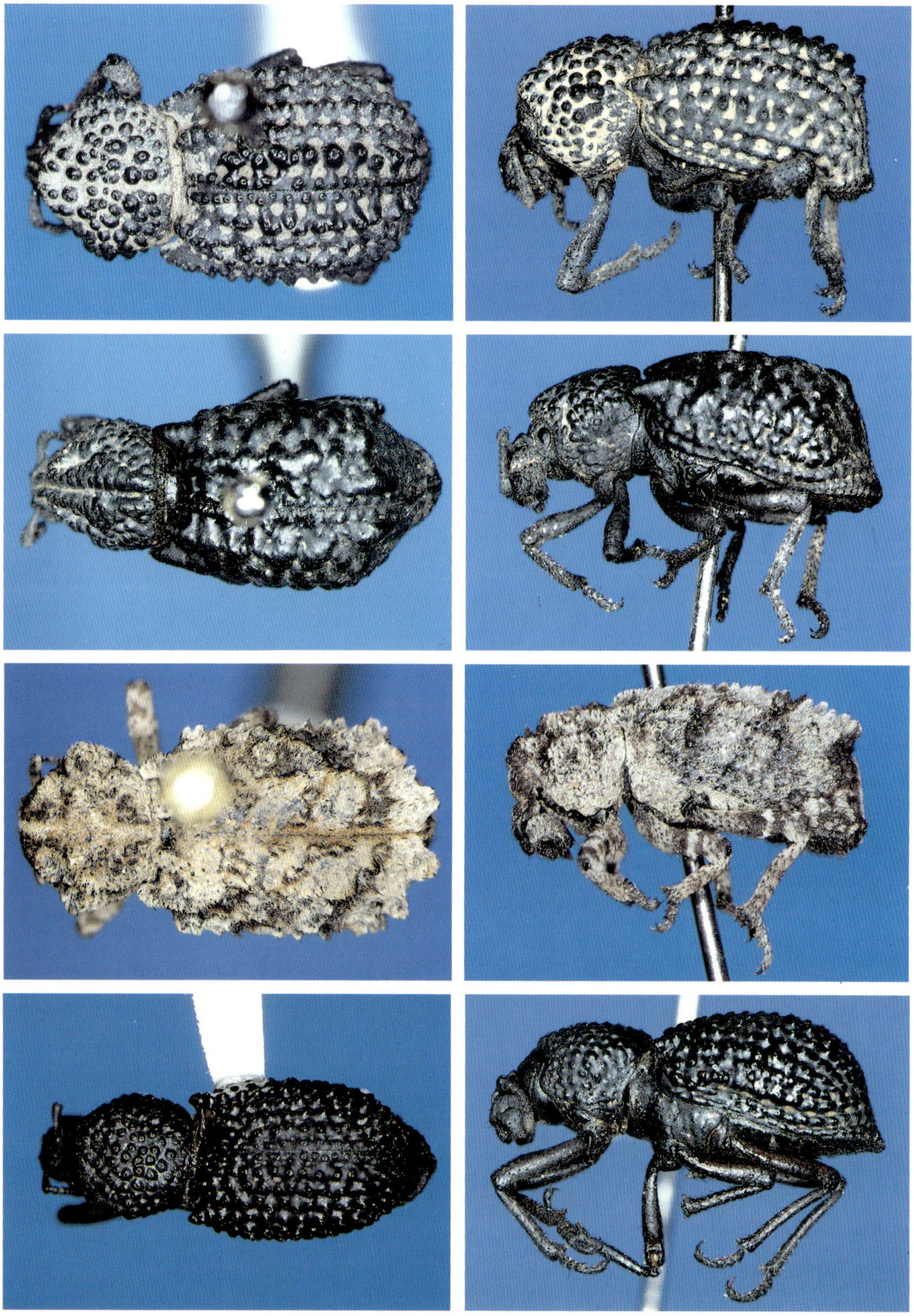

Plate 157 Amycterinae

1, 2 *Talaurinus acromialis* Ferguson
Brisbane, Queensland
Holotype male; 14 mm

3, 4 *Talaurinus acutipennis* Ferguson
Mount Macedon, Victoria
Holotype female; 17 mm

5, 6 *Talaurinus alternans* Macleay
Clyde River, New South Wales
Holotype male; 17 mm

7, 8 *Talaurinus alternatus* Macleay
New Holland
Holotype female; 23 mm

1	2
3	4
5	6
7	8

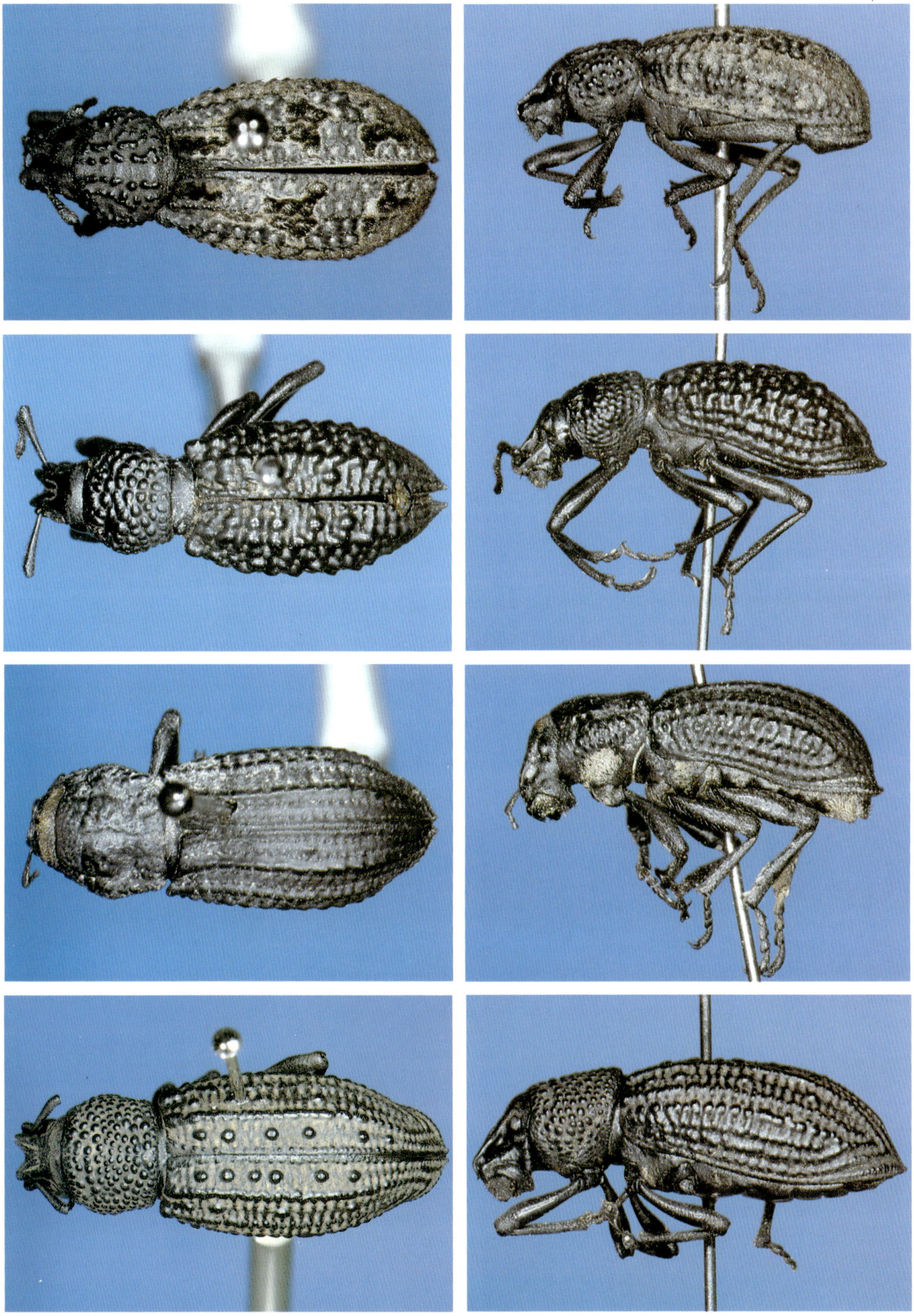

Plate 158 Amycterinae

1, 2 *Talaurinus angularis* Ferguson
Queensland
Holotype male; 16 mm

3, 4 *Talaurinus angustatus* Macleay
King George Sound, Western Australia
Syntype male; 15 mm

5, 6 *Talaurinus angustus* Ferguson
Glenample, Victoria
Holotype male; 15.5 mm

7, 8 *Talaurinus brevior* Ferguson
No locality data
Holotype male; 12.5 mm

1	2
3	4
5	6
7	8

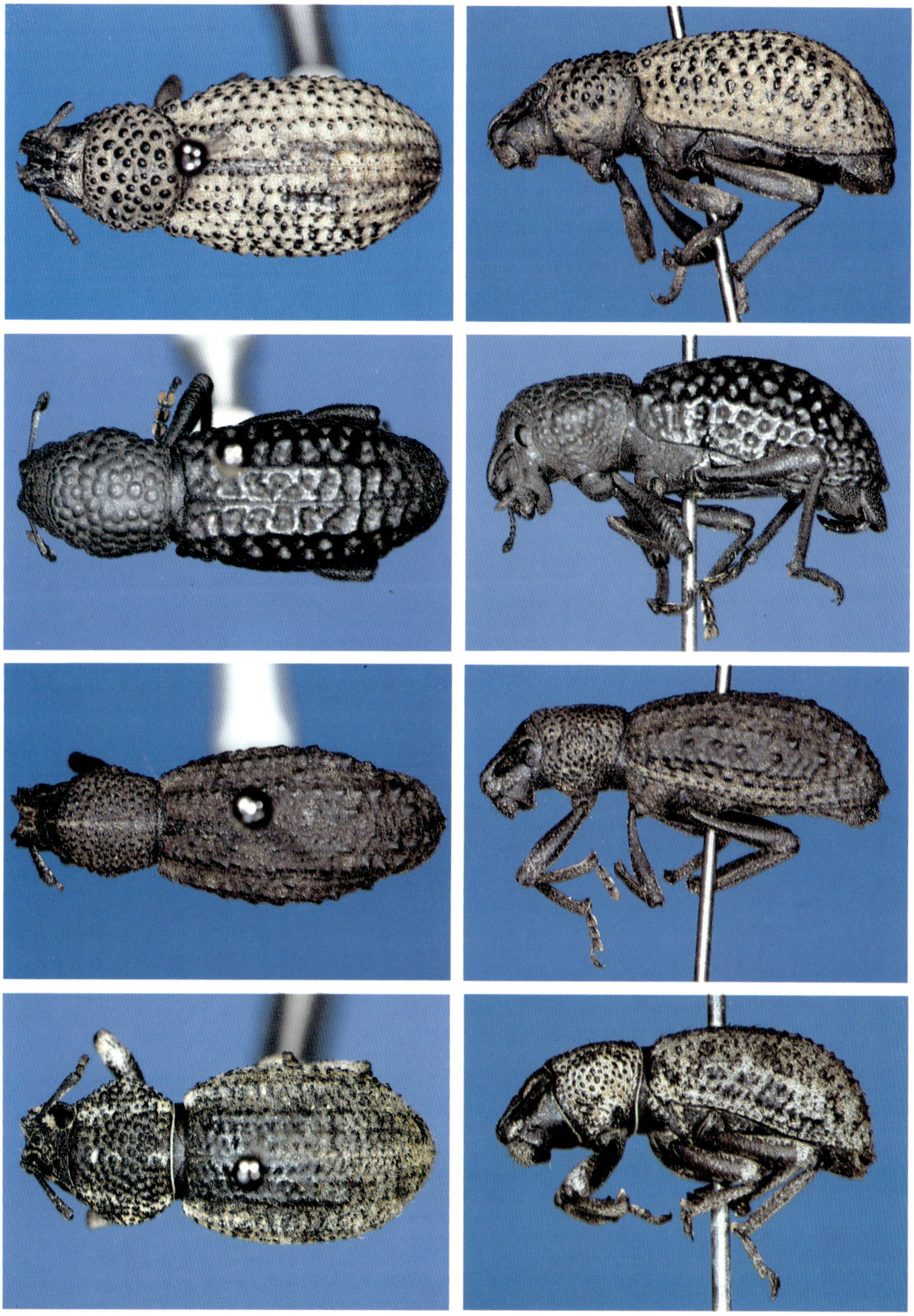

Plate 159 Amycterinae

1, 2 *Talaurinus bubaroides* Ferguson
Fowlers Bay, South Australia
Paratype female; 12.5 mm

3, 4 *Talaurinus bucephalus* (Olivier)
Blackheath, New South Wales
Male; 19 mm

5, 6 *Talaurinus capito* Pascoe
Champion Bay, Western Australia
Holotype female; 19 mm

7, 8 *Talaurinus carbonarius* Pascoe
Swan River, Western Australia
Male, compared with holotype; 14 mm

1	2
3	4
5	6
7	8

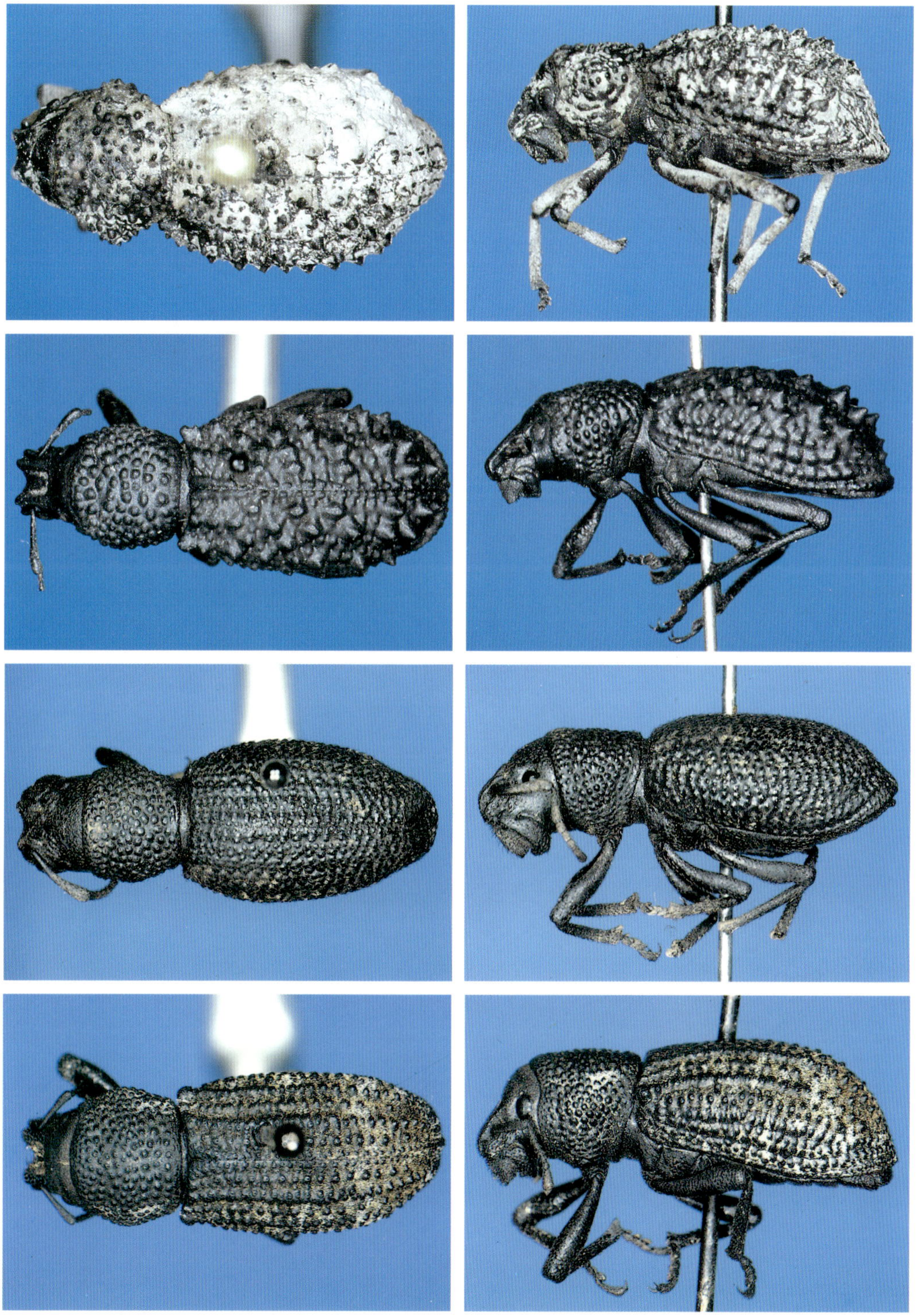

Plate 160 Amycterinae

1, 2 *Talaurinus carinatifrons* Ferguson
South Australia
Holotype male; 21 mm

3, 4 *Talaurinus carinatior* Ferguson
Coramba, New South Wales
Holotype male; 24 mm

5, 6 *Talaurinus carinatus* Ferguson
Victoria
Holotype male; 16 mm

7, 8 *Talaurinus carpentariae* Ferguson
North Queensland
Holotype male; 15 mm

1	2
3	4
5	6
7	8

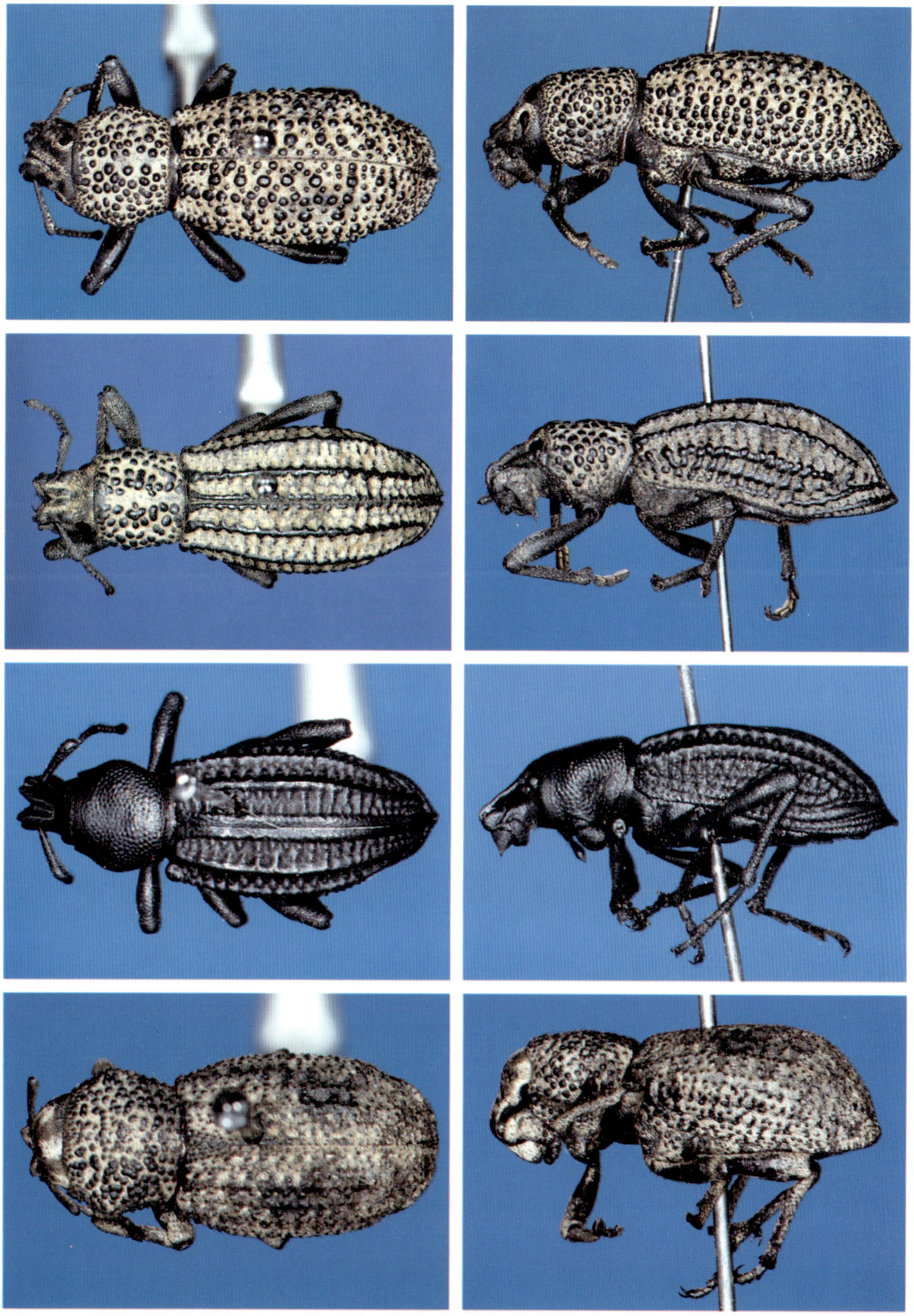

Plate 161 Amycterinae

1, 2 *Talaurinus carteri* Ferguson
Fletcher, Queensland
Male, compared with holotype; 26 mm

3, 4 *Talaurinus catenulatus* Macleay
New Holland
Holotype male; 21.5 mm

5, 6 *Talaurinus caviceps* Macleay
Port Lincoln, South Australia
Syntype female; 24.5 mm

7, 8 *Talaurinus confusus* Ferguson
Sarah, New South Wales
Holotype male; 16 mm

1	2
3	4
5	6
7	8

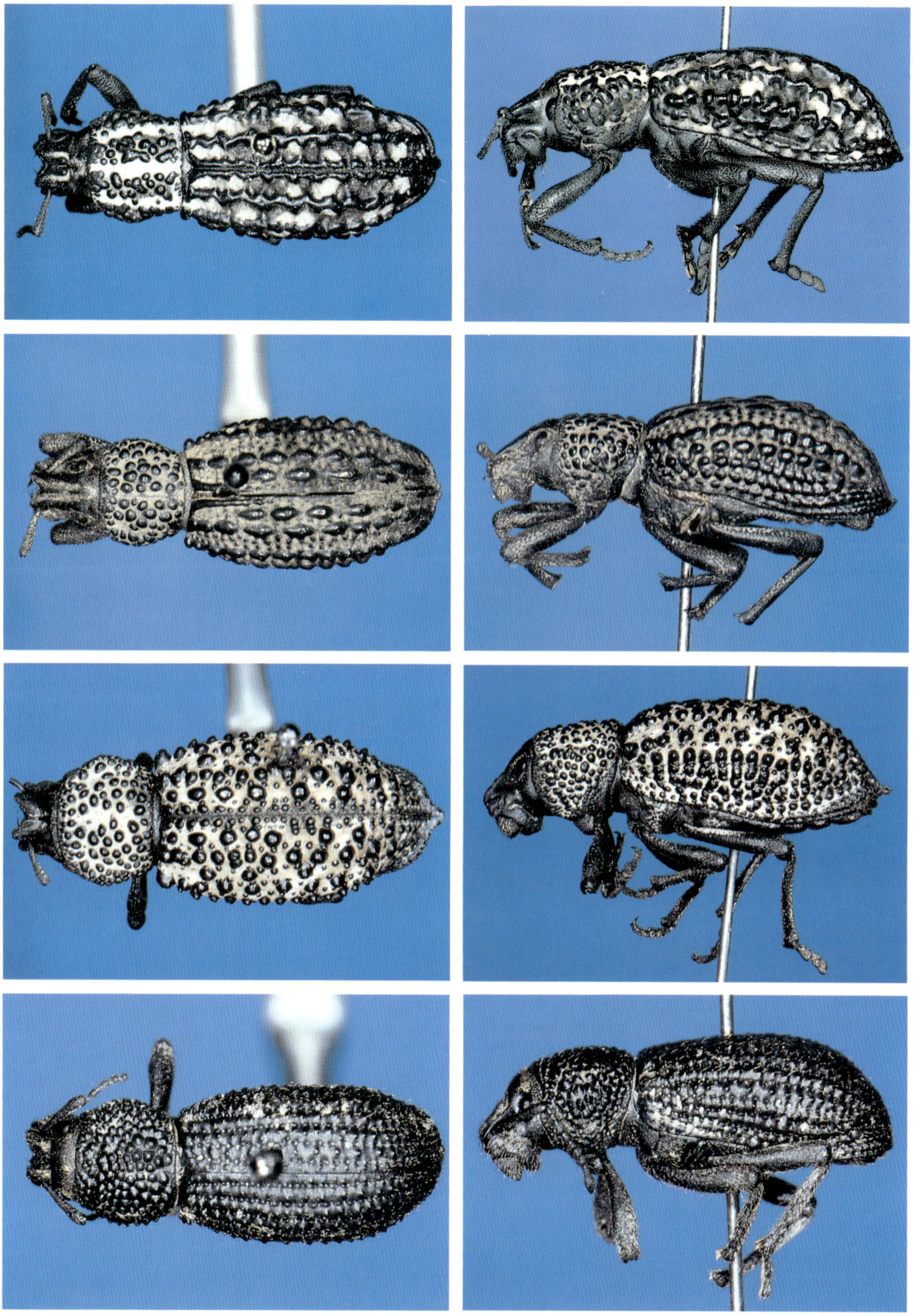

Plate 162 Amycterinae

1, 2 *Talaurinus costipennis* Ferguson
Victoria
Male; 15 mm

3, 4 *Talaurinus crassiceps* Sloane
Barrow Range, Western Australia
Holotype male; 15 mm

5, 6 *Talaurinus crenulatus* Ferguson
Tenterfield, New South Wales
Holotype male; 16 mm

7, 8 *Talaurinus euomoides* Macleay
Moreton Bay, Queensland
Holotype female; 15 mm

1	2
3	4
5	6
7	8

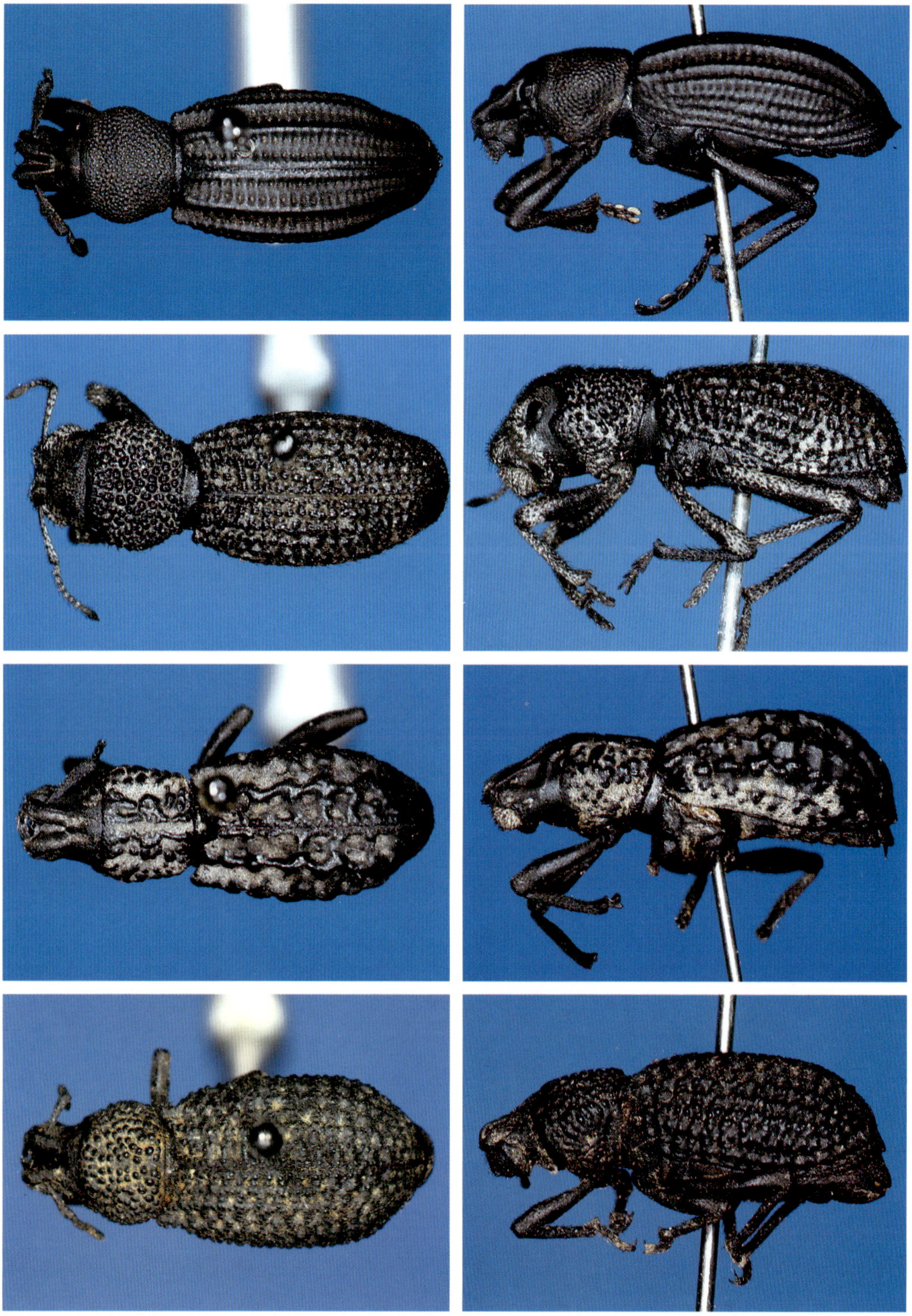

Plate 163 Amycterinae

1, 2 *Talaurinus fergusoni* Carter
Wyberba, Queensland
Paratype male; 22 mm

3, 4 *Talaurinus flaveolus* Ferguson
Naretha, Western Australia
Male; 11 mm
The dorsum is abraded.

4, 5 *Talaurinus flaveolus* Ferguson
Eucla, Western Australia
Allotype female; 11 mm
This may not be the same species as the holotype.

7, 8 *Talaurinus fossulatus* Ferguson
New South Wales
Holotype male; 21 mm

1	2
3	4
5	6
7	8

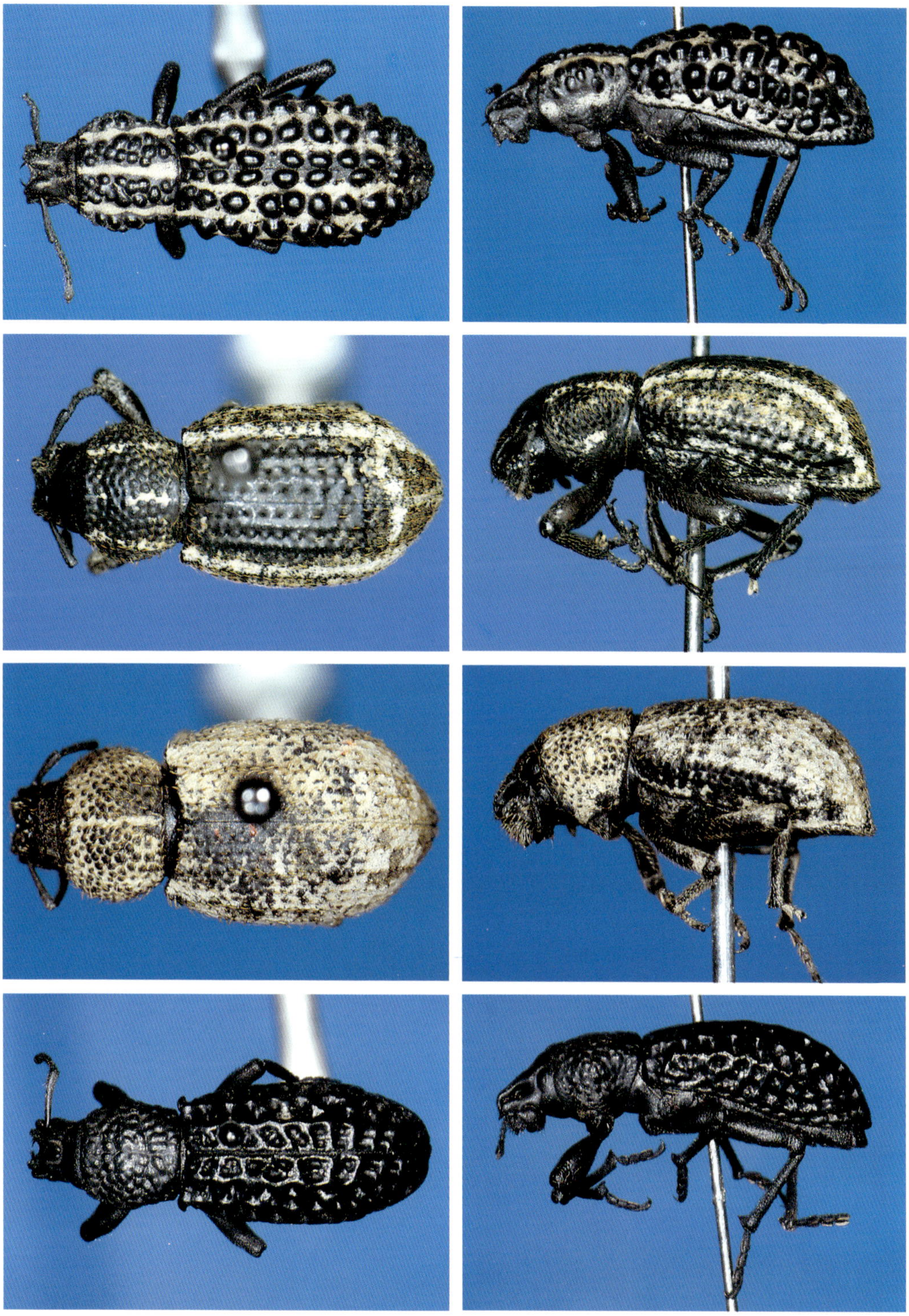

Plate 164 Amycterinae

1, 2 *Talaurinus foveatus* "variety" *montanus* Ferguson
Mount Irvine, Blue Mountains, New South Wales
Male; 25 mm

3, 4 *Talaurinus foveipennis* Ferguson
Wyberba near Stanthorpe, Queensland
Male, compared with holotype; 16 mm

5, 6 *Talaurinus gayndahensis* Ferguson
Gayndah, Queensland
Holotype male; 16 mm

7, 8 *Talaurinus griseus* Macleay
Dawson River, Queensland
Male, compared with holotype; 17 mm

1	2
3	4
5	6
7	8

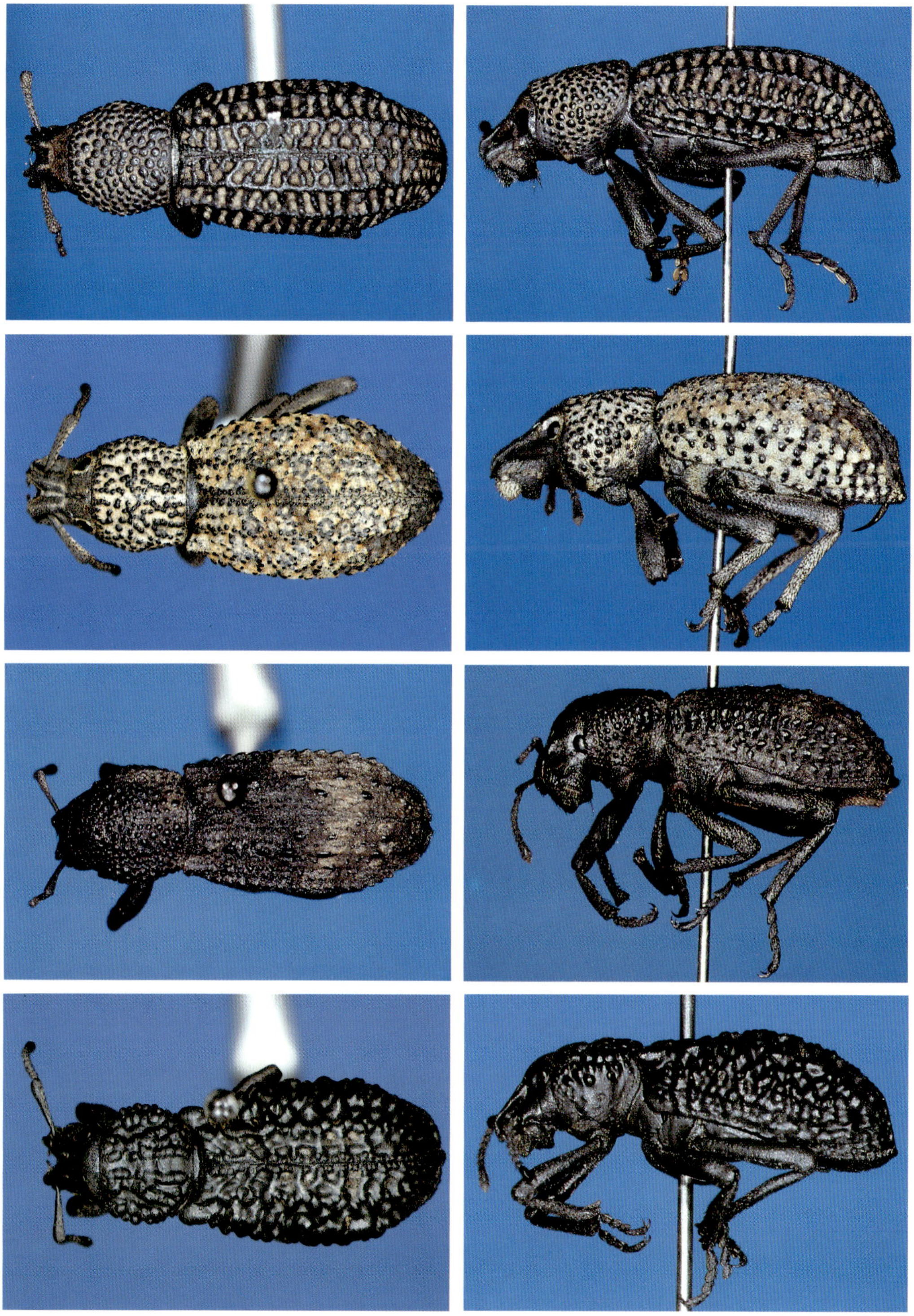

Plate 165 Amycterinae

1, 2 *Talaurinus halmaturinus* Ferguson
Kangaroo Island, South Australia
Paratype male; 20 mm

3, 4 *Talaurinus howittii* Macleay
Victoria
Holotype male; 12 mm

5, 6 *Talaurinus humeralis* Macleay
Victoria River, Queensland
Holotype male; 17 mm

7, 8 *Talaurinus hystrix* Ferguson
Portland, Victoria
Paratype male; 16.5 mm

1	2
3	4
5	6
7	8

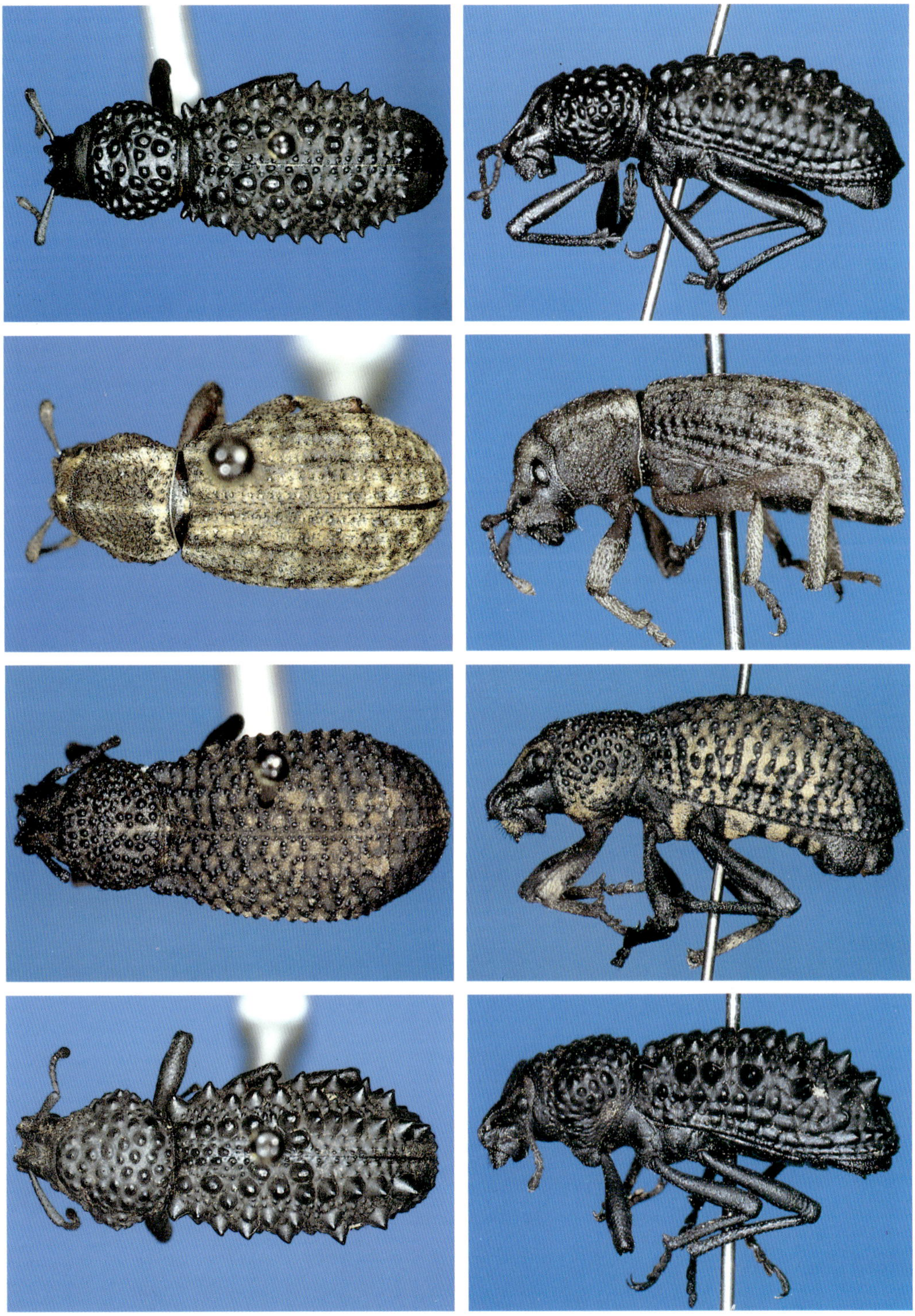

Plate 166 Amycterinae

1, 2 *Talaurinus illidgei* Ferguson
Brisbane, Queensland
Holotype male; 15 mm

3, 4 *Talaurinus imitator* Blackburn
Arnhemland, Northern Territory
Male; 13 mm

5, 6 *Talaurinus impressicollis* Macleay
Victoria
Syntype male; 21.5 mm

7, 8 *Talaurinus incanescens* Macleay
King George Sound, Western Australia
Syntype female; 15 mm

1	2
3	4
5	6
7	8

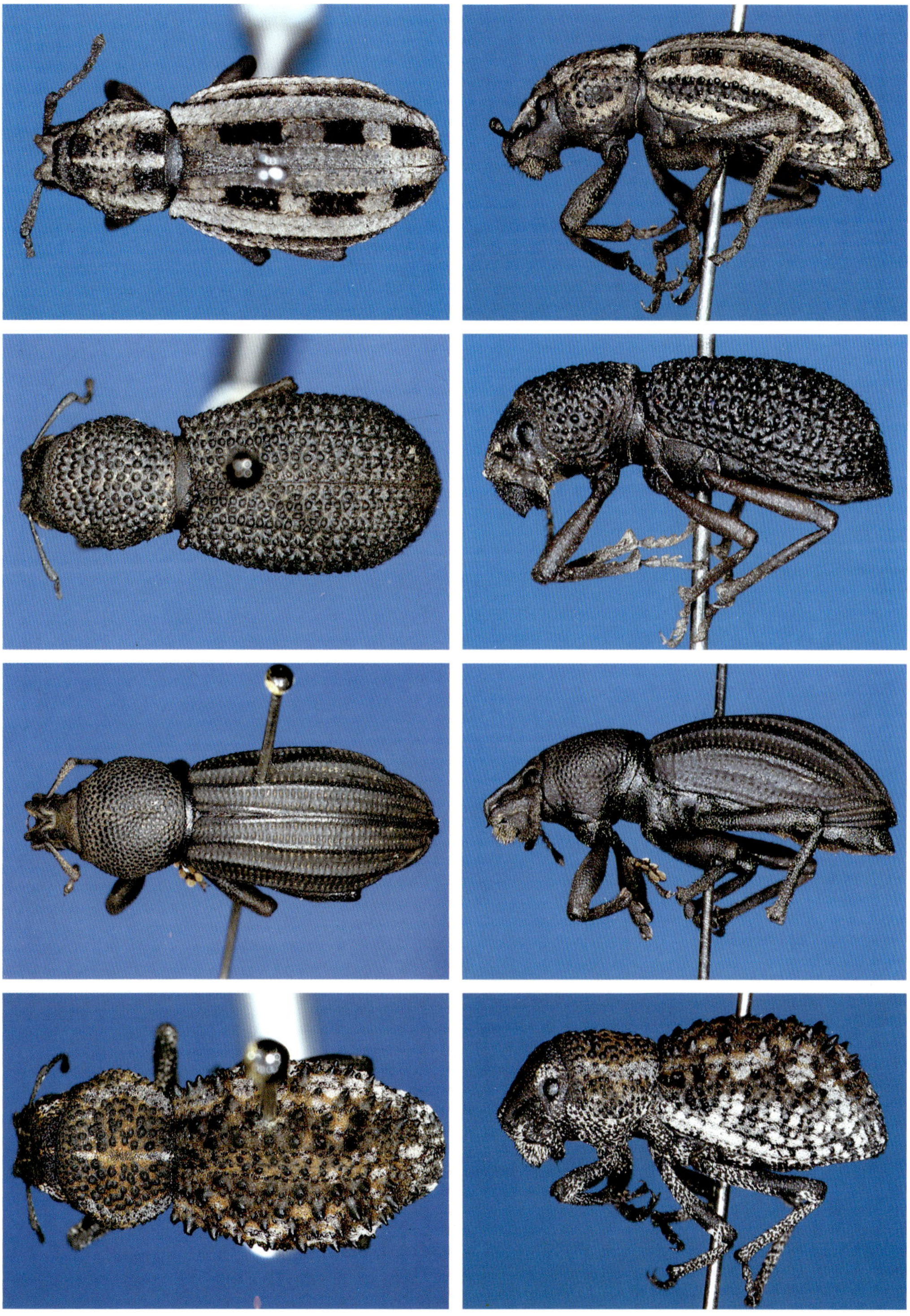

Plate 167 Amycterinae

1, 2 *Talaurinus inconspicuus* Ferguson
Brisbane, Queensland
Holotype male; 14 mm

3, 4 *Talaurinus irroratus* Ferguson
Blackheath, New South Wales
Male; 21.5 mm

5, 6 *Talaurinus kirbii* Macleay
New Holland
Holotype male; 16.5 mm

7, 8 *Talaurinus lacunosus* Macleay
Manning River, New South Wales
Holotype male; 21 mm

1	2
3	4
5	6
7	8

Plate 168 Amycterinae

1, 2 *Talaurinus leai* Ferguson
Swan River, Western Australia
Holotype male; 16 mm

3, 4 *Talaurinus longipes* Ferguson
New South Wales
Holotype male; 20.5 mm

5, 6 *Talaurinus maculipennis* Lea
Bardoc, Western Australia
Paratype male; 14.5 mm

7, 8 *Talaurinus megalongensis* Ferguson
Blue Mountains, New South Wales
Allotype female; 14.5 mm

1	2
3	4
5	6
7	8

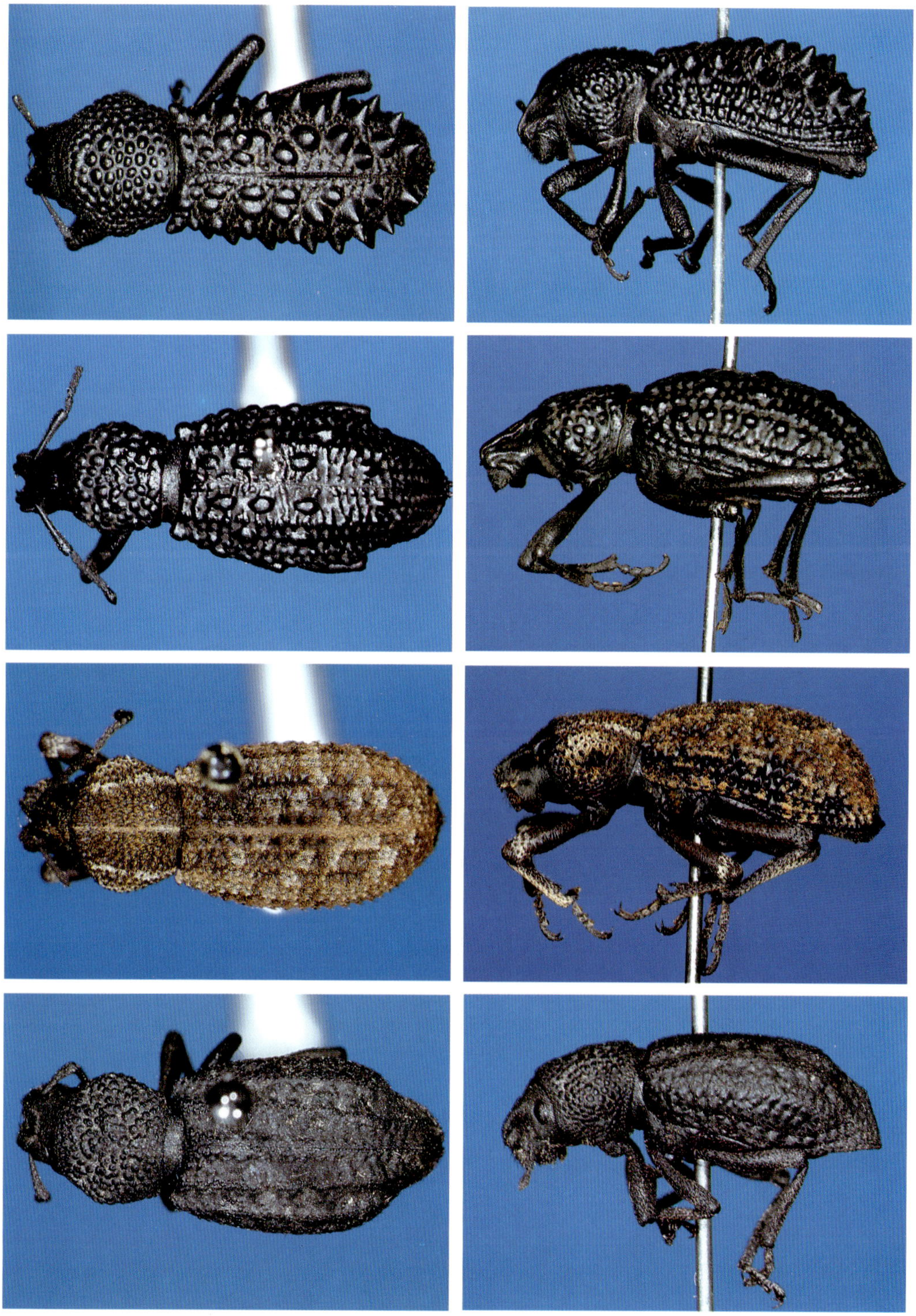

Plate 169 Amycterinae

1, 2 *Talaurinus miliaris* Ferguson
Mittagong, New South Wales
Holotype male; 21 mm

3, 4 *Talaurinus mitchellii* Macleay
Victoria River, Queensland
Holotype female; 19 mm

5, 6 *Talaurinus multigranulatus* (Lea)
Queensland?
Holotype female; 17 mm

7, 8 *Talaurinus mythitoides* Ferguson
Coonabarabran, New South Wales
Holotype male; 16.5 mm

1	2
3	4
5	6
7	8

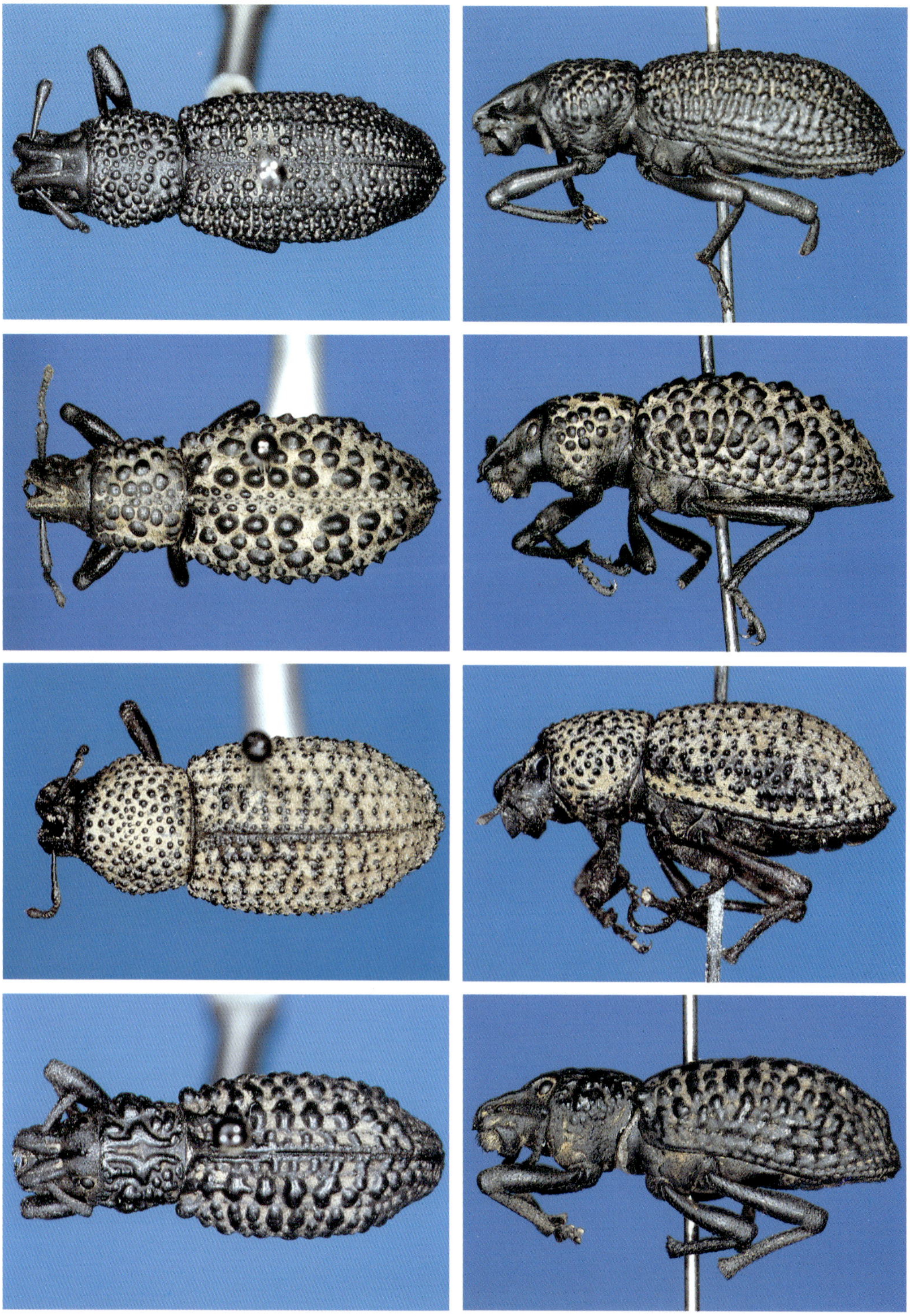

Plate 170 Amycterinae

1, 2 *Talaurinus niveovittatus* Ferguson
Inglewood, Queensland
Paratype male; 23.5 mm

3, 4 *Talaurinus pallidus* Macleay
Dawson River, Queensland
Male; 17 mm

5, 6 *Talaurinus papulosus* Macleay
New Holland
Holotype male; 17 mm

7, 8 *Talaurinus parallelus* Macleay
New Holland
Holotype male; 19.6 mm; abdomen missing.

1	2
3	4
5	6
7	8

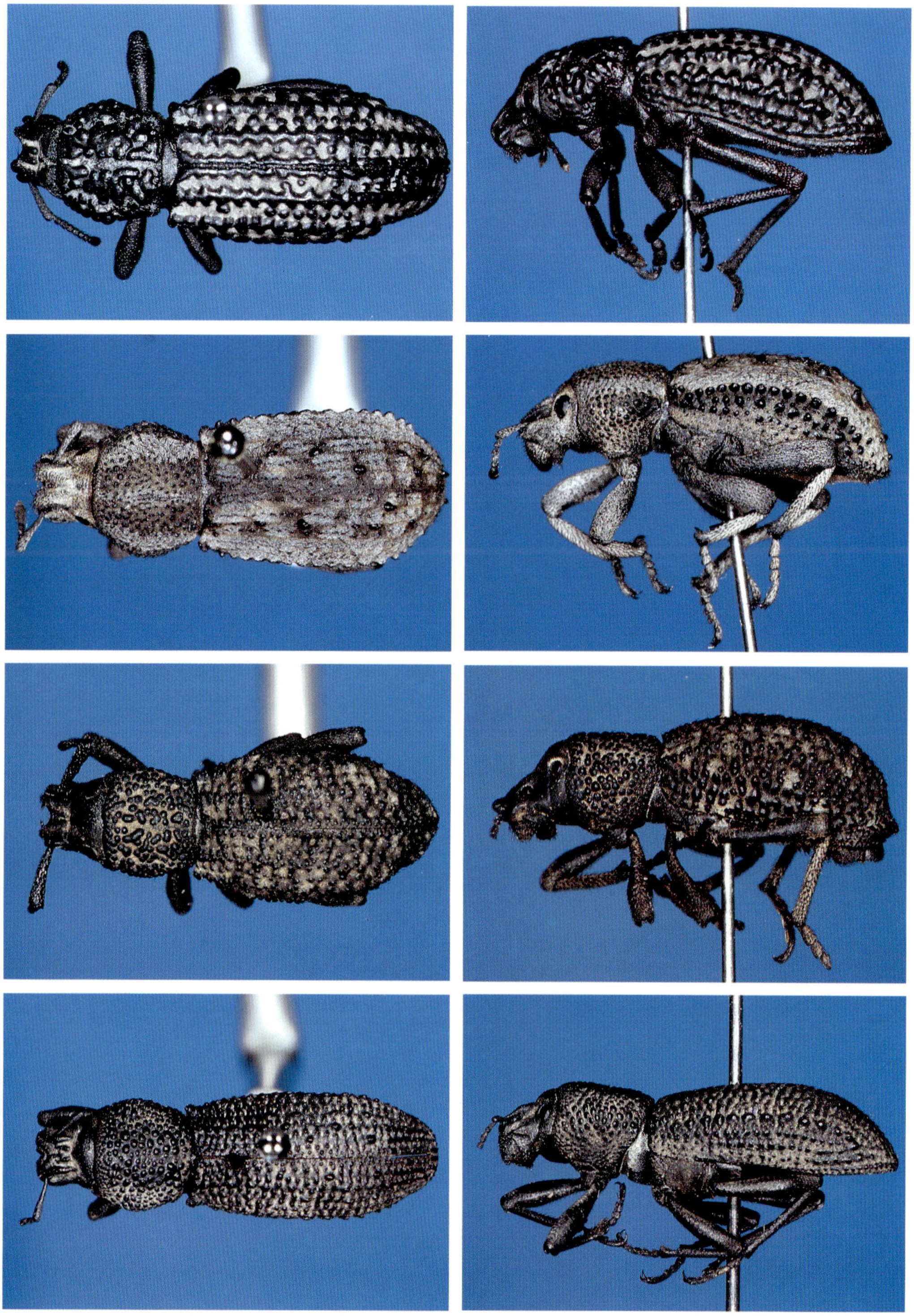

Plate 171 Amycterinae

1, 2 *Talaurinus parvus* Ferguson
Queensland
Holotype male; 11 mm

3, 4 *Talaurinus perplexus* Ferguson
Portland, Victoria
Holotype male; 18 mm

5, 6 *Talaurinus plagiatus* Ferguson
Coen, Queensland
Male; 10.5 mm

7, 8 *Talaurinus posticalis* Ferguson
Coramba, New South Wales
Holotype male; 20 mm

1	2
3	4
5	6
7	8

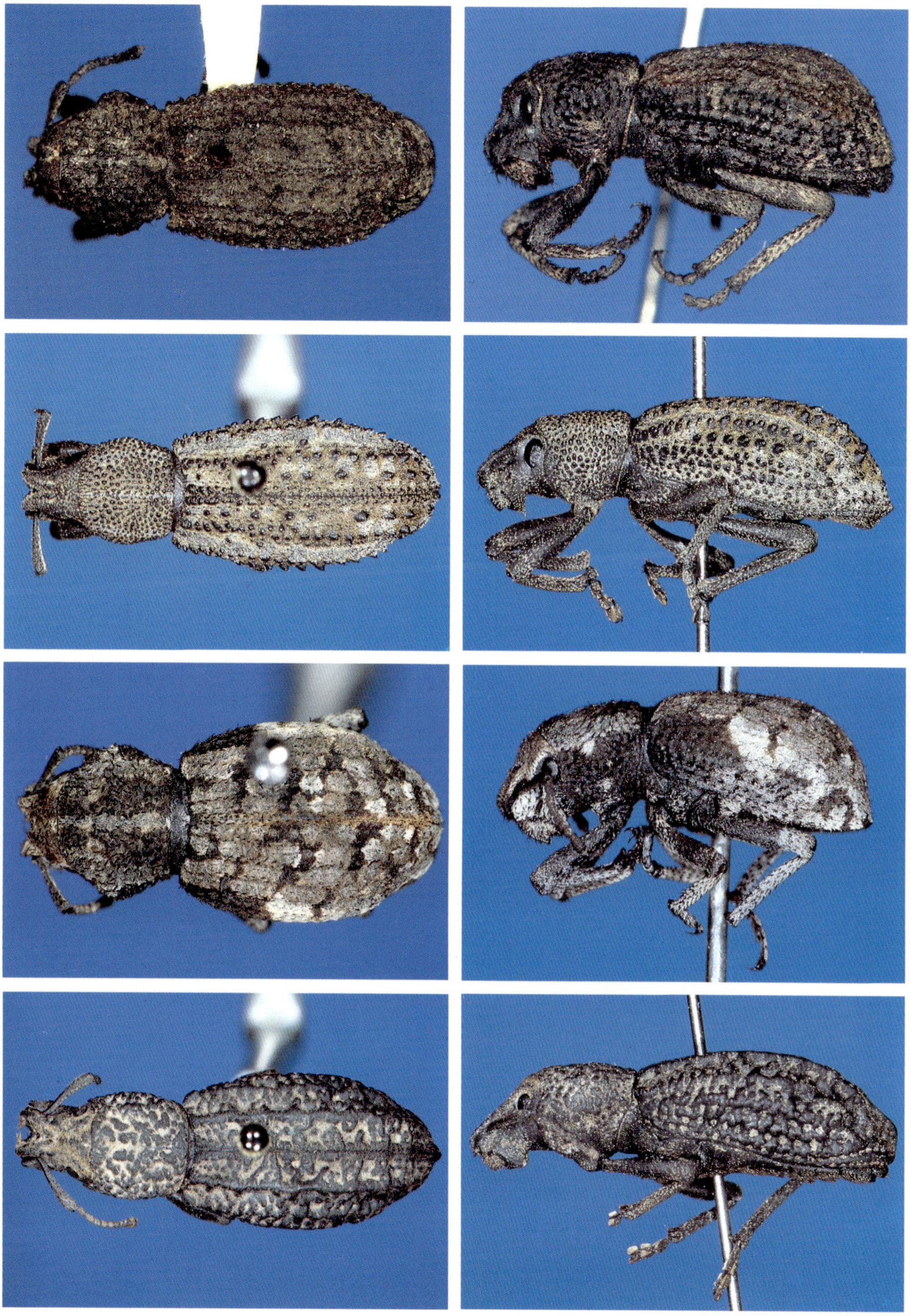

Plate 172 Amycterinae

1, 2 *Talaurinus prypnoides* Ferguson
South Australia
Holotype male; 11.5 mm

3, 4 *Talaurinus regularis* Sloane
Central Australia
Holotype male; 17 mm

5, 6 *Talaurinus roei* (Boheman)
King George Sound, Western Australia
Male; 20 mm

7, 8 *Talaurinus rugicollis* Macleay
Singleton, New South Wales
Holotype female; 21.5 mm

1	2
3	4
5	6
7	8

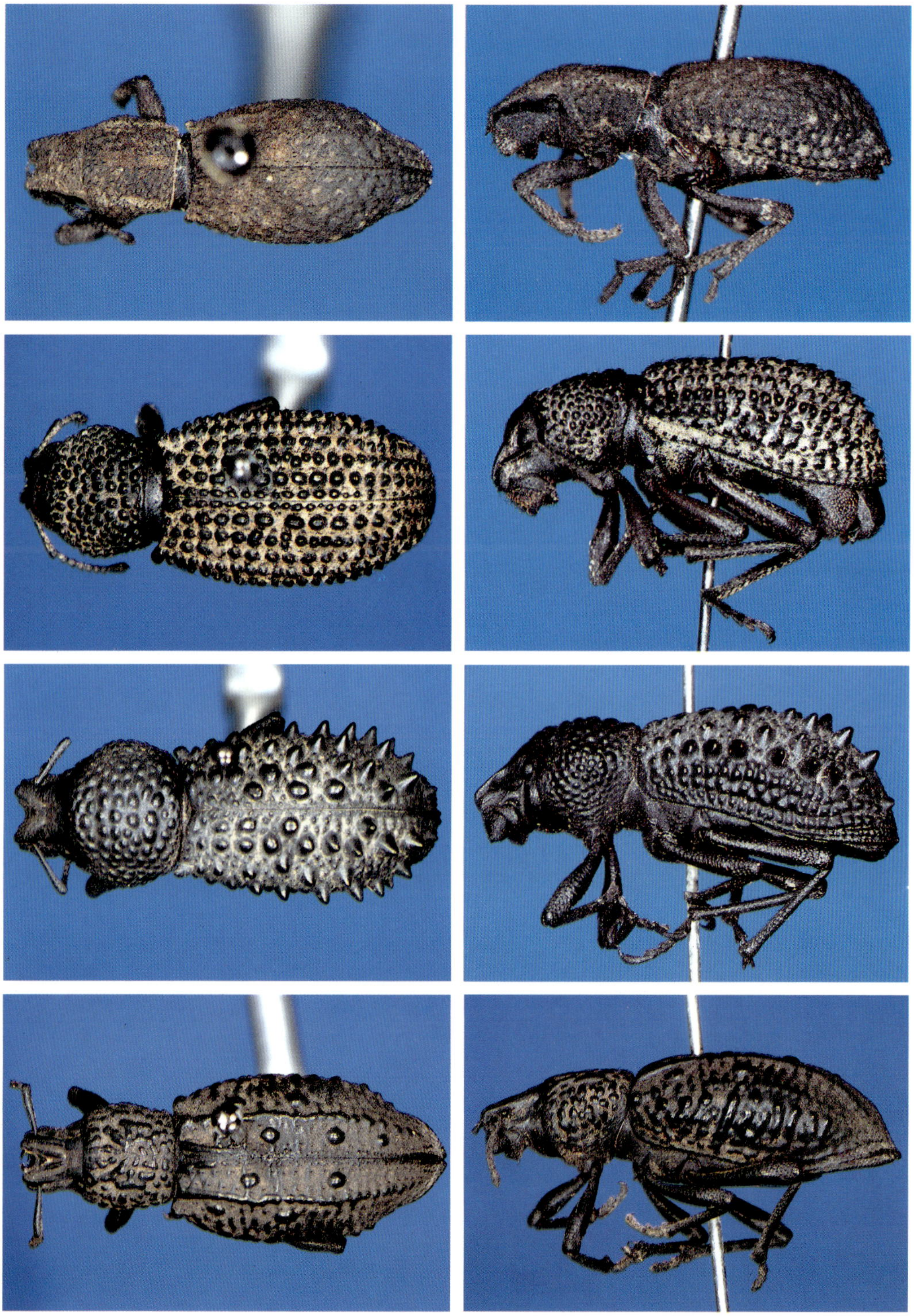

Plate 173 Amycterinae

1, 2 *Talaurinus rugifer* (Boisduval)
Swan River, Western Australia
Male; 24 mm

3, 4 *Talaurinus scaber* (Boisduval)
Victoria
Male; 12 mm; from a syntype of *aberrans* Macleay.

5, 6 *Talaurinus scabricollis* Ferguson
Swan River, Western Australia
Female; 24 mm; from the holotype of *scaber* Macleay, not Boisduval.

7, 8 *Talaurinus scabriosus* Macleay
Mitchell River, Queensland
Lectotype male; 24 mm

1	2
3	4
5	6
7	8

Plate 174 Amycterinae

1, 2 *Talaurinus scapularis* Ferguson
Queensland
Lectotype female; 14 mm

3, 4 *Talaurinus semispinosus* (Boheman)
King Goerge Sound, Western Australia
Male; 16.5 mm

5, 6 *Talaurinus septentrionalis* Ferguson
Mount Garnet, Queensland
Holotype male; 11.5 mm

7, 8 *Talaurinus simplex* Ferguson
Ouldea, South Australia
Male; 15 mm

1	2
3	4
5	6
7	8

Plate 175 Amycterinae

1, 2 *Talaurinus sobrinus* Ferguson
Queensland
Holotype male; 17 mm

3, 4 *Talaurinus sphaerulatus* Macleay
No locality data.
Holotype male; 18 mm

5, 6 *Talaurinus spiniger* Ferguson
South Australia
Paratype male; 15 mm

7, 8 *Talaurinus squamosus* Macleay
Murrumbidgee, New South Wales
Male; 13 mm

1	2
3	4
5	6
7	8

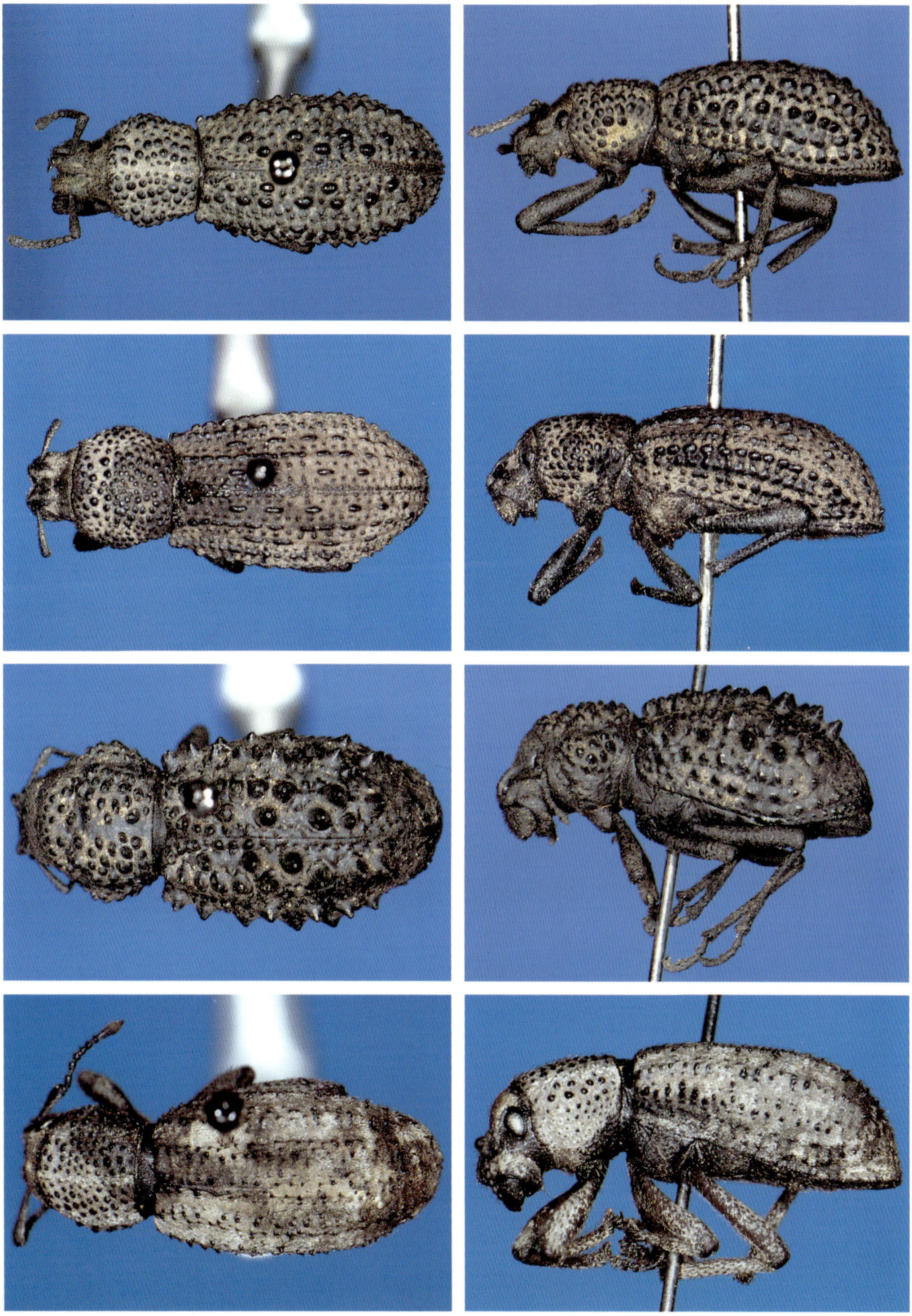

Plate 176 Amycterinae

1, 2 *Talaurinus strangulatus* Blackburn
Oodnadatta, South Australia
Paratype male; 16 mm

3, 4 *Talaurinus subvittatus* Ferguson
North Queensland
Holotype male; 17 mm

5, 6 *Talaurinus suttoni* Carter
Fletcher, south Queensland
Female; 25 mm

7, 8 *Talaurinus tuberculosus* Zimmerman
Victoria
Holotype male; 17.5 mm

1	2
3	4
5	6
7	8

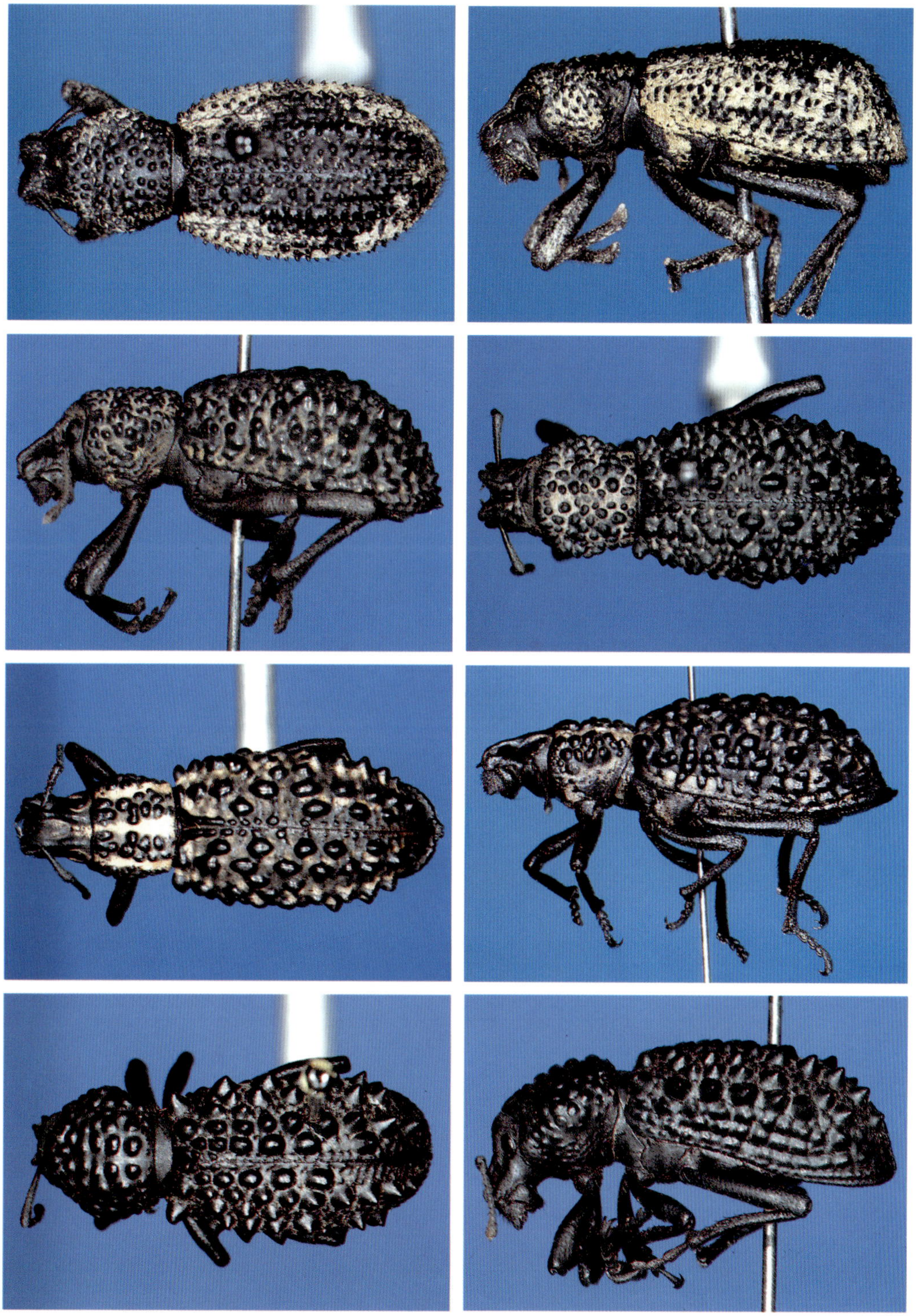

Plate 177 Amycterinae

1, 2 *Talaurinus tumulosus* Ferguson
Inverell, New South Wales
Holotype female; 17.5 mm

3, 4 *Talaurinus typicus* Macleay
Argyle, New South Wales
Syntype male; 18 mm

5, 6 *Talaurinus vermicollis* Ferguson
Mudgee, New South Wales
Holotype male; 14 mm

7, 8 *Talaurinus verrucosus* (Boisduval)
Blue Mountains, New South Wales
Female; 25.5 mm

1	**2**
3	**4**
5	**6**
7	**8**

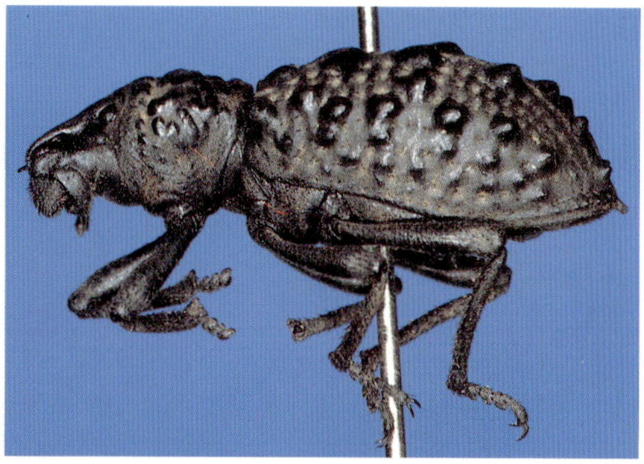

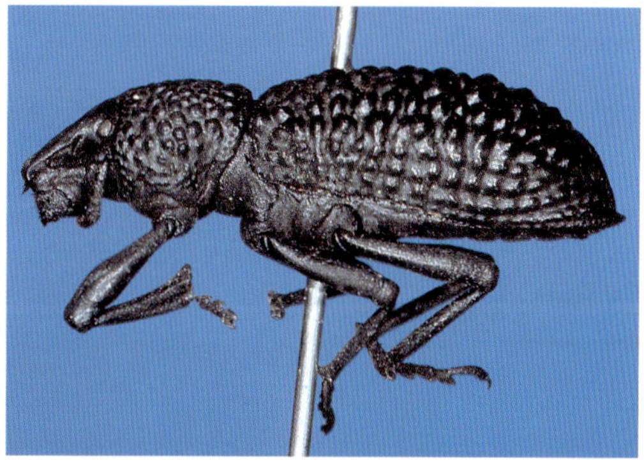

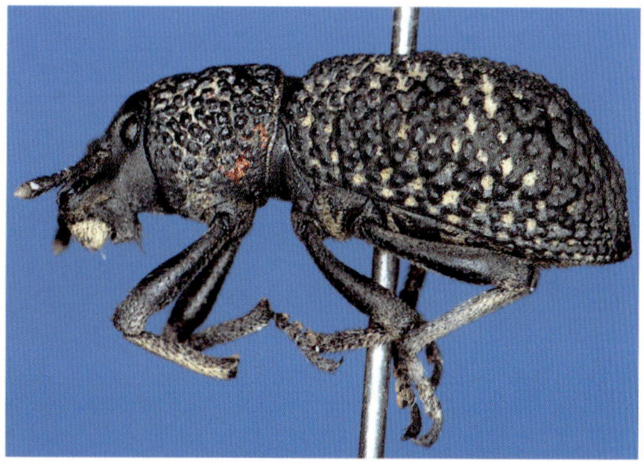

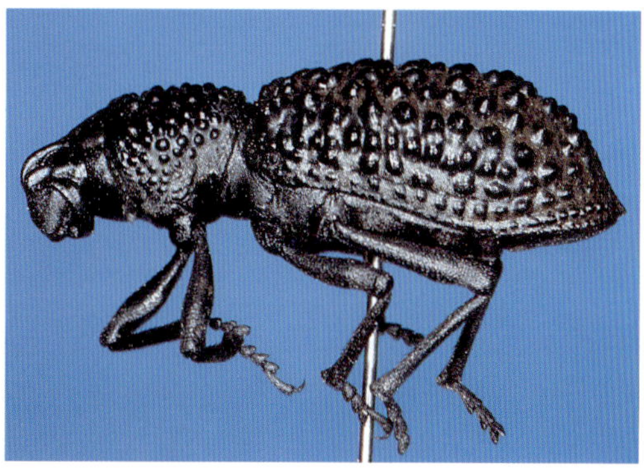

Plate 178 Amycterinae

1, 2 *Talaurinus vitticollis* Ferguson
Queensland
Holotype female; 17 mm

3, 4 *Sclerorinus subcarinatus* Ferguson
Grenfell, New South Wales
Holotype male; 17 mm

5, 6 *Talaurinus laevicollis* Pascoe
Victoria
Male; 17 mm

7, 8 *Talaurinus simplicipes* Lea
South Australia
Holotype male; 12.5 mm

1	2
3	4
5	6
7	8

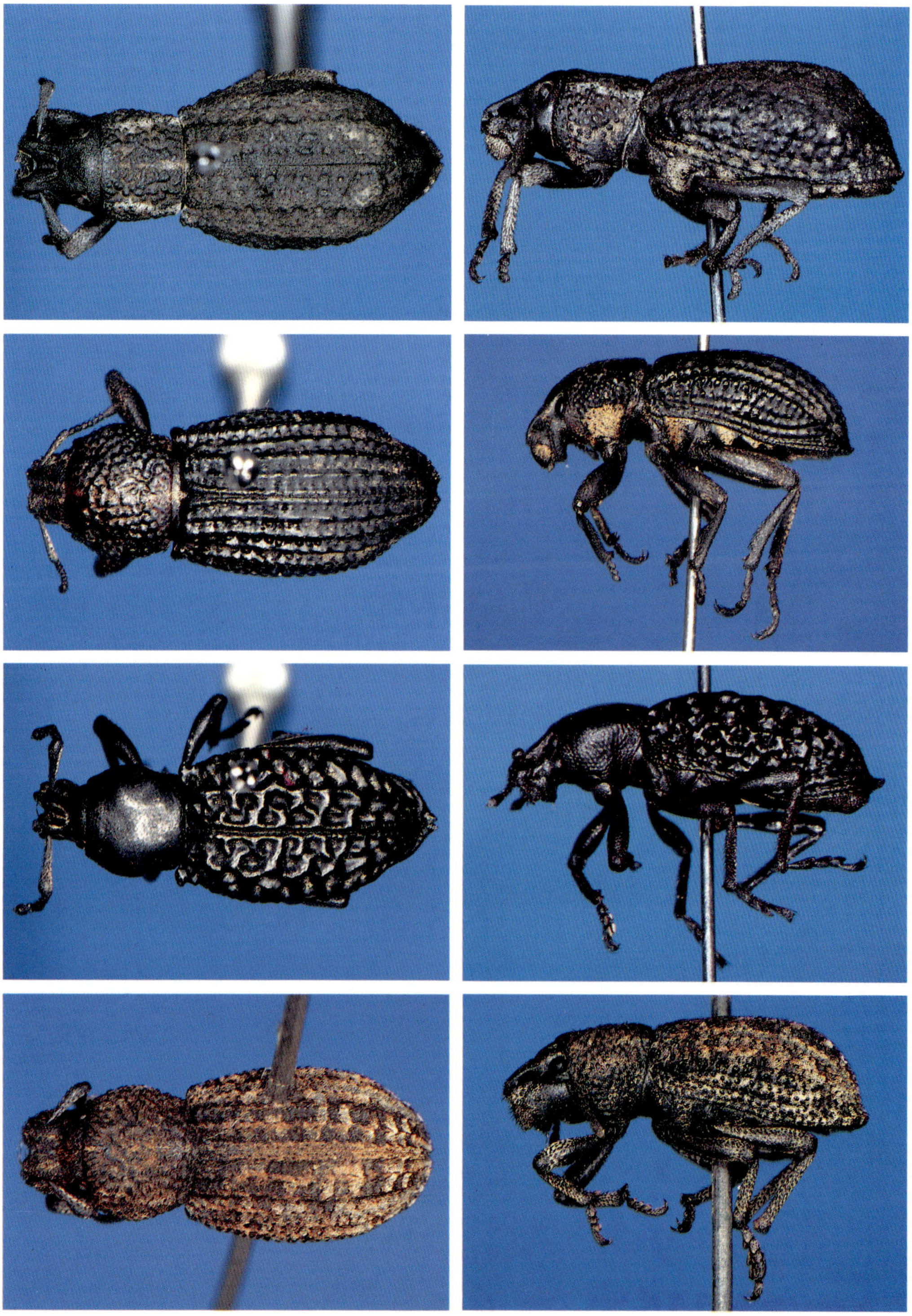

Plate 179 Amycterinae

1, 2 *Sosyteles carteri* (Ferguson)
Dalveen, Queensland
Holotype male; 9 mm

3, 4 *Tetralophus sculpturatus* Waterhouse
Victoria
Male; 12.5 mm

5 *Phalidura reticulata* (Boisduval)
Mount Tamborine, Queensland
Male; 22.5 mm
From a specimen of the synonym *costipennis* (Ferguson) with abdomen extended in precopulatory position. Note the co-called "gular" horn.

6 The same; enlarged side view of the caudal abdominal segments.

7 The same; dorsal view of tergites 7 and 8 (forceps protruding from sternite 8).

8 The same; ventral view of head to show the so-called "gular" horn, which is not on the gular area. See also Plate 138:7, 8.

Coloured illustrations of Amycterinae are continued in Volume VI.

1	2
3	4
5	6
7	8

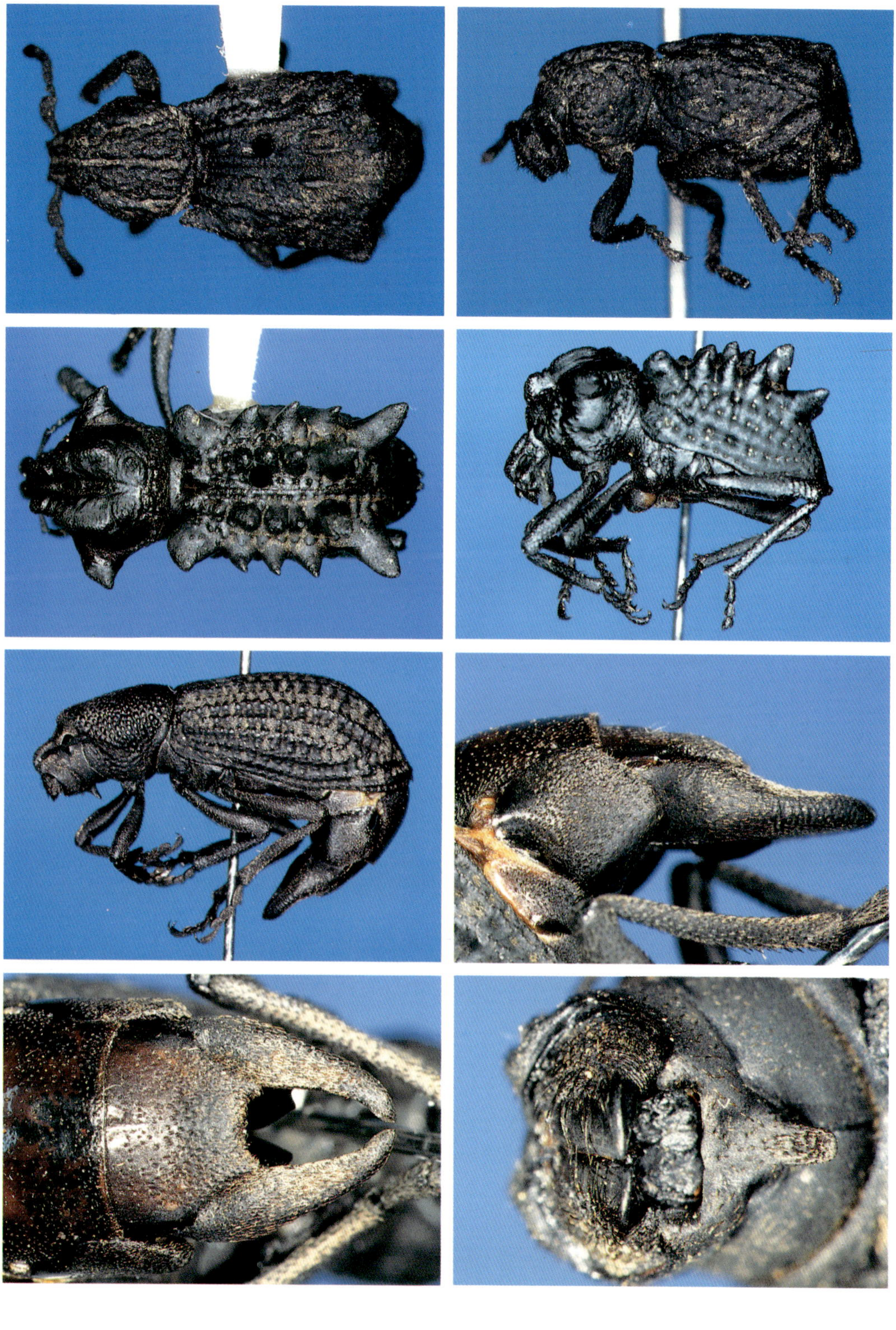

Plate 180 Adelognatha

1, 2 *Afurius sublineatus* (Lea)
Forest Reefs, New South Wales
Paratype female; 4.8 mm

3, 4 *Allotimareta setistriata* (Lea)
Perth, Western Australia
Holotype female; 3 mm

5, 6 *Amandalotus cordipennis* (Lea)
Beaconsfield, Victoria
Paratype male; 2.3 mm

7, 8 *Asceparnodes duplicatus* (Lea)
Mittagong, New South Wales
Male; 6.3 mm

1	2
3	4
5	6
7	8

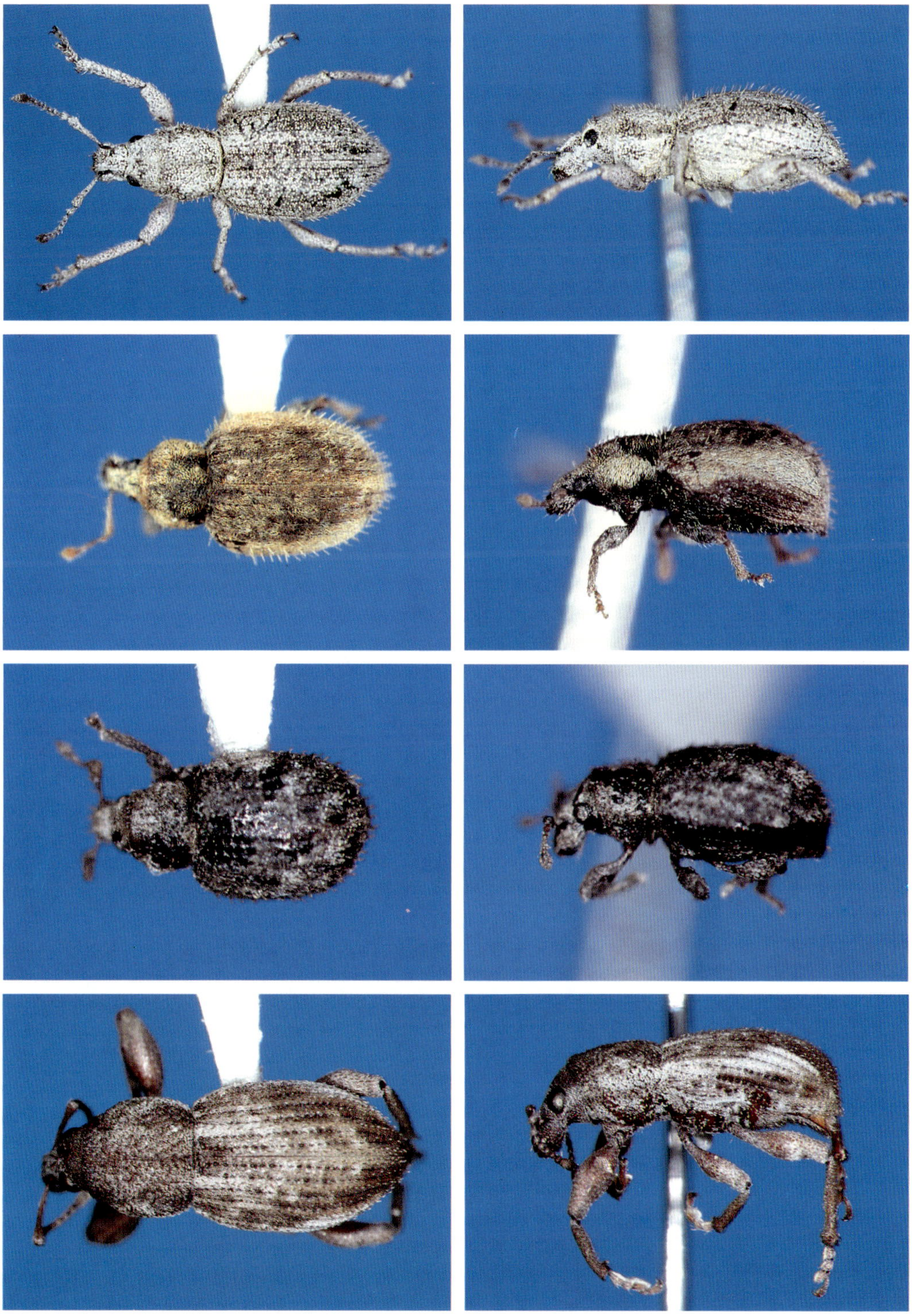

Plate 181 Adelognatha

1, 2 *Asceparnus nodipennis* (Lea)
Mount Barker, Western Australia
Holotype female; 7.5 mm

3, 4 *Asceparnus subfasciatus* (Blackburn)
Western Australia
Female; 6 mm

5, 6 *Asceparnus xanthorrhoeae* (Lea)
Darling Ranges, Western Australia
Paratype female; 4.5 mm

7, 8 *Catasarcus carbo* Pascoe
Sharks Bay, Western Australia
Male; 14 mm

1	2
3	4
5	6
7	8

Plate 182 Adelognatha

1, 2 *Catasarcus griseus* Pascoe
Pinjarra, Western Australia
Female; 13.5 mm

3, 4 *Catasarcus impressipennis* Boisduval
Torbay, Western Australia
Male; 13.5 mm

5, 6 *Catasarcus obesus* Thompson
Transmitter Tower, east of Yellowstone, Western Australia
Female; 14 mm

7, 8 *Catasarcus pallidiventris* Thompson
Hill River district, Western Australia
Paratype female; 17 mm

1	2
3	4
5	6
7	8

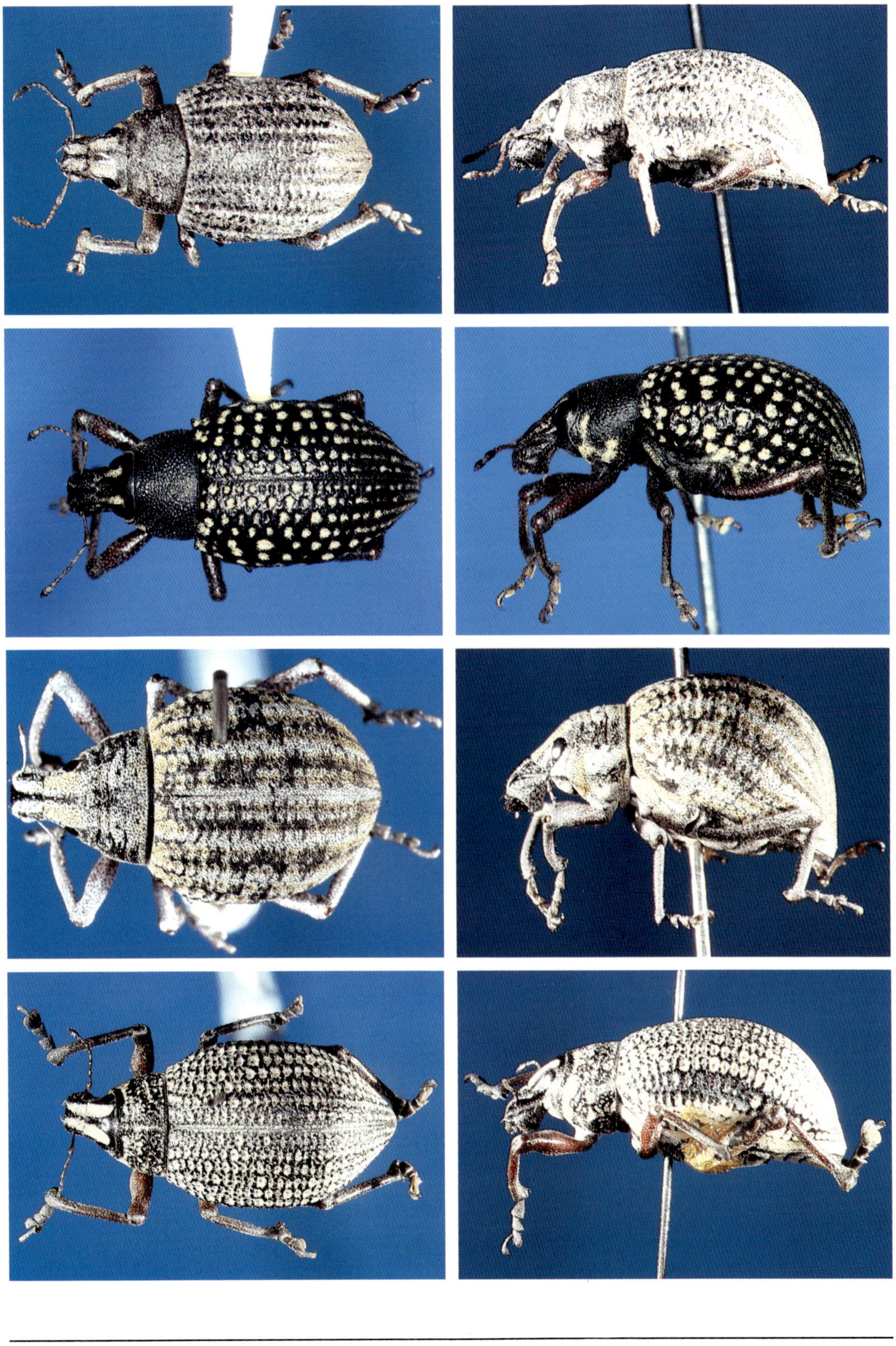

Plate 183 Adelognatha

1, 2 *Catasarcus rugosus* Boheman
Albany, Western Australia
Female; 9.5 mm

3, 4 *Catasarcus varus* Thompson
Porongorup, Western Australia
Male; 14.5 mm

5, 6 *Catastygnus elegans* Lea
Stanthorpe, Queensland
Male; 15 mm

7, 8 *Catastygnus rivularis* Pascoe
New South Wales
Female; 18 mm; from the holotype of *amplipennis* Lea, a synonym.

For *Catastignus stigmus* Pascoe, see Plate 217:7, 8.

1	2
3	4
5	6
7	8

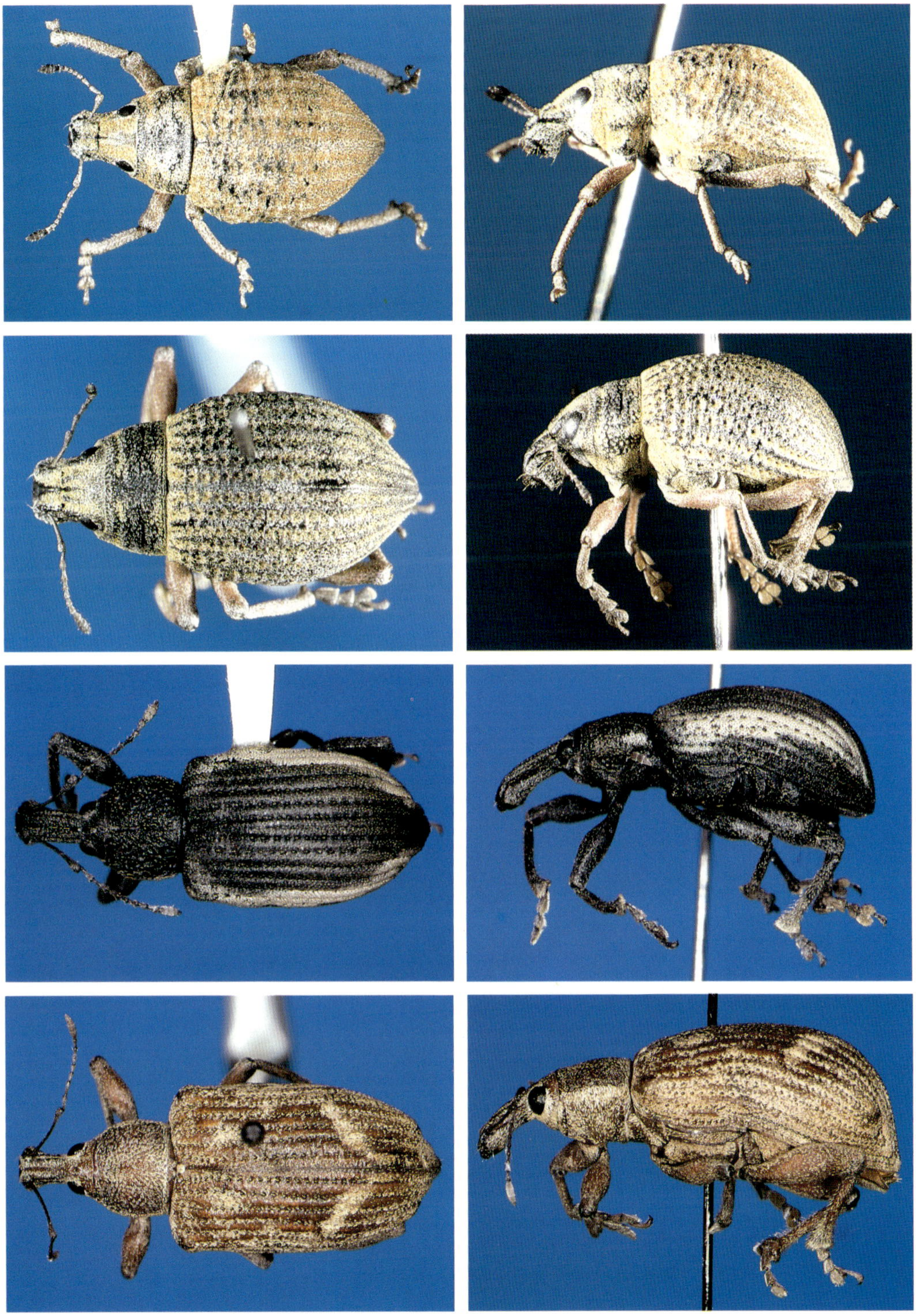

Plate 184 Adelognatha

1, 2 *Catastygnus textilis* Pascoe
Rockhampton, Queensland
Female; 15 mm

3, 4 *Catastygnus v-album* Lea
New South Wales
Female, compared with holotype; 15 mm

5, 6 *Catastygnus variabilis* Lea
Cairns, Queensland
Paratype female; 14.5 mm

7, 8 *Ecrizothis boviei* Lea
Dividing Range, Victoria
Female; 12.5 mm

1	2
3	4
5	6
7	8

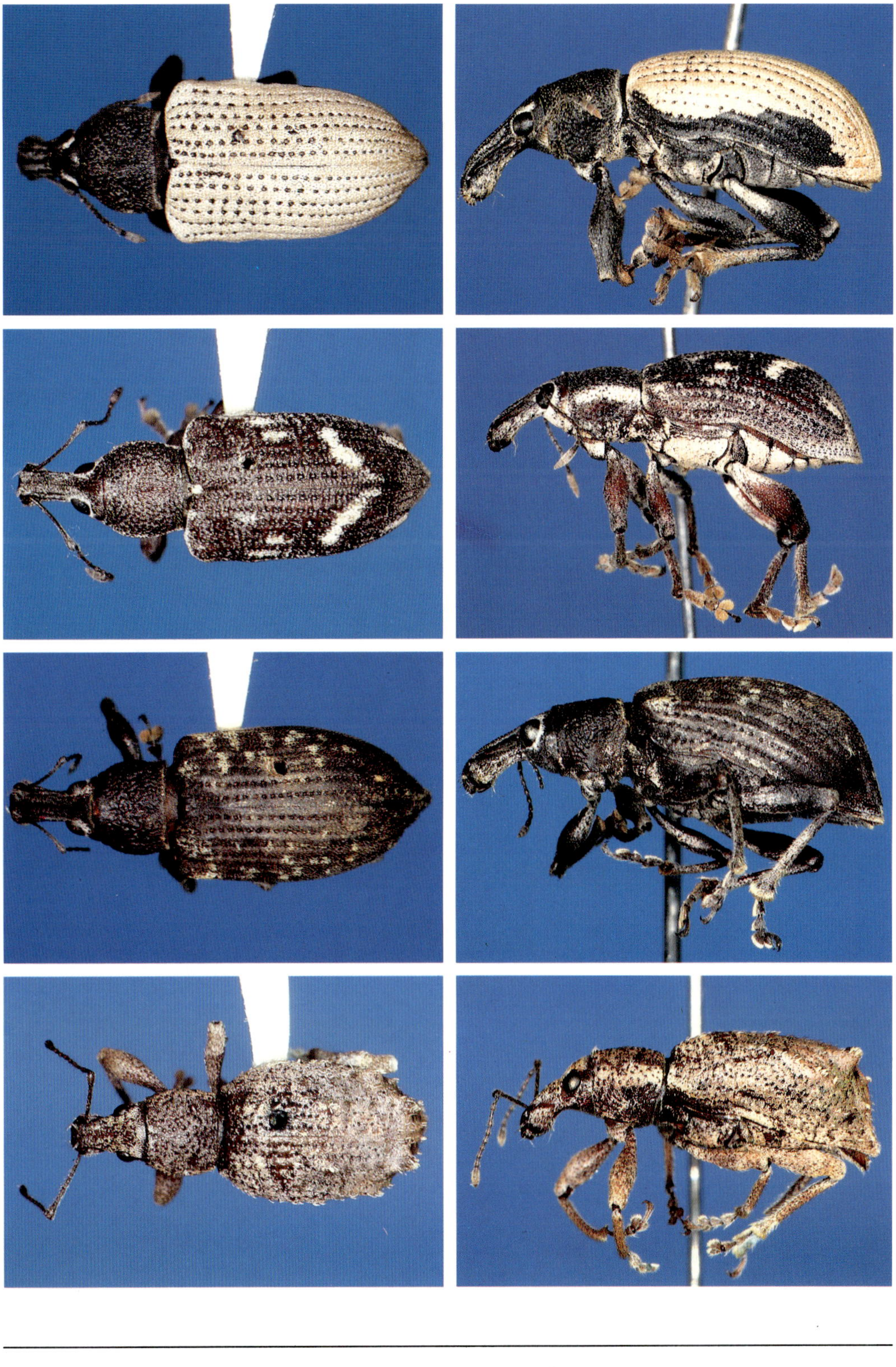

Plate 185 Adelognatha

1, 2 *Ecrizothis terminalis* Oke
Apollo Bay, Victoria
Female, compared with paratype; 7.8 mm

3, 4 *Essolithna cordipennis* Lea
Geraldton, Western Australia
Holotype male; 4.1 mm

5, 6 *Essolithna jonesi* Lea
Ooldea, South Australia
Holotype male; 4.2 mm

7,8 *Essolithna kingiae* Lea
Pinjarrah, Western Australia
Holotype male; 3.8 mm

1	2
3	4
5	6
7	8

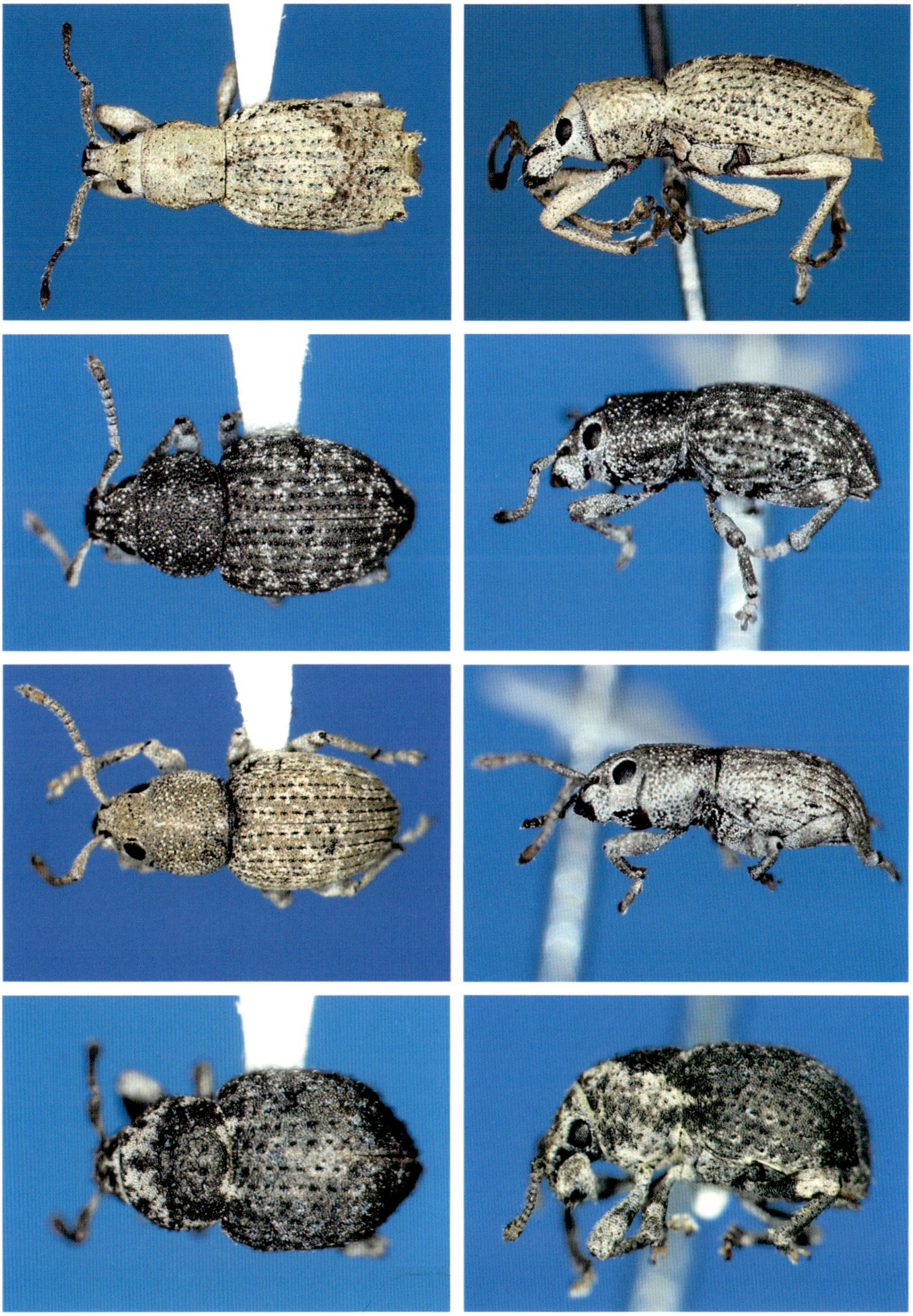

Plate 186 Adelognatha

1, 2 *Essolithna maculata* Lea
Western Australia
Paratype male; 4 mm

3, 4 *Essolithna mediofusca* Lea
Alexandria, Northern Territory
Paratype male; 5.7 mm

5, 6 *Essolithna seriata* Blackburn
Ooldea, South Australia
Female; 5.8 mm

7, 8 *Essolithna terrena* Lea
Swan River, Western Australia
Holotype male; 3.7 mm

1	2
3	4
5	6
7	8

Plate 187 Adelognatha

1,2 *Eurymetopus birabeni* Kuschel
Canberra, Australian Capital Territory
Female; 8 mm

3, 4 *Eutinophaea bicristata* Lea
Cairns District, Queensland
Male; 2.9 mm

5, 6 *Eutinophaea nana* Pascoe
Sydney. New South Wales
From a female paratype of *Eutinophaea dispar* Lea; 2.5 mm

7, 8 *Eutinophaea falcata* Lea
Dunk Island, Queensland
Female; 3.1 mm

1	2
3	4
5	6
7	8

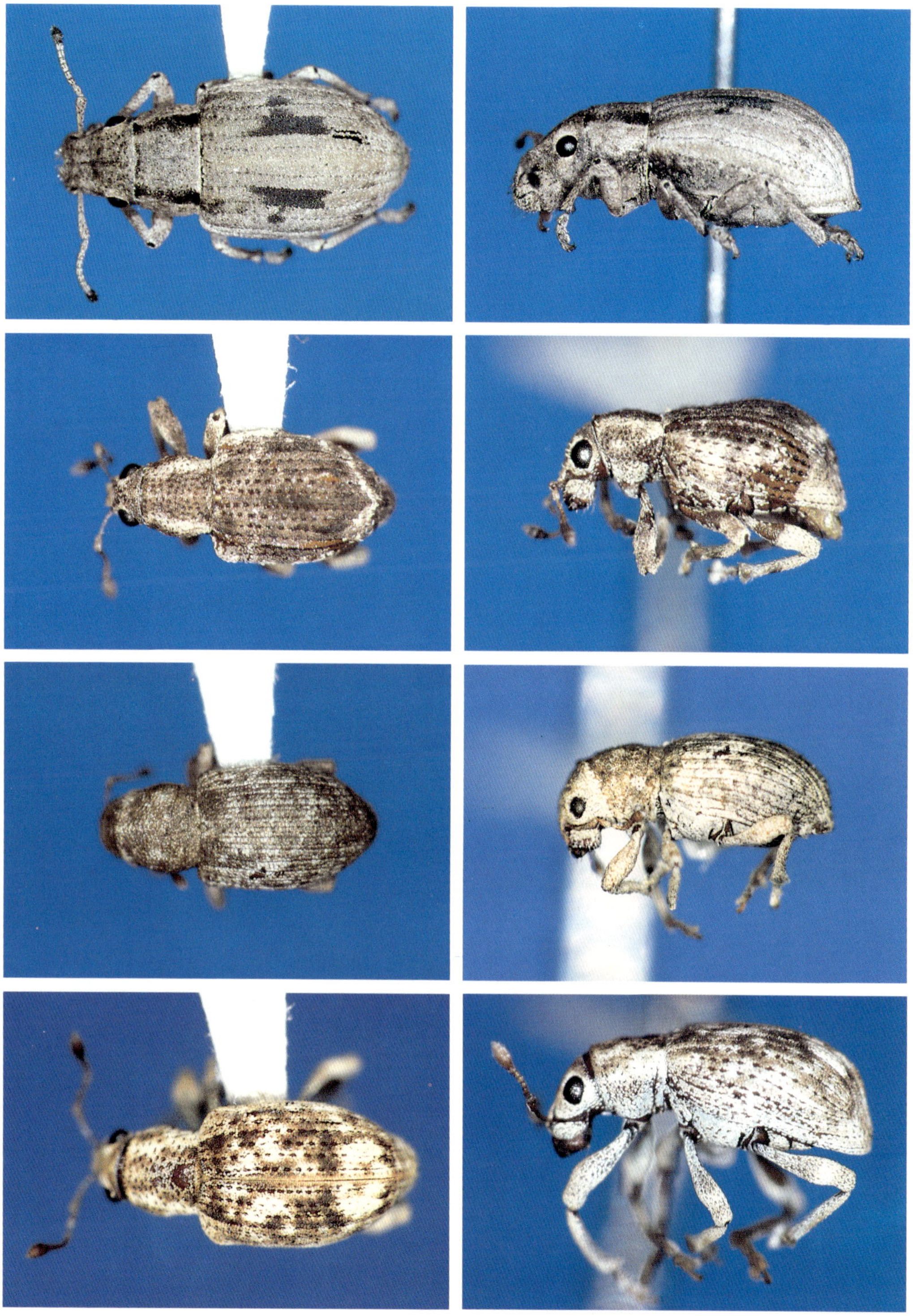

Plate 188 Adelognatha

1, 2 *Eutinophaea fasciculata* Lea
Cairns District, Queensland
Paratype female; 3.3 mm

3, 4 *Eutinophaea setistriata* Lea
Mount Tamborine, Queensland
Male; 2.8 mm

5, 6 *Eutinophaea subviridis* Lea
Cairns District, Queensland
Paratype male; 3 mm

7, 8 *Eutinophaea suturalis* Lea
Mount Tamborine, Queensland
Paratype male; 3.4 mm

1	2
3	4
5	6
7	8

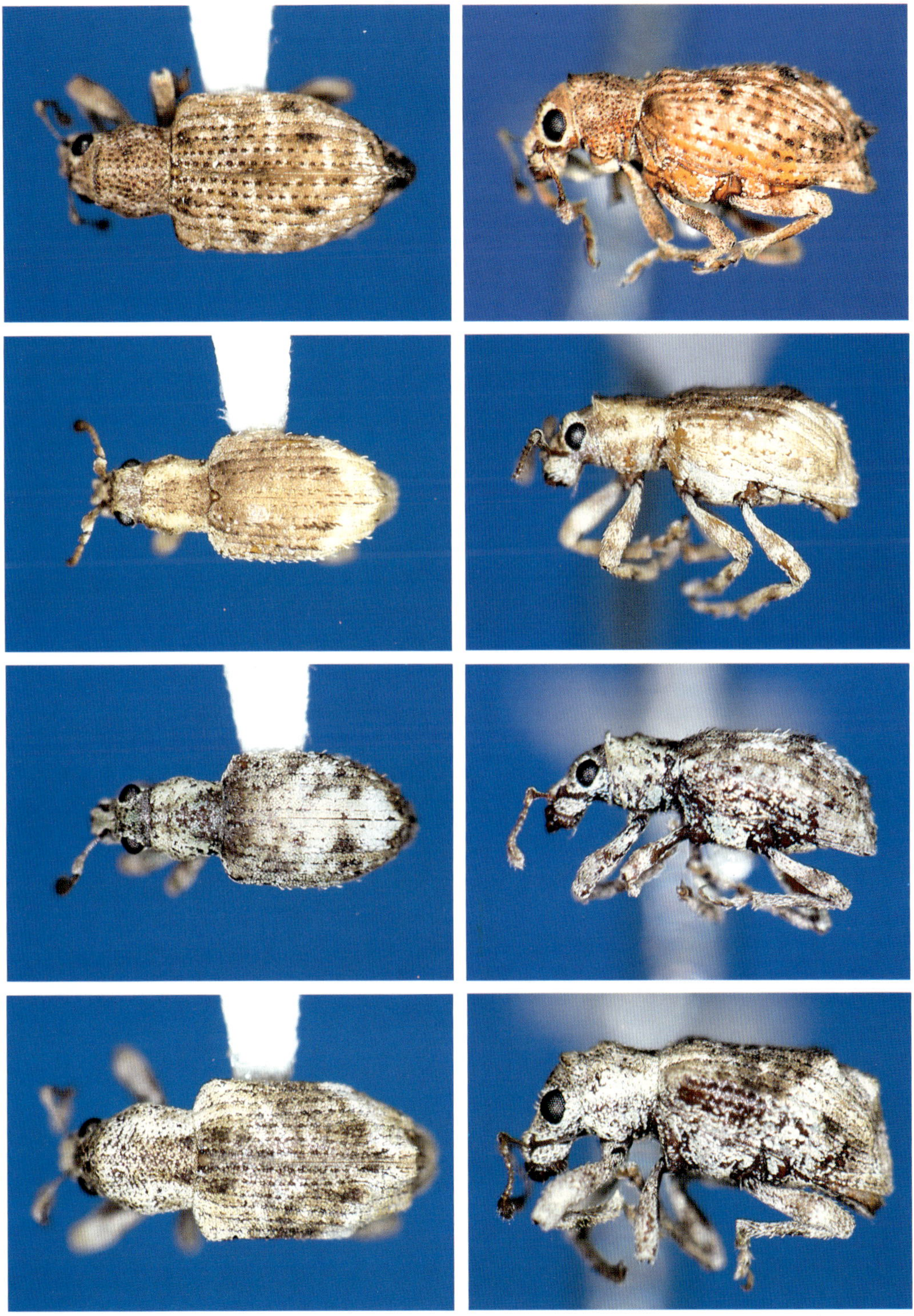

Plate 189 Adelognatha

1, 2 *Eutinophaea variegata* Lea
Kuranda, Queensland
Holotype male; 2 mm

3, 4 *Eutinophaea murina* Lea
Groote Eylandt, Northern Territory
Female; 5.9 mm

5, 6 *Florisianus sordidus* Hustache
Murrumbateman, New South Wales
Female; 5.8 mm (excluding ovipositor)

7, 8 *Furius intermedius* (Blackburn)
Kangaroo Island, South Australia
Paratype female; 5.9 mm

1	2
3	4
5	6
7	8

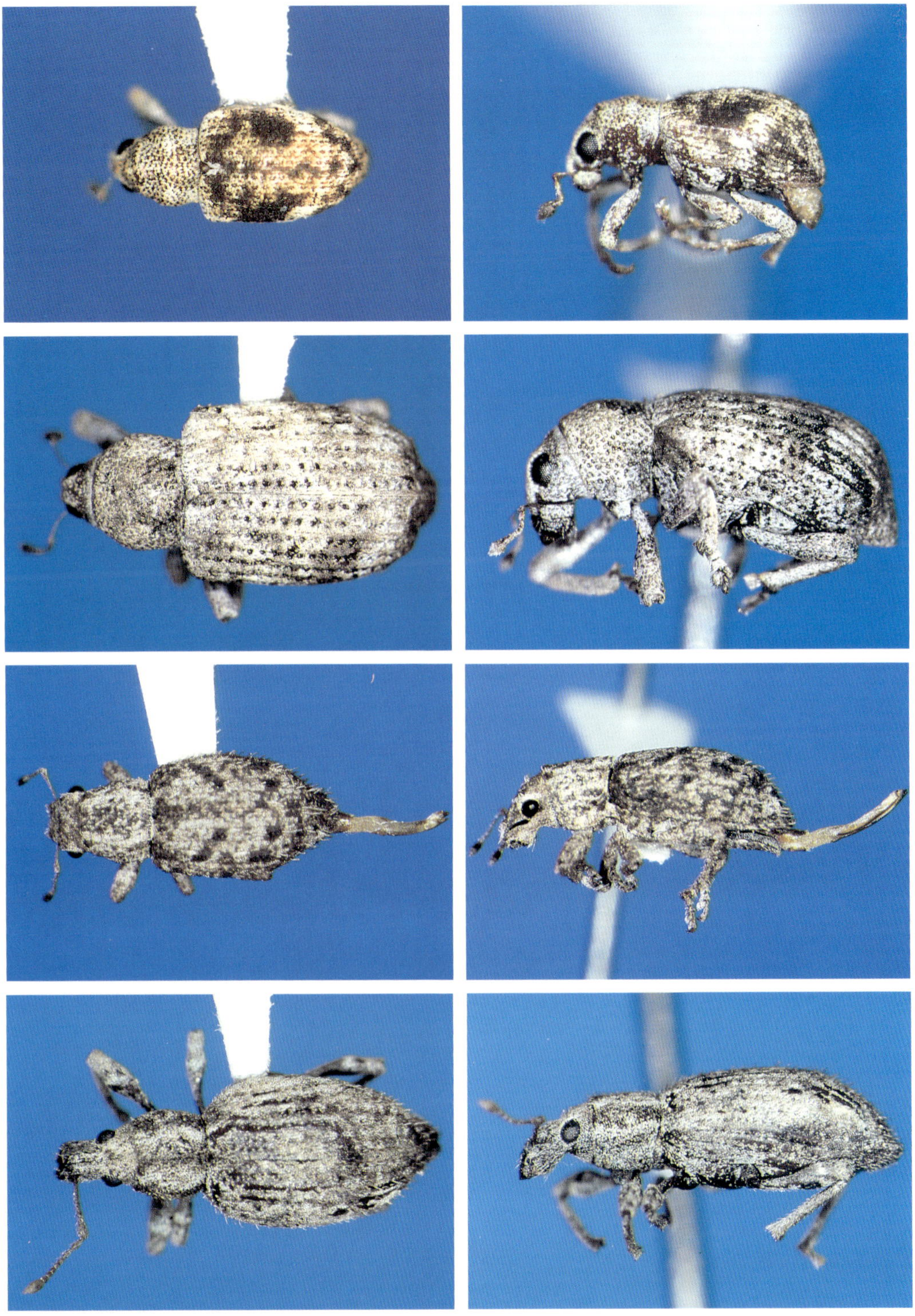

Plate 190 Adelognatha

1, 2 *Furius subfasciatus* (Blackburn)
South Australia
Syntype female; 3.7 mm

3, 4 *Graphognathus leucoloma* (Boheman)
Brooklyn Park, South Australia
Female; 1.5 mm

5, 6 *Graphognathus peregrinus* Buchanan
Jerrys Plain, New South Wales
Female; 10.5 mm

7, 8 *Hackeria chrysidea* Pascoe
Somerset, Queensland
Paratype male; 5.6 mm

1	2
3	4
5	6
7	8

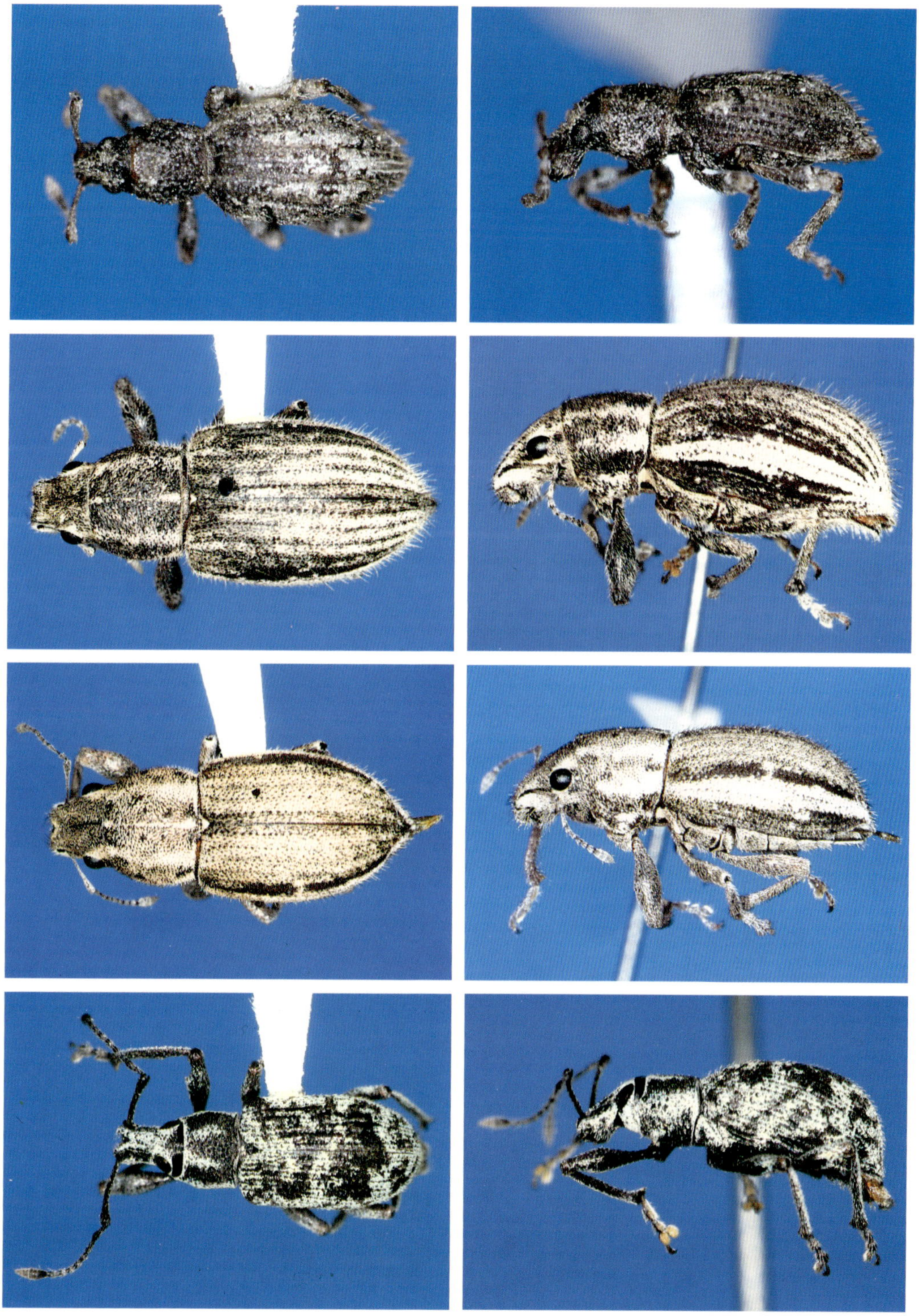

Plate 191 Adelognatha

1, 2 *Hackeria gnopholota* (Lea)
Daly River, Northern Territory
Female; 5.2 mm

3, 4 *Hackeria hardcastlei* (Lea)
Cunnamulla, Queensland
Holotype male; 5 mm

5, 6 *Hackeria nigrovaria* (Lea)
Coen River, Queensland
Paratype female; 4.9 mm

7, 8 *Hackeria viridivaria* Lea
Coen River, Queensland
Female; 6 mm

1	2
3	4
5	6
7	8

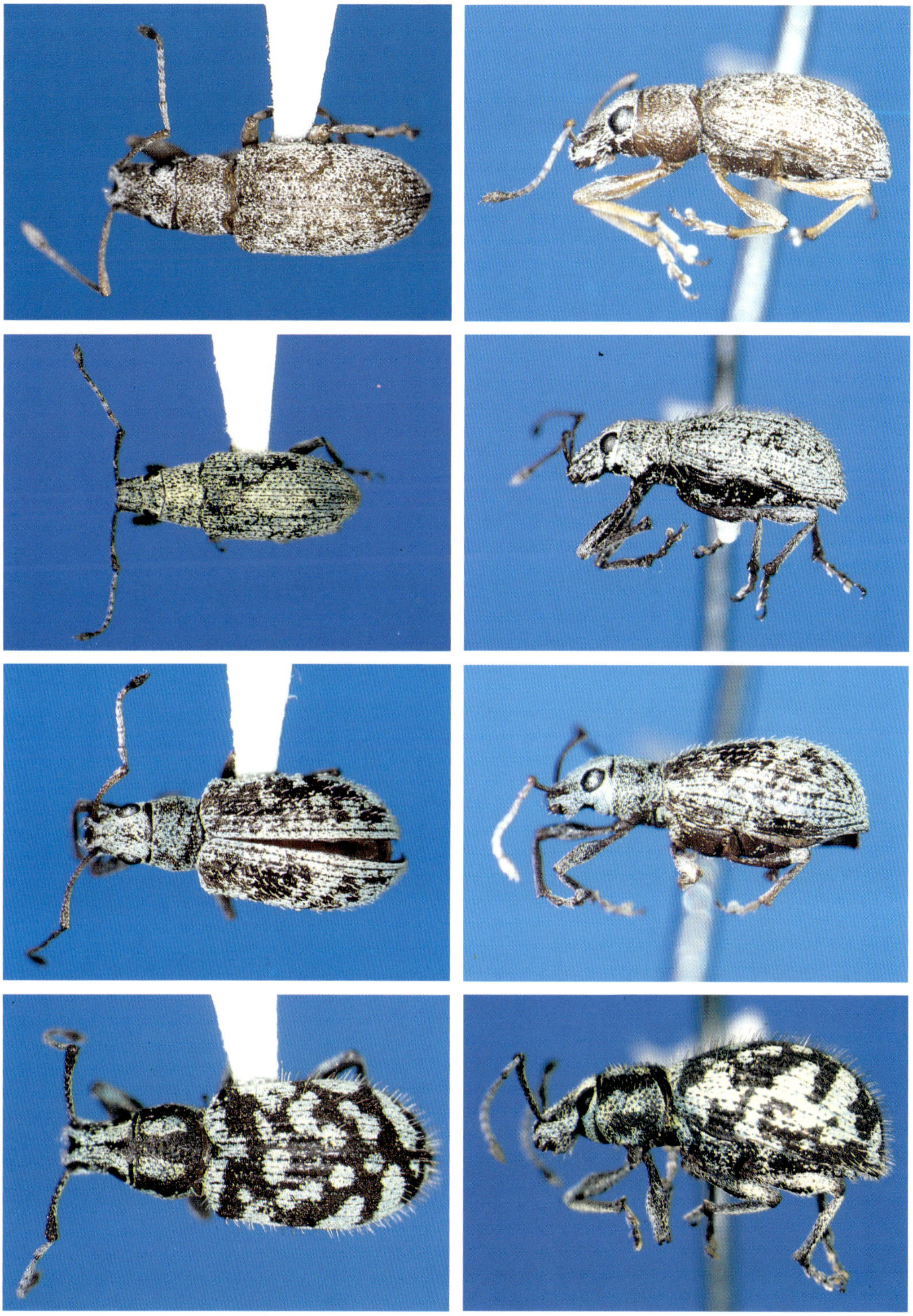

Plate 192 Adelognatha

1, 2 *Hadrorhinus basirostris* (Lea)
Melville Island, Northern Territory
Paratype male; 5.6 mm

3, 4 *Hadrorhinus concinnus* (Lea)
Kimberley District, Western Australia
Male; 4.5 mm

5, 6 *Hadrorhinus collaris* (Lea)
Batchelor, Northern Territory
Paratype female; 6.3 mm

7, 8 *Homoeotrachelus hadromerus* Lea
Geraldton, Western Australia
Holotype female; 4.8 mm

1	2
3	4
5	6
7	8

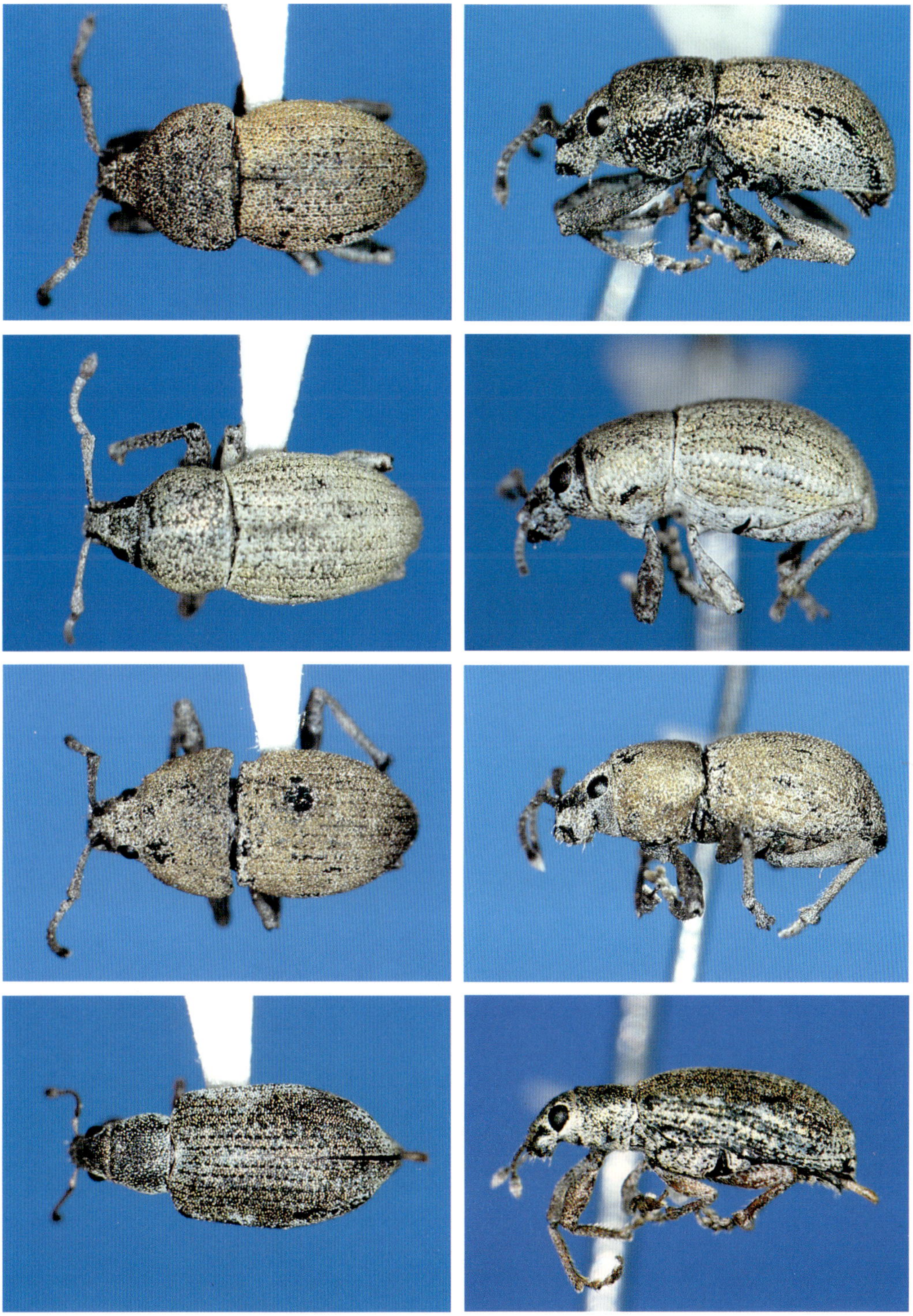

Plate 193 Adelognatha

1, 2 *Homoeotrachelus tricarinatus* Lea
Port Denison, Queensland
Holotype male; 6.1 mm

3, 4 *Idiopsodes ocularis* (Lea)
Darnley Island, Torres Strait, Queensland
Female; 5.4 mm

5, 6 *Leptopius areolatus* (Blackburn)
Yuendumu, Northern Territory
Female, compared with holotype; 29.5 mm

7, 8 *Leptopius biordinatus* (Blackburn)
40 km northwest of Mount Ragged, Western Australia
Male; 29 mm

1	2
3	4
5	6
7	8

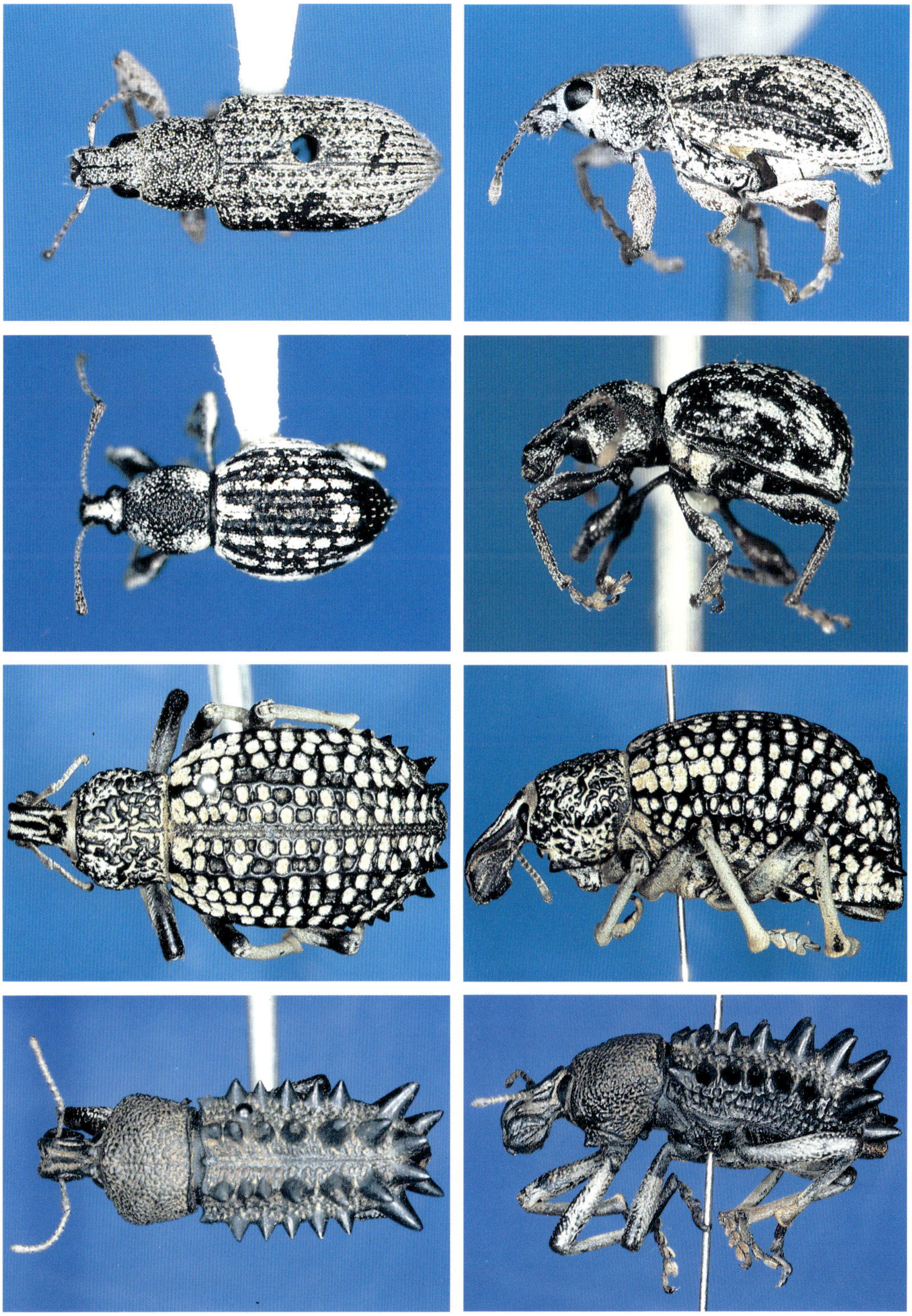

Plate 194 Adelognatha

1, 2 *Leptopius etheridgei* (Olliff)
Lord Howe Island
Holotype male; 19.5 mm

3, 4 *Leptopius fasciculatus* (Lea)
Queensland
Paratype female; 10.5 mm

5, 6 *Leptopius glaucus* (Pascoe)
Victoria
Female; 13 mm

7, 8 *Leptopius humeralis* (Germar)
Mount Lofty, South Australia
Female; 16 mm
The humeral tubercles are abraded.

1	2
3	4
5	6
7	8

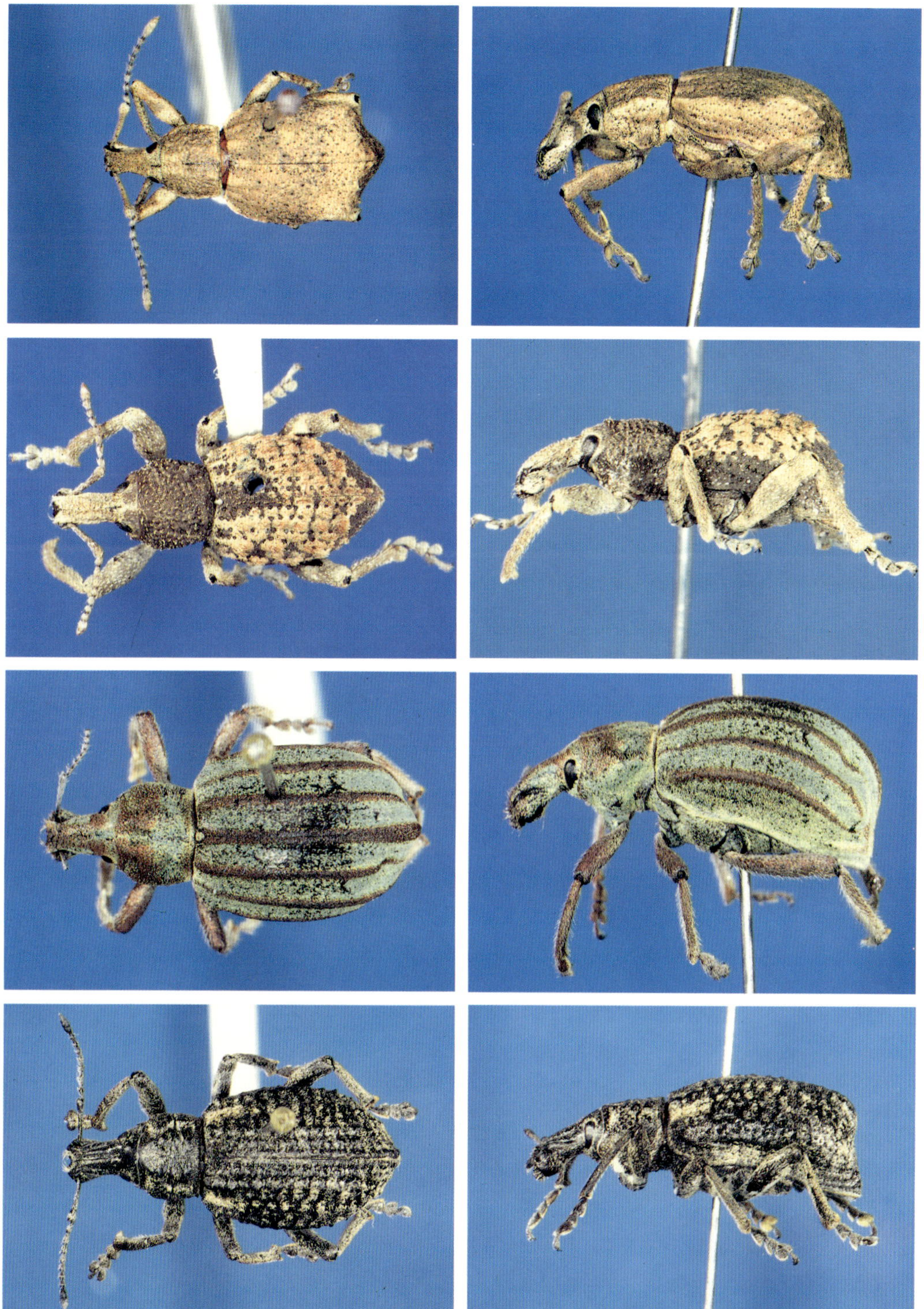

Plate 195 Adelognatha

1, 2 *Leptopius insularis* (Lea)
Baudin Island, Western Australia
Holotype female; 13.5 mm

3, 4 *Leptopius interruptus* (Lea)
Quorn, South Australia
Male; 6.5 mm

5, 6 *Leptopius maleficus* (Lea)
Coen River, Queensland
Male; 15.5 mm

7, 8 *Leptopius maleficus* (Lea)
Coen River, Queensland
Female; 21.5 mm

1	2
3	4
5	6
7	8

Plate 196 Adelognatha

1, 2 *Leptopius mirabilis* (Lea)
Lord Howe Island
Female; 20.5 mm

3, 4 *Leptopius nigropunctatus* (Lea)
Richmond River, New South Wales
Paratype female; 9.5 mm

5, 6 *Leptopius nodicollis* (Lea)
Cairns, Queensland
Holotype female; 19 mm

7, 8 *Leptopius puellaris* (Pascoe)
Tamborine Mountains, Queensland
Male; 6.2 mm

1	2
3	4
5	6
7	8

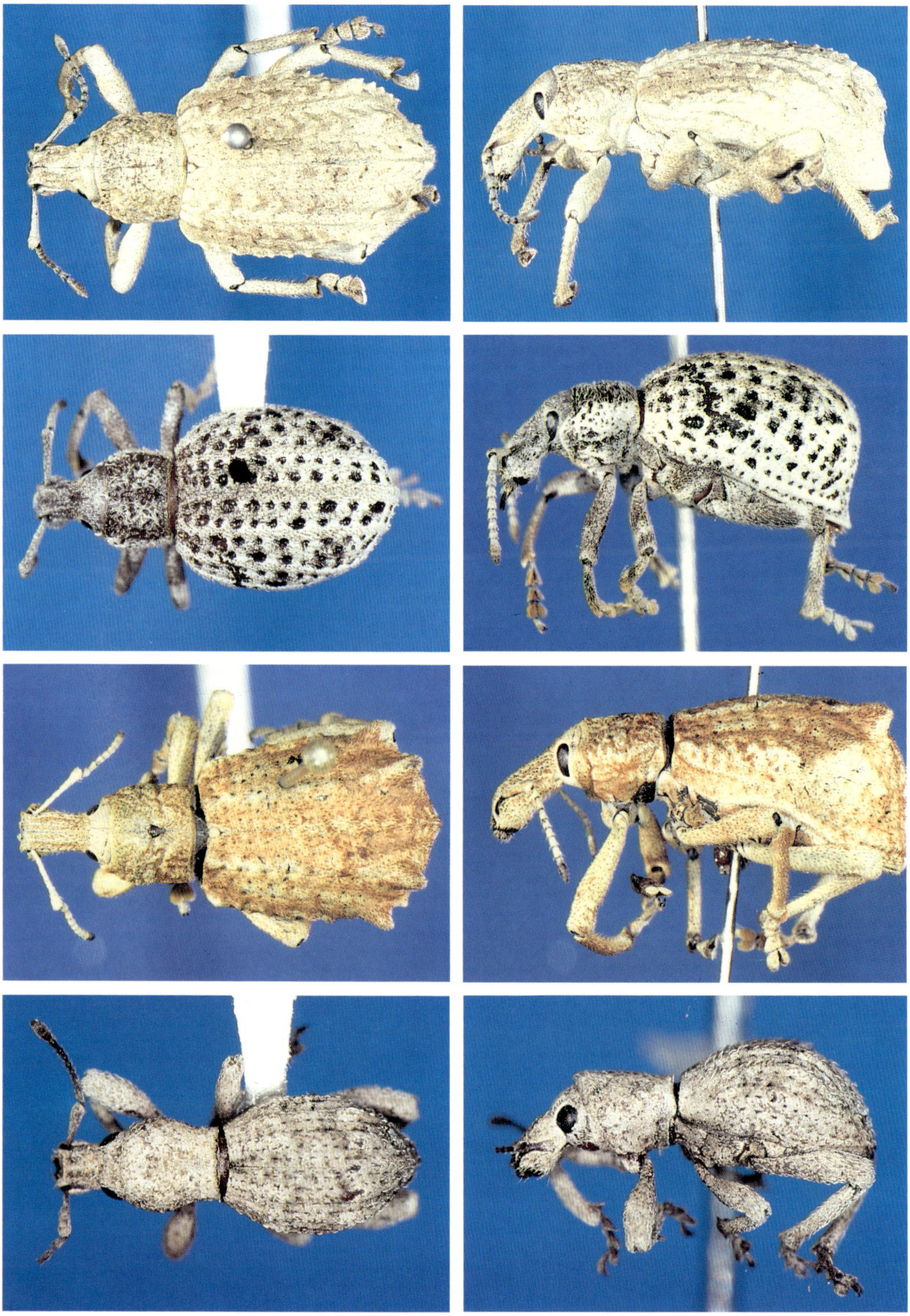

Plate 197 Adelognatha

1, 2 *Leptopius rhizophagus* (Lea)
Wirrabarra, South Australia
Paratype female; 12.5 mm

3, 4 *Leptopius robustus* (Olivier)
New South Wales
Male; 16 mm

5, 6 Leptopius robustus (Olivier)
Illawarra, New South Wales
Female; 19.5 mm

7, 8 *Leptopius spinipennis* (Fahraeus)
Wembley, South Australia
Female, compared with holotype; 13.5 mm

1	2
3	4
5	6
7	8

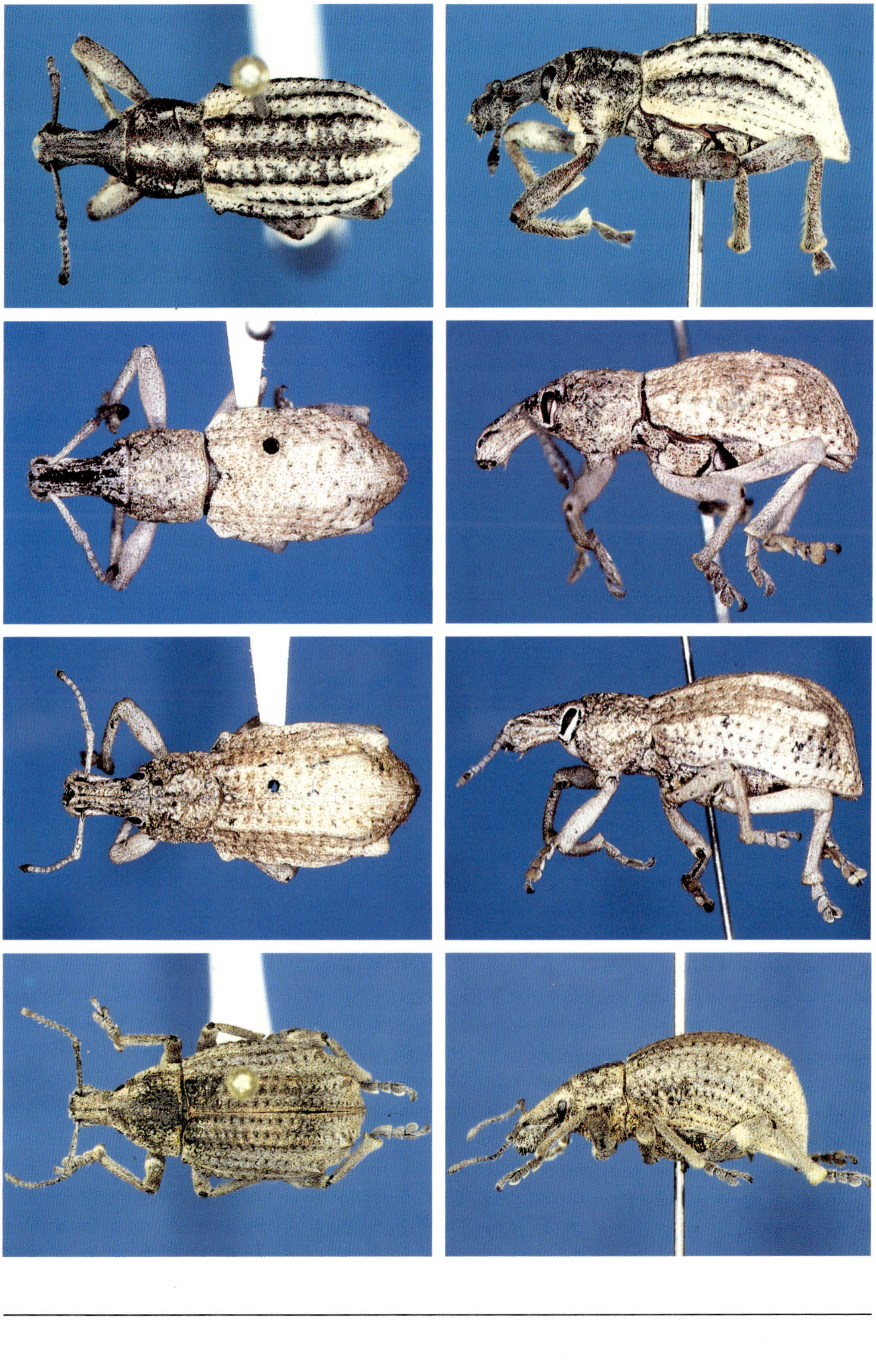

Plate 198 Adelognatha

1, 2 *Lipothyrea arrowi* Lea
Tolga, North Queensland
Female; 7.5 mm

3, 4 *Lipothyrea chloris* Pascoe
Port Denison, Queensland
Male; 8.8 mm

5, 6 *Mandalotus* (*Mandalotus*) *abdominalis* Lea
Mount Tamborine, Queensland
Male; 3.7 mm

7, 8 *Mandalotus* (*Mandalotus*) *albonotatus* Lea
Forrest, Victoria
Male; 4.8 mm

1	2
3	4
5	6
7	8

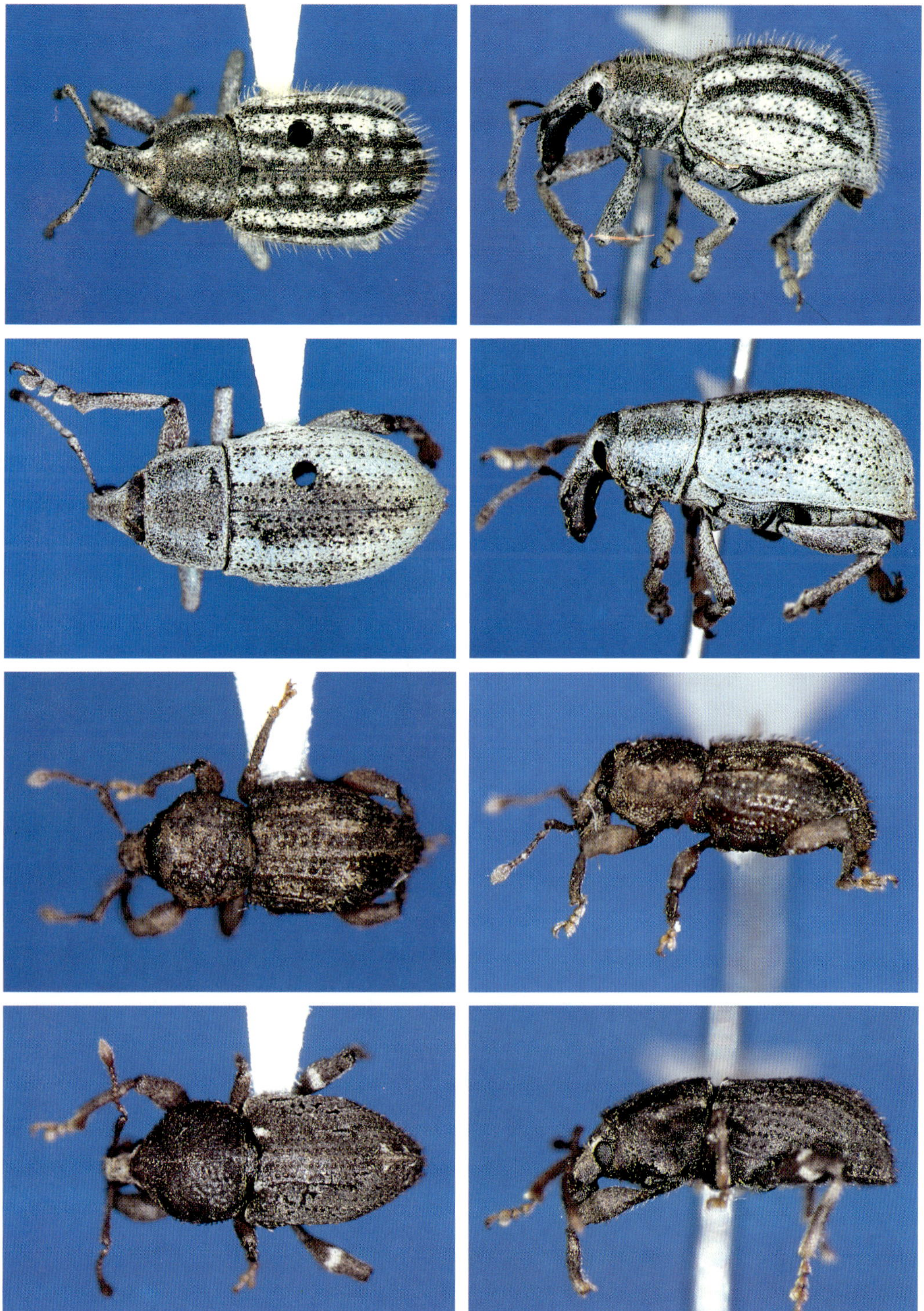

Plate 199 Adelognatha

1, 2 *Mandalotus* (*Mandalotus*) *angustipennis* Lea
Tasmania
Paratype male; 5.2 mm

3, 4 *Mandalotus* (*Mandalotus*) *angustus* Lea
Toowoomba, Queensland
Paratype male; 4.6 mm

5, 6 *Mandalotus* (*Mandalotus*) *arciferus* Lea
Portland, Victoria
Male; 6.2 mm

7, 8 *Mandalotus* (*Mandalotus*) *arcuatus* Lea
Rooty Hill, New South Wales
Holotype male; 3.2 mm

1	2
3	4
5	6
7	8

Plate 200 Adelognatha

1, 2 *Mandalotus* (*Mandalotus*) *armicoxis* Lea
Victoria
Paratype male; 5.2 mm

3, 4 *Mandalotus* (*Mandalotus*) *armipectus* Lea
Shoalhaven, New South Wales
Paratype male; 4.6 mm

5, 6 *Mandalotus* (*Mandalotus*) *armivarius* Lea
Belgrave, Victoria
Paratype male; 4.1 mm
The tubercle on ventrite 1 can be seen near the metacoxa.

7, 8 *Mandalotus* (*Mandalotus*) *auchmeresthes* Lea
Eccleston, New South Wales
Holotype male; 5.2 mm

1	2
3	4
5	6
7	8

Plate 201 Adelognatha

1, 2 *Mandalotus* (*Mandalotus*) *avenaceus* Lea
Daylesford, Victoria
Holotype male; 3.4 mm

3, 4 *Mandalotus* (*Mandalotus*) *bilobicollis* Lea
Lord Howe Island
Male, compared with holotype; 4.6 mm

5, 6 *Mandalotus* (*Mandalotus*) *bivitticollis* Lea
Melbourne, Victoria
Holotype male; 3.6 mm

7, 8 *Mandalotus* (*Mandalotus*) *blackmorei* Lea
Forest Reefs, New South Wales
Holotype male; 5 mm

1	2
3	4
5	6
7	8

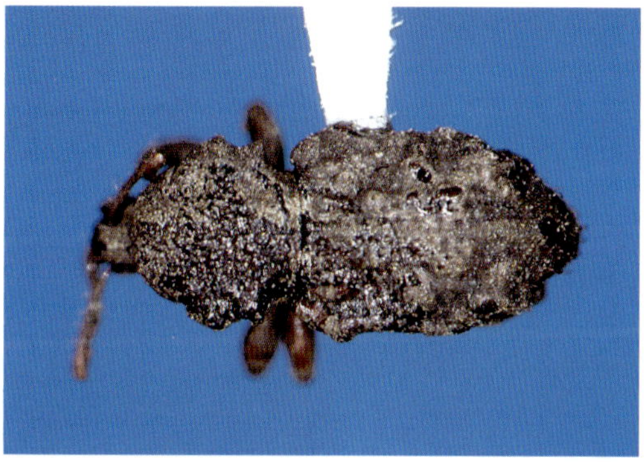

Plate 202 Adelognatha

1, 2 *Mandalotus* (*Mandalotus*) *bryophilus* Oke
Lorne, Victoria
Holotype male; 4.7 mm

3, 4 *Mandalotus* (*Mandalotus*) *carteri* Lea
Mount Hotham, Victoria
Male, compared with holotype; 6.5 mm

5, 6 *Mandalotus* (*Mandalotus*) *cellaris* Pascoe
Berry, New South Wales
Male; 4.5 mm

7, 8 *Mandalotus* (*Mandalotus*) *ciliatus* Lea
Nelson, Victoria
Holotype male; 6 mm

1	2
3	4
5	6
7	8

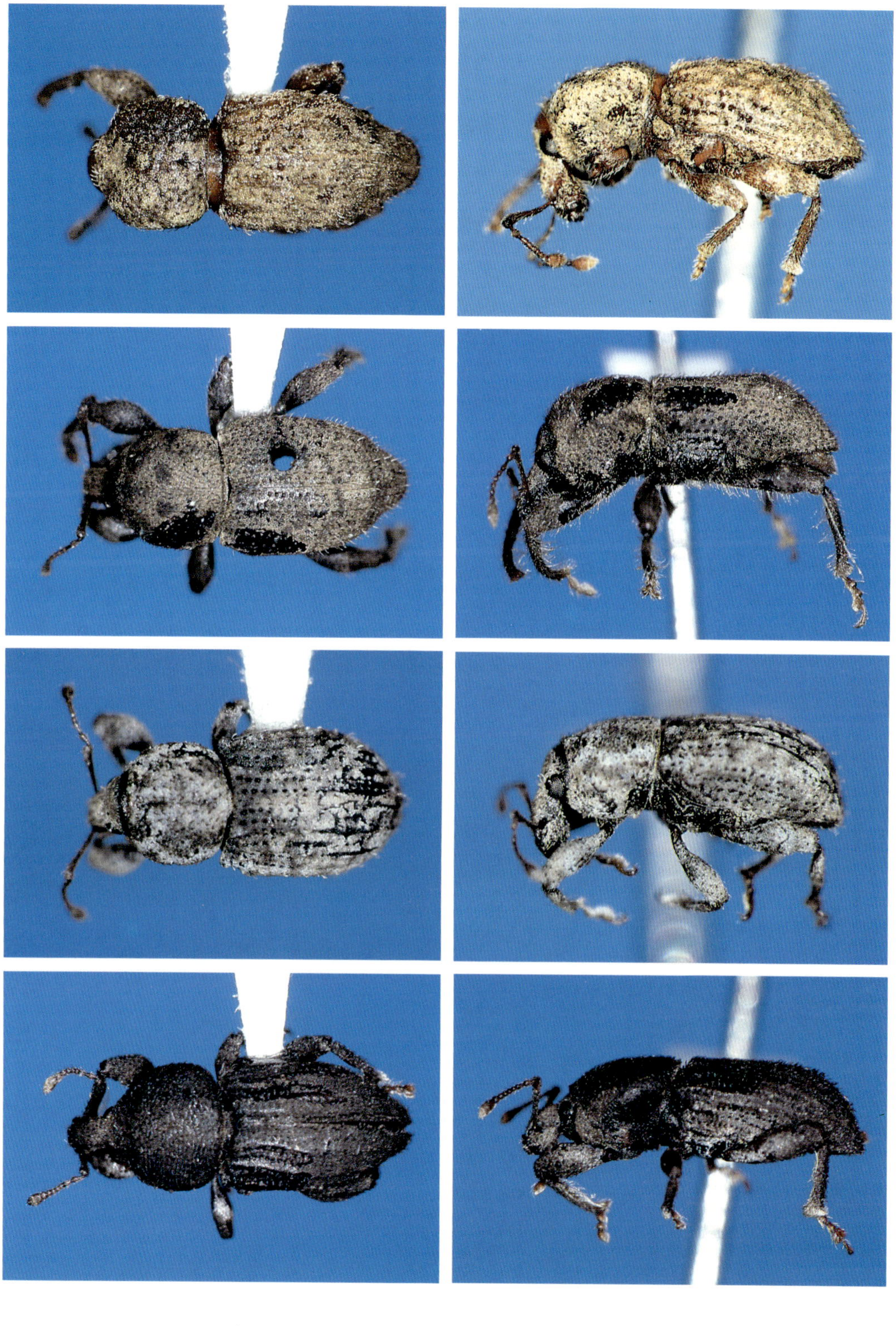

Plate 203 Adelognatha

1, 2 *Mandalotus* (*Mandalotus*) *coatesi* Lea
Richmond River, New South Wales
Type male; 4.9 mm

3, 4 *Mandalotus* (*Mandalotus*) *collaris* Lea
Lord Howe Island
Paratype male; 5.3 mm

5, 6 *Mandalotus* (*Mandalotus*) *contortus* Lea
Ebor, New South Wales
Holotype male; 9 mm

7, 8 *Mandalotus* (*Mandalotus*) *crudus* Erichson
Victoria or Tasmania (no data on specimen)
Male; 7.5 mm
From the holotype of the synonym *Mandalotus rigidus* Marshall; see details in Volume IV.

1	2
3	4
5	6
7	8

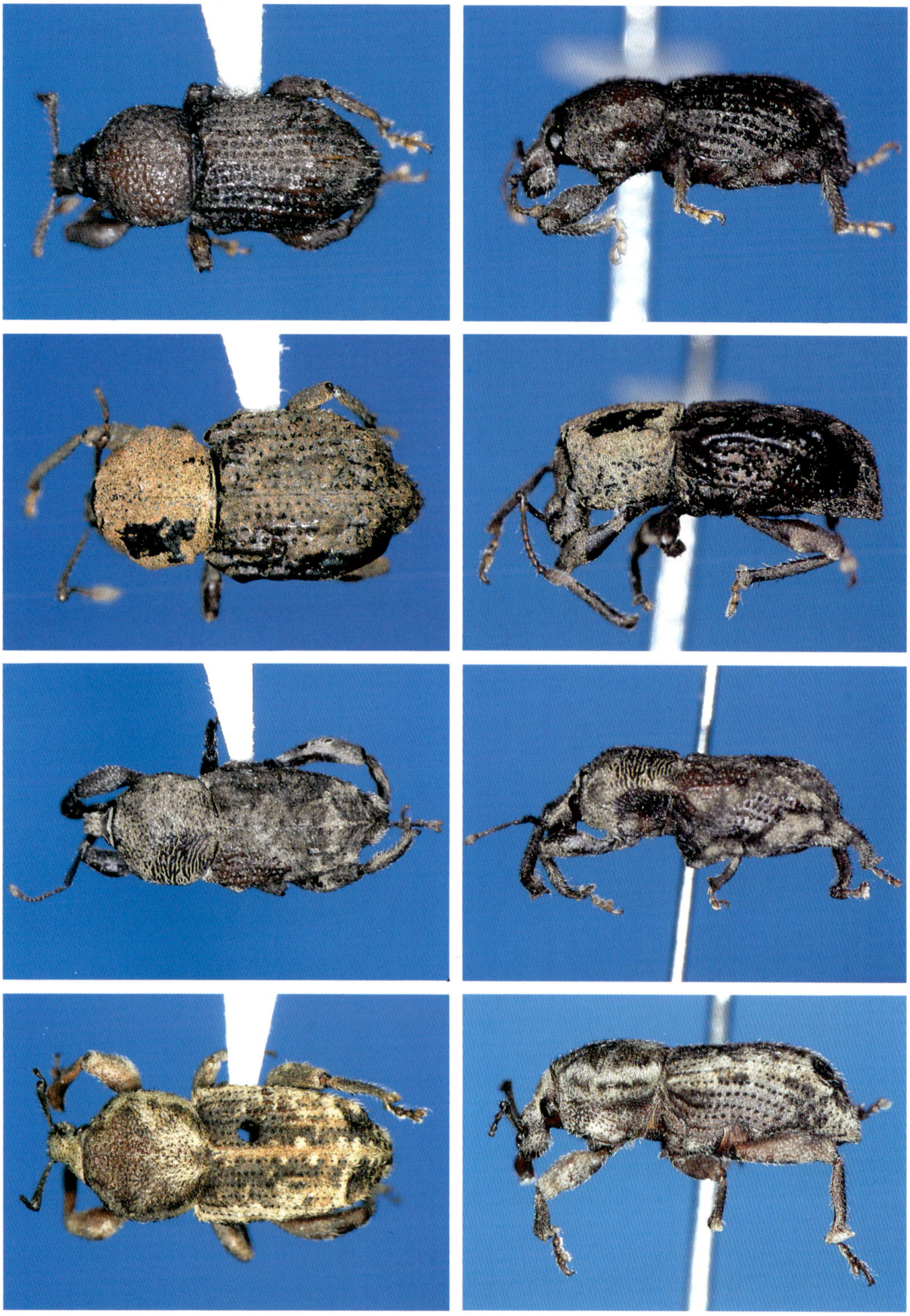

Plate 204 Adelognatha

1, 2 *Mandalotus (Mandalotus) decipiens* Blackburn
Mount Blowhard, Victoria, 1,220 metres
Male; 5.1 mm

3, 4 *Mandalotus (Mandalotus) exavatus* Lea
Brisbane, Queensland
Holotype male; 4.7 mm

5, 6 *Mandalotus (Mandalotus) explanicollis* Oke
Lorne, Victoria
Holotype male; 3.9 mm

7, 8 *Mandalotus (Mandalotus) fergusoni* Lea
Muswellbrook, New South Wales
Paratype female; 5.1 mm

1	2
3	4
5	6
7	8

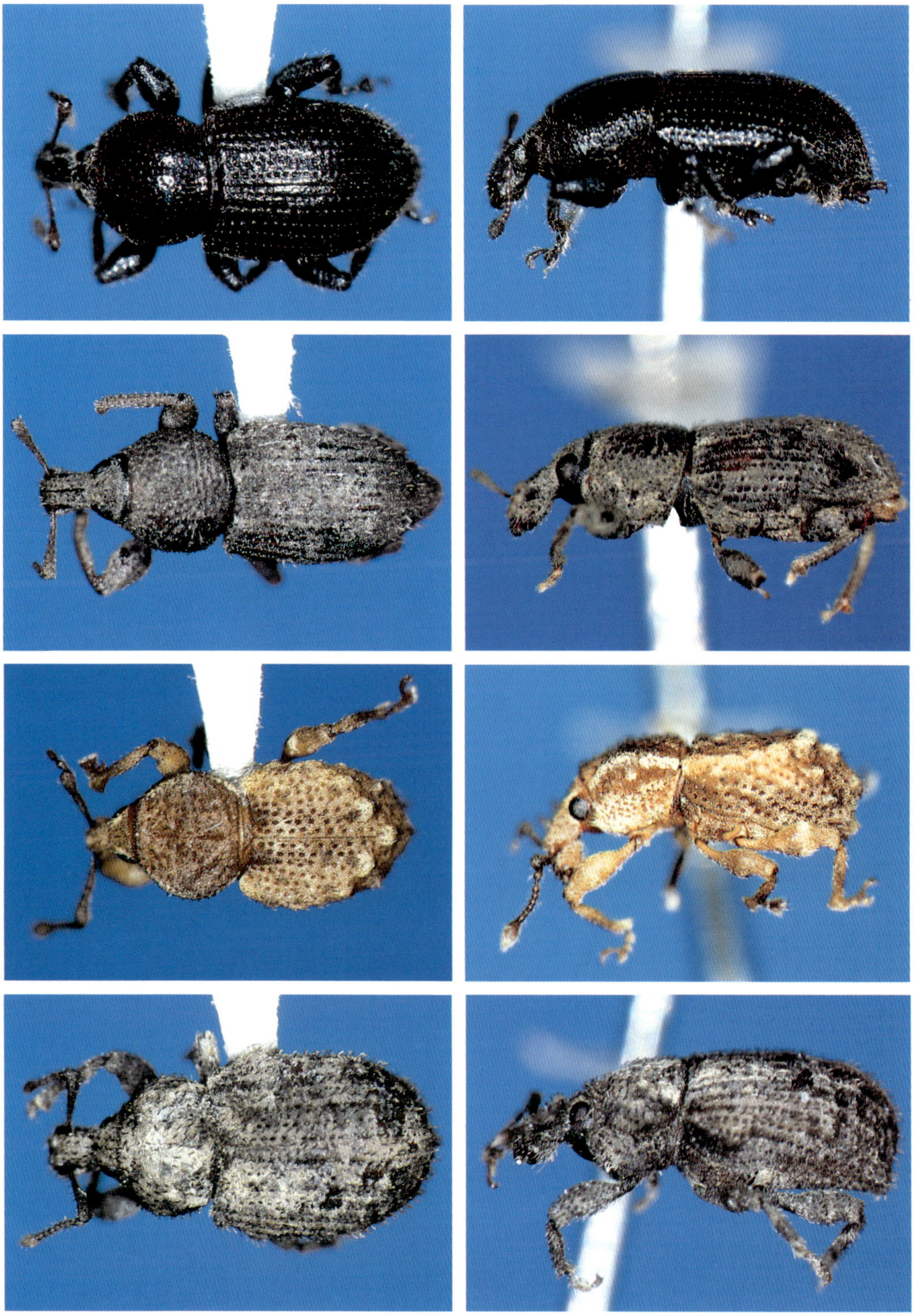

Plate 205 Adelognatha

1, 2 *Mandalotus* (*Mandalotus*) *fimbriatus* Lea
Bogong Plateau, Victoria
Paratype male; 7.5 mm

3, 4 *Mandalotus* (*Mandalotus*) *foveatus* Lea
Guyra, New South Wales
Holotype male; 5.8 mm

5, 6 *Mandalotus* (*Mandalotus*) *funereus* Lea
Victorian Alps, Victoria
Holotype male; 8.3 mm

7, 8 *Mandalotus* (*Mandalotus*) *glaber* (Blackburn)
Alps, Victoria
Paratype male; 6.5 mm

1	2
3	4
5	6
7	8

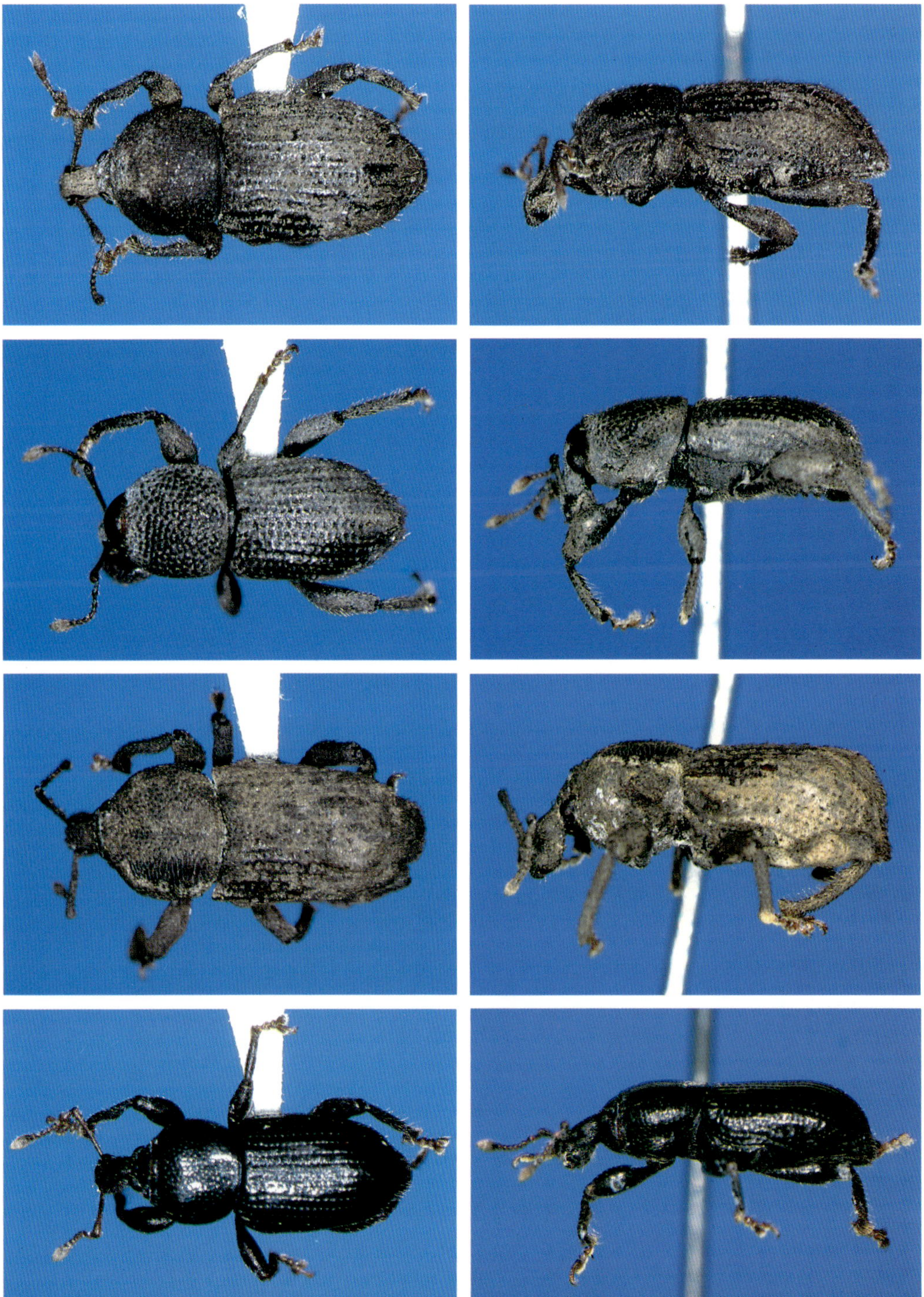

Plate 206 Adelognatha

1, 2 *Mandalotus* (*Mandalotus*) *graminicola* Oke
Belgrave, Victoria
Paratype male; 2.9 mm

3, 4 *Mandalotus* (*Mandalotus*) *hoplocnemus* Lea
Mount Lofty, South Australia
Holotype male; 3.2 mm

5, 6 *Mandalotus* (*Mandalotus*) *hoplosternus* Lea
Melbourne, Victoria
Holotype male; 6.6 mm

7, 8 *Mandalotus* (*Mandalotus*) *hoplostethus* (Pascoe)
Tasmania
Male; 4.9 mm; identified by Lea

1	2
3	4
5	6
7	8

Plate 207 Adelognatha

1, 2 *Mandalotus* (*Mandalotus*) *humeralis* Lea
National Park, Sydney, New South Wales
Holotype male; 3.6 mm

3, 4 *Mandalotus* (*Mandalotus*) *hystricosus* Lea
Corinna, Tasmania
Holotype male; 7 mm

5, 6 *Mandalotus* (*Mandalotus?*) *imponderosus* Lea
Mount Tamborine, Queensland
Holotype male; 1.7 mm

7, 8 *Mandalotus* (*Mandalotus*) *impressicollis* Oke
Emerald, Victoria
Paratype male; 4.1 mm

1	2
3	4
5	6
7	8

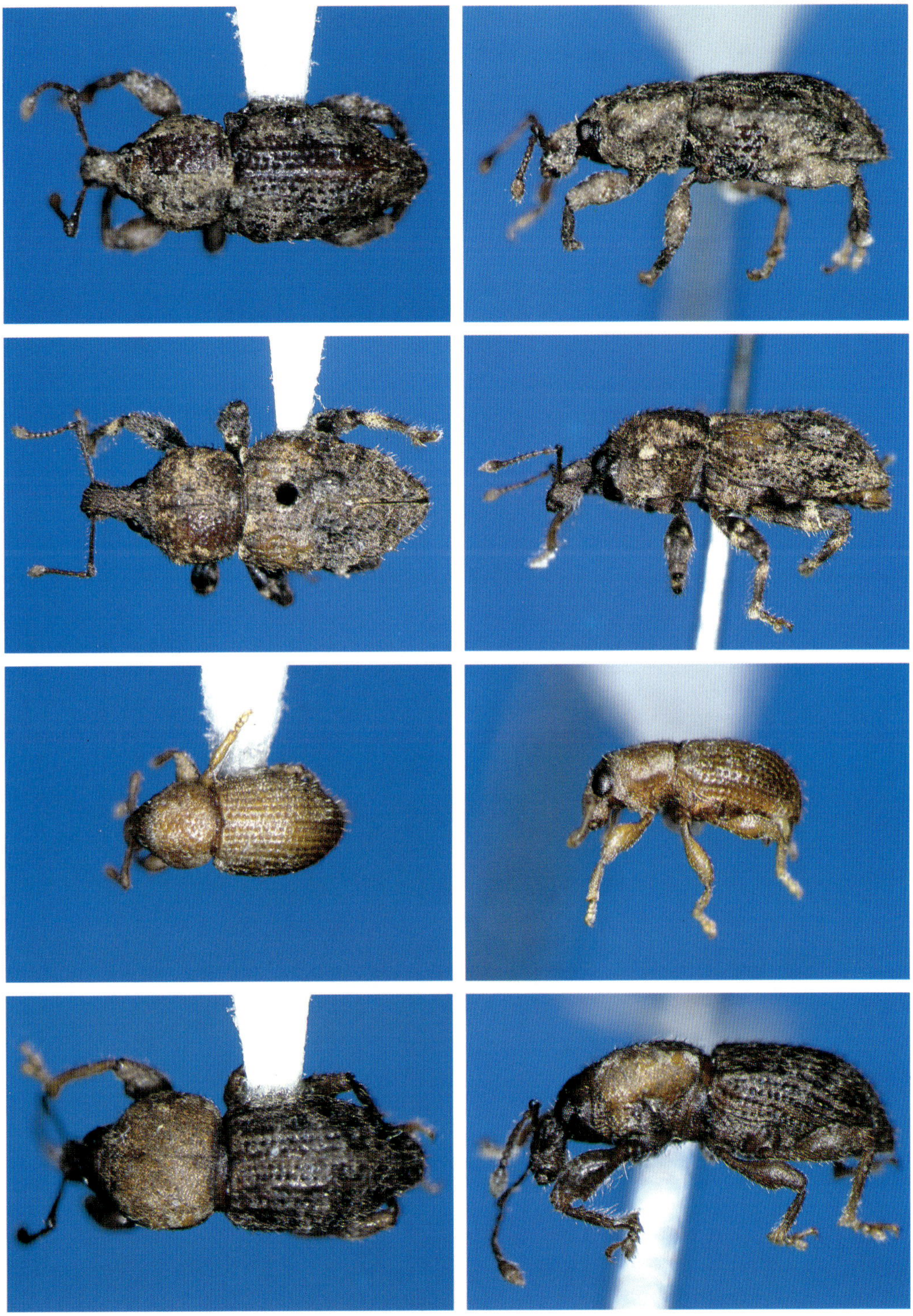

Plate 208 Adelognatha

1, 2 *Mandalotus* (*Mandalotus*) *incisipes* Lea
Mittagong, New South Wales
Holotype male; 6.5 mm
Note the formation of the apical area of the protobia.

3, 4 *Mandalotus* (*Mandalotus*) *inconspicuus* Lea
Mount Lofty Ranges, South Australia
Paratype male; 2.5 mm

5, 6 *Mandalotus* (*Mandalotus*) *insignis* Lea
4 km northeast of Mount Wog Wog, near Bombala, New South Wales
Male; 11 mm

7, 8 Female of the same species from the same locality; 7.5 mm

1	2
3	4
5	6
7	8

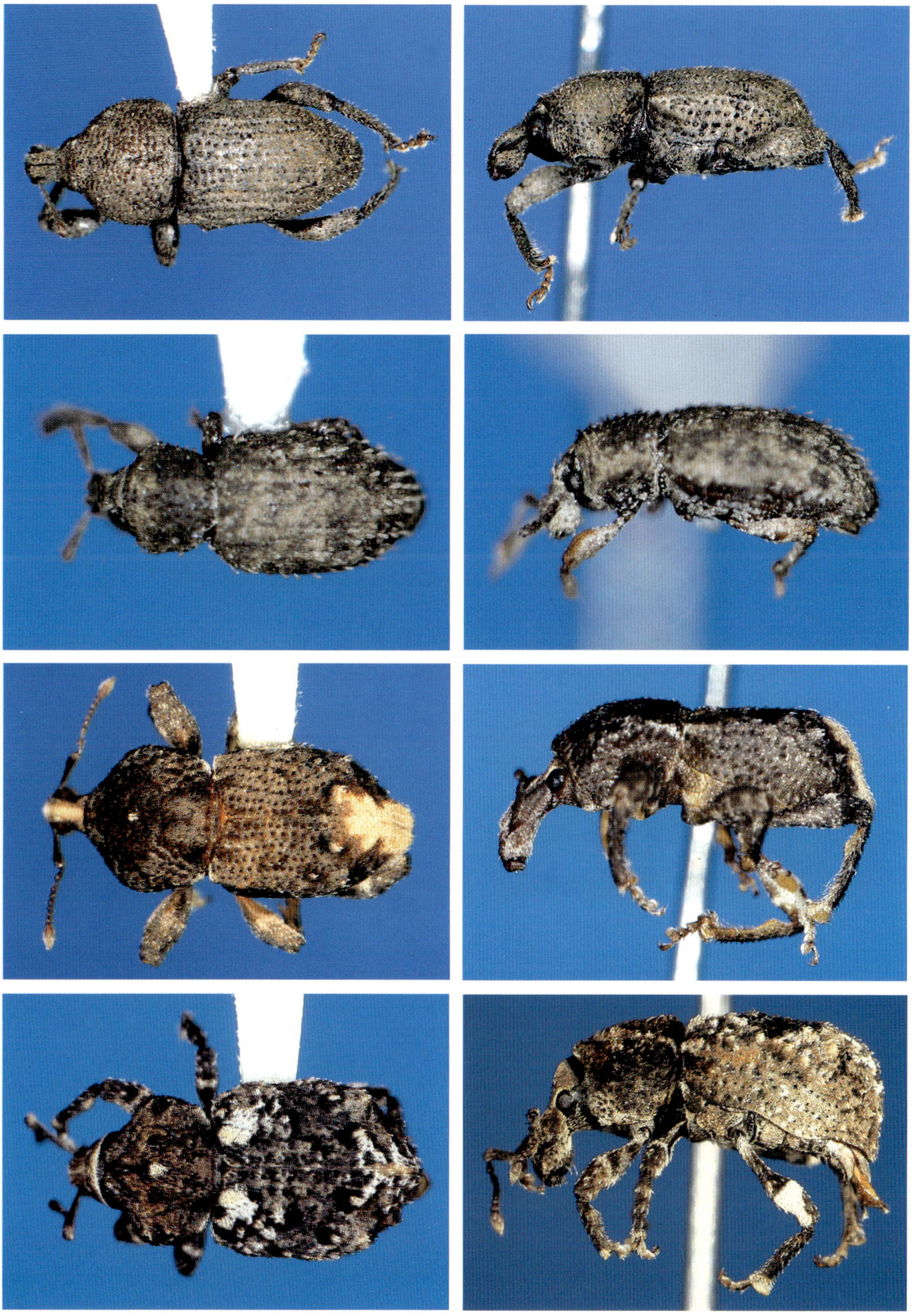

Plate 209 Adelognatha

1, 2 *Mandalotus* (*Mandalotus*) *insularis* Lea
Stradbroke Island, Queensland
Holotype male; 4.2 mm

3, 4 *Mandalotus* (*Mandalotus*) *intercoxalis* Lea
Near Gibraltar Falls, Australian Capital Territory
Male, compared with the holotype; 5.6 mm

5, 6 *Mandalotus* (*Mandalotus*) *interocularis* Lea
Stanley, Tasmania
Paratype male; 6.2 mm

7, 8 *Mandalotus* (*Mandalotus*) *irrasus* Lea
New South Wales
Holotype male; 5.1 mm

1	2
3	4
5	6
7	8

Plate 210 Adelognatha

1, 2 *Mandalotus* (*Mandalotus*) *laminatipes* Lea
Nar Nar Goon, Victoria
Holotype male; 4.8 mm

3, 4 *Mandalotus* (*Mandalotus*) *laminipectus* Lea
Glenelg River, Victoria
Holotype male; 5.2 mm

5, 6 *Mandalotus* (*Mandalotus*) *latebricola* Lea
Ringwood, Victoria
Holotype male; 3.7 mm

7, 8 *Mandalotus* (*Mandalotus*) *latens* Lea
Tamborine, Queensland
Paratype male; 2.3 mm

1	2
3	4
5	6
7	8

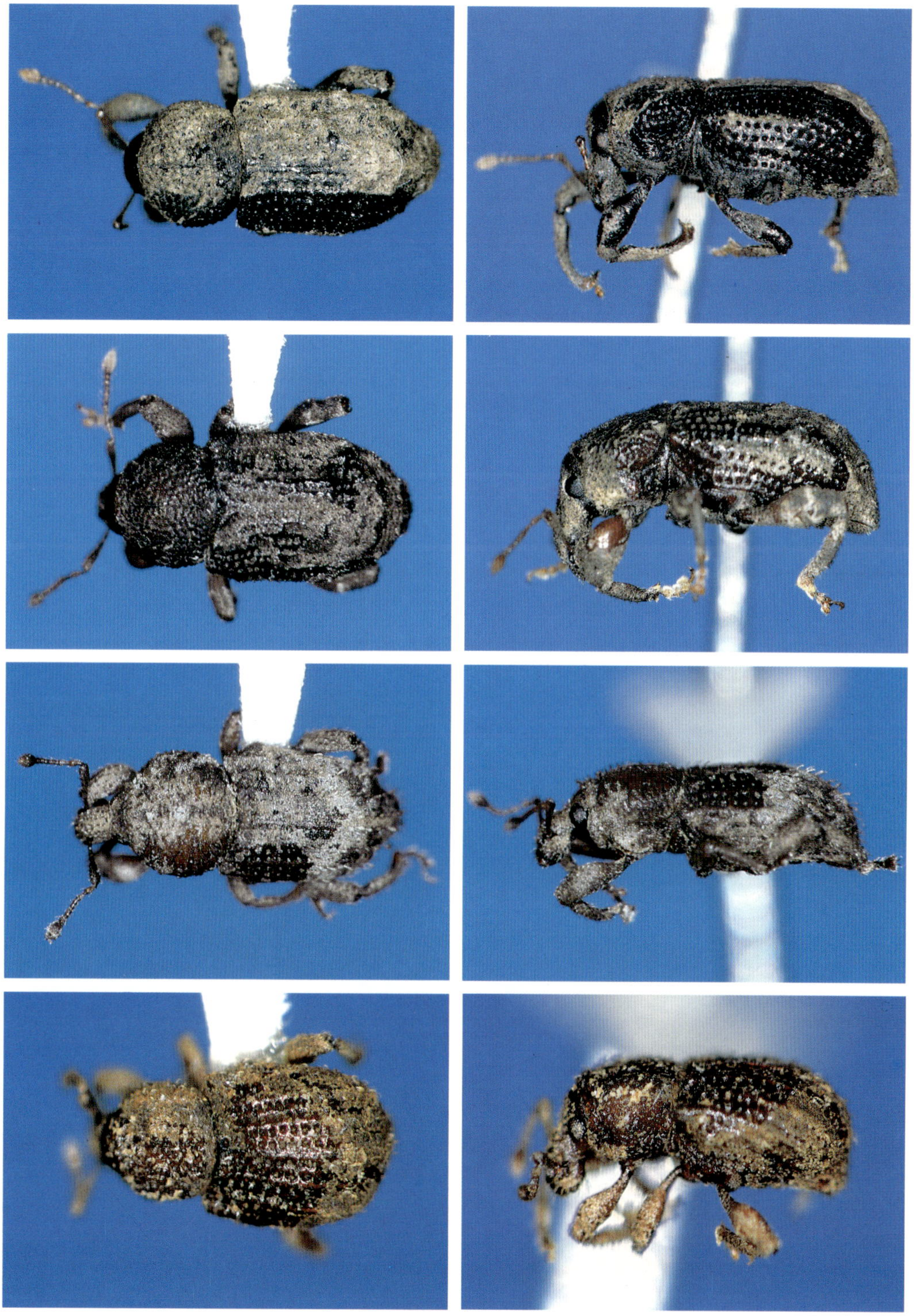

Plate 211 Adelognatha

1, 2 *Mandalotus* (*Mandalotus*) *leai* Oke
Evelyn, Victoria
Male; 7.5 mm

3, 4 *Mandalotus* (*Mandalotus*) *longicollis* Lea
Greenmirk Rocks Reserve, Victoria
Male, compared with holotype; 6.7 mm

5, 6 *Mandalotus* (*Mandalotus*) *luciphilus* Oke
Paterson, New South Wales
Holotype male; 4.9 mm

7, 8 *Mandalotus* (*Mandalotus*) *lutosus* Lea
Australia
Holotype male; 3.2 mm

1	2
3	4
5	6
7	8

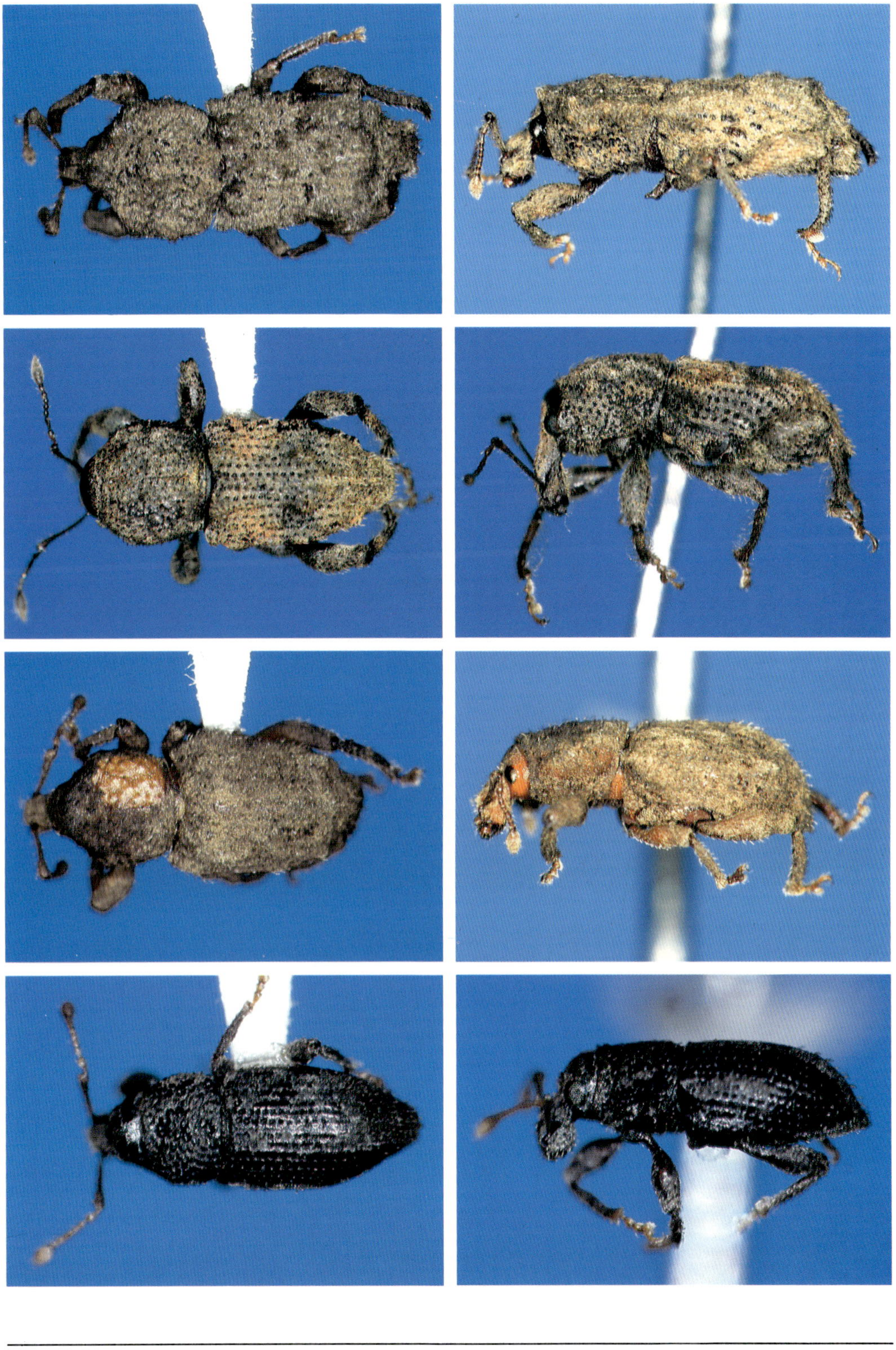

Plate 212 Adelognatha

1, 2 *Mandalotus* (*Mandalotus*) *magnicollis* Lea
Taralga, New South Wales
Holotype male; 5.5 mm

3, 4 *Mandalotus* (*Mandalotus*) *medcoxalis* Lea
Dorrigo, New South Wales
Paratype male; 7.1 mm

5, 6 *Mandalotus* (*Mandalotus*) *medianus* Lea
New South Wales
Holotype male; 6.3 mm

7, 8 *Mandalotus* (*Mandalotus*) *mesosternalis* Lea
National Park, Sydney, New South Wales
Paratype male; 4.7 mm

1	2
3	4
5	6
7	8

Plate 213 Adelognatha

1, 2 *Mandalotus* (*Mandalotus*) *metasternalis* Lea
Gosford, New South Wales
Holotype male; 4.5 mm

3, 4 *Mandalotus* (*Mandalotus*) *microscopicus* Lea
Muswellbrook, New South Wales
Holotype male; 2.3 mm

5, 6 *Mandalotus* (*Mandalotus*) *minusculus* Oke
Carrum, Victoria
Paratype male; 2.5 mm

7, 8 *Mandalotus* (*Mandalotus*) *modicus* Lea
Maryborough, Queensland
Paratype male; 5.9 mm
The specimen is clean and teneral, and its appearance may give a false impression.

1	2
3	4
5	6
7	8

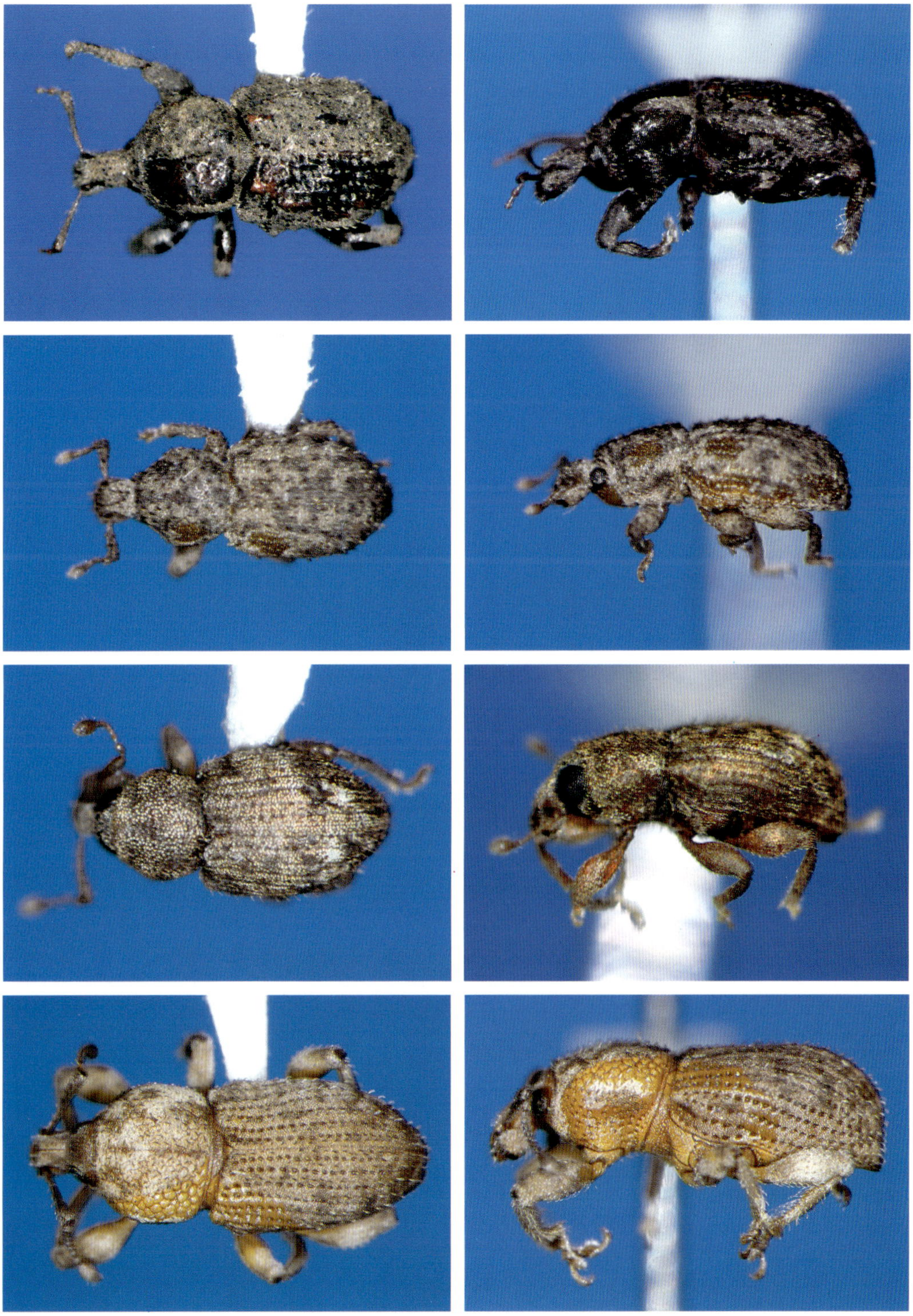

Plate 214 Adelognatha

1, 2 *Mandalotus* (*Mandalotus*) *nodicollis* Lea
Mackay, Queensland
Paratype male; 5 mm

3, 4 *Mandalotus* (*Mandalotus*) *norfolcensis* Lea
Norfolk Island
Holotype male; 2.8 mm

5, 6 *Mandalotus* (*Mandalotus*) *obliquus* Oke
Dorrigo, New South Wales
Holotype male; 5.8 mm

7, 8 *Mandalotus* (*Mandalotus*) *ochreonotatus* Lea
Muswellbrook, New South Wales
Holotype male; 4.7 mm

1	2
3	4
5	6
7	8

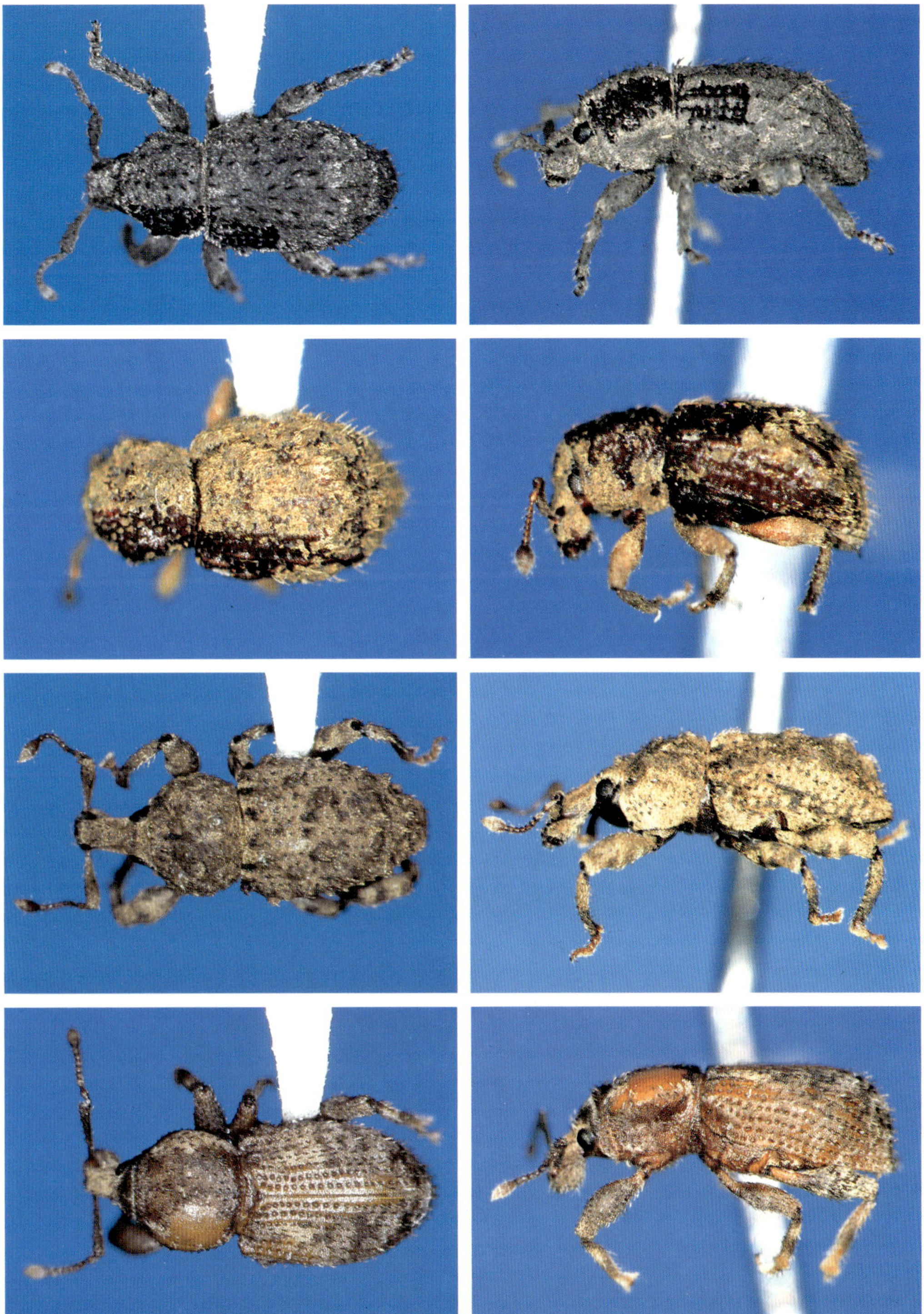

Plate 215 Adelognatha

1, 2 *Mandalotus* (*Mandalotus*) *octagonalis* Oke
Warburton, Victoria
Holotype male; 5.1 mm

3, 4 *Mandalotus* (*Mandalotus*) *oxyomus* Lea
Victoria
Male; 4.8 mm

5, 6 *Mandalotus* (*Mandalotus*) *pallidus* Lea
Sydney, New South Wales
Holotype male; 5.2 mm

7, 8 *Mandalotus* (*Mandalotus*) *parentheticus* Lea
Australia
Holotype male; 3.8 mm

1	2
3	4
5	6
7	8

Plate 216 Adelognatha

1, 2 *Mandalotus* (*Mandalotus*) *pentagonalis* Lea
Black Rock, Victoria
Male; 4.6 mm

3, 4 *Mandalotus* (*Mandalotus*) *piliventris* Lea
New South Wales
Paratype male; 6.3 mm

5, 6 *Mandalotus* (*Mandalotus*) *postcoxalis* Lea
Ringwood, Victoria
Paratype male; 5.3 mm

7, 8 *Mandalotus* (*Mandalotus*) *puncticollis* Lea
Kinchega National Park, New South Wales
Female; 3.2 mm

1	2
3	4
5	6
7	8

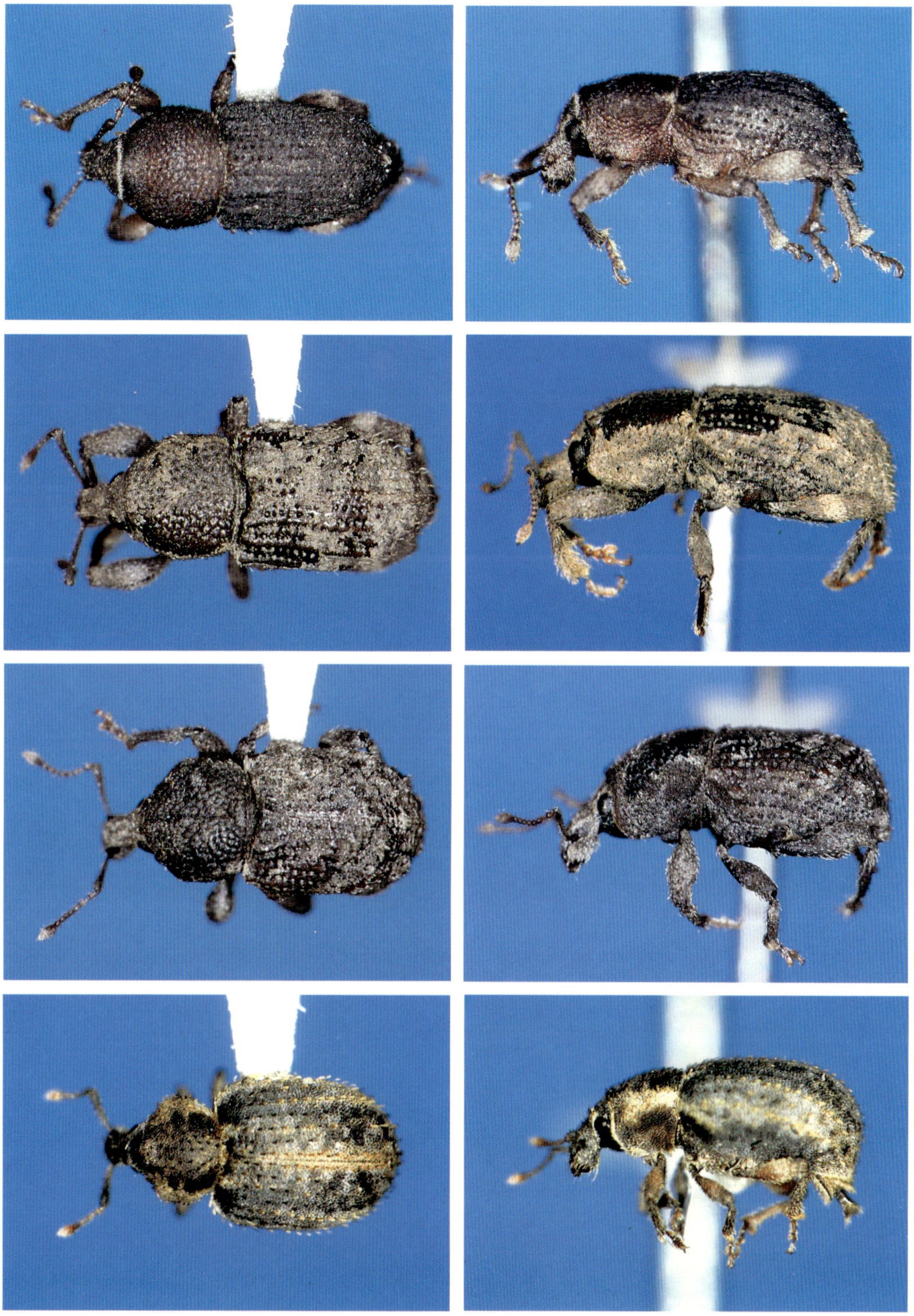

Plate 217 Adelognatha

1, 2 *Mandalotus* (*Mandalotus*) *raui* Lea
Adelaide, South Australia
Holotype male; 4 mm

3, 4 *Mandalotus* (*Mandalotus*) *recticornis* Lea
Myponga, South Australia
Paratype male; 2.9 mm

5, 6 *Mandalotus* (*Mandalotus*) *reticulatus* Lea
Inverell, New South Wales
Holotype male; 3.4 mm

7, 8 *Catastygnus stigmus* (Pascoe)
Bellinger, New South Wales
Female; 12 mm

1	2
3	4
5	6
7	8

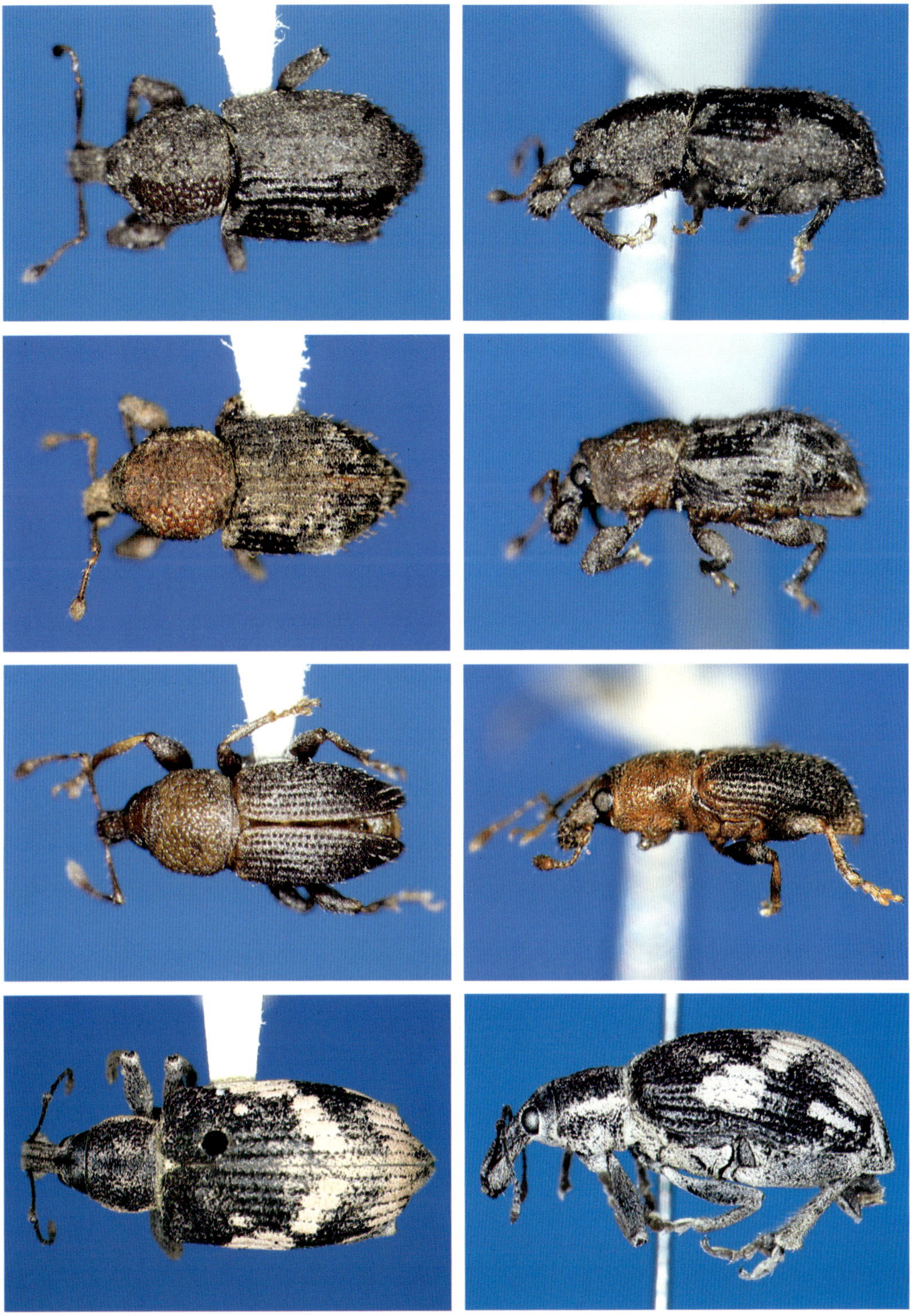

Plate 218 Adelognatha

1,2 *Mandalotus* (*Mandalotus*) *rufimanus* Lea
Adelaide, South Australia
Paratype male; 3.9 mm

3, 4 *Mandalotus* (*Mandalotus*) *scaber* Lea
Tweed River, New South Wales
Remains of holotype male; elytra 2.8 mm

5, 6 *Mandalotus* (*Mandalotus*) *seticollis* Lea
Stanley, Tasmania
Paratype male; 5.8 mm

7, 8 *Mandalotus* (*Mandalotus*) *setistriatus* Lea
New South Wales
Holotype male; 3.5 mm

1	2
3	4
5	6
7	8

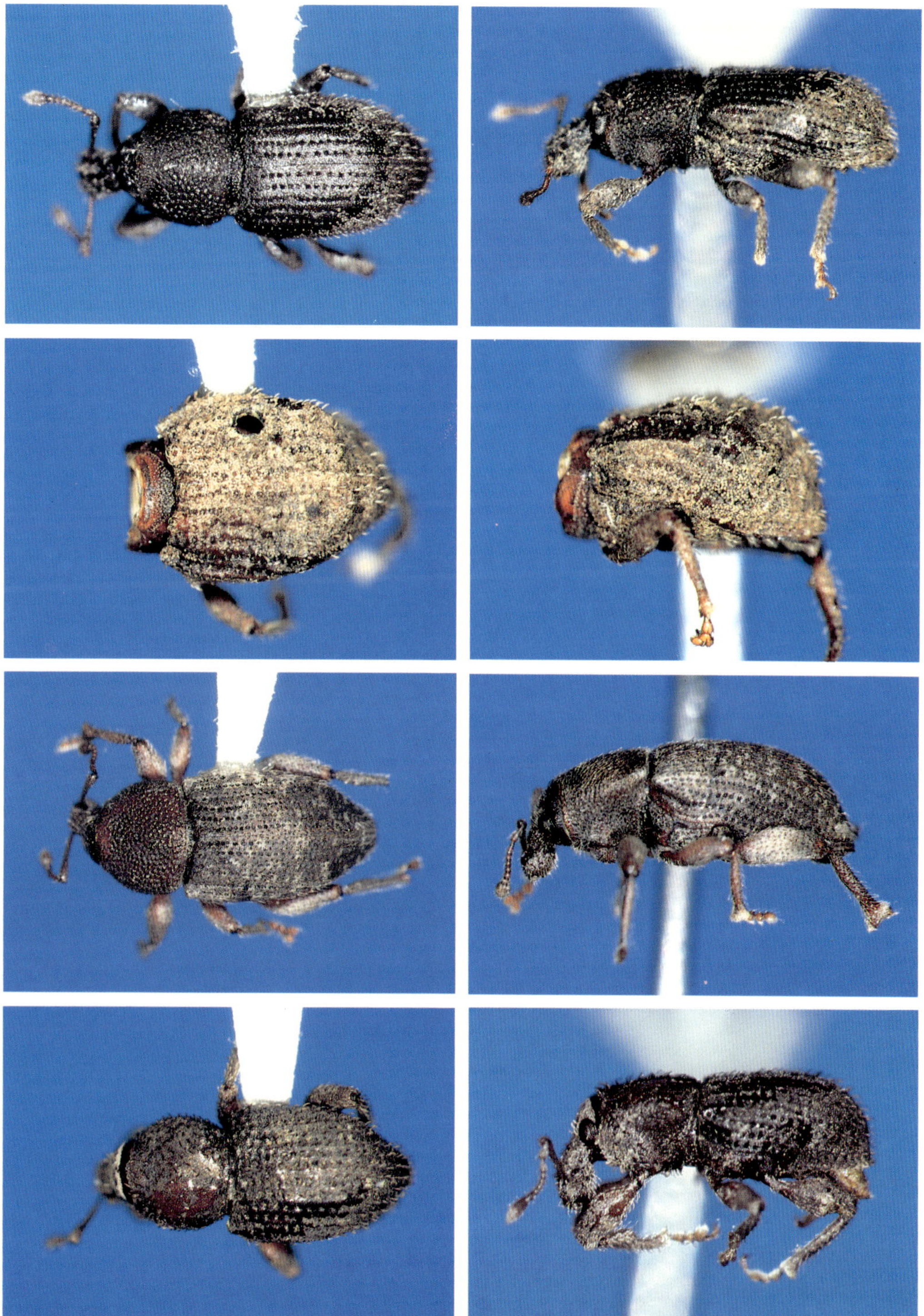

Plate 219 Adelognatha

1, 2 *Mandalotus* (*Mandalotus*) *simulator* Lea
Blue Mountains, New South Wales
Male; 6.9 mm

3, 4 *Mandalotus* (*Mandalotus*) *spurcus* Lea
Sydney, New South Wales
Paratype male; 4.8 mm

5, 6 *Mandalotus* (*Mandalotus*) *squalidus* Lea
Adelaide, South Australia
Holotype male; 3.6 mm

7, 8 *Mandalotus* (*Mandalotus*) *squamibundus* Lea
Port Denison, Queensland
Holotype male; 3 mm

1	2
3	4
5	6
7	8

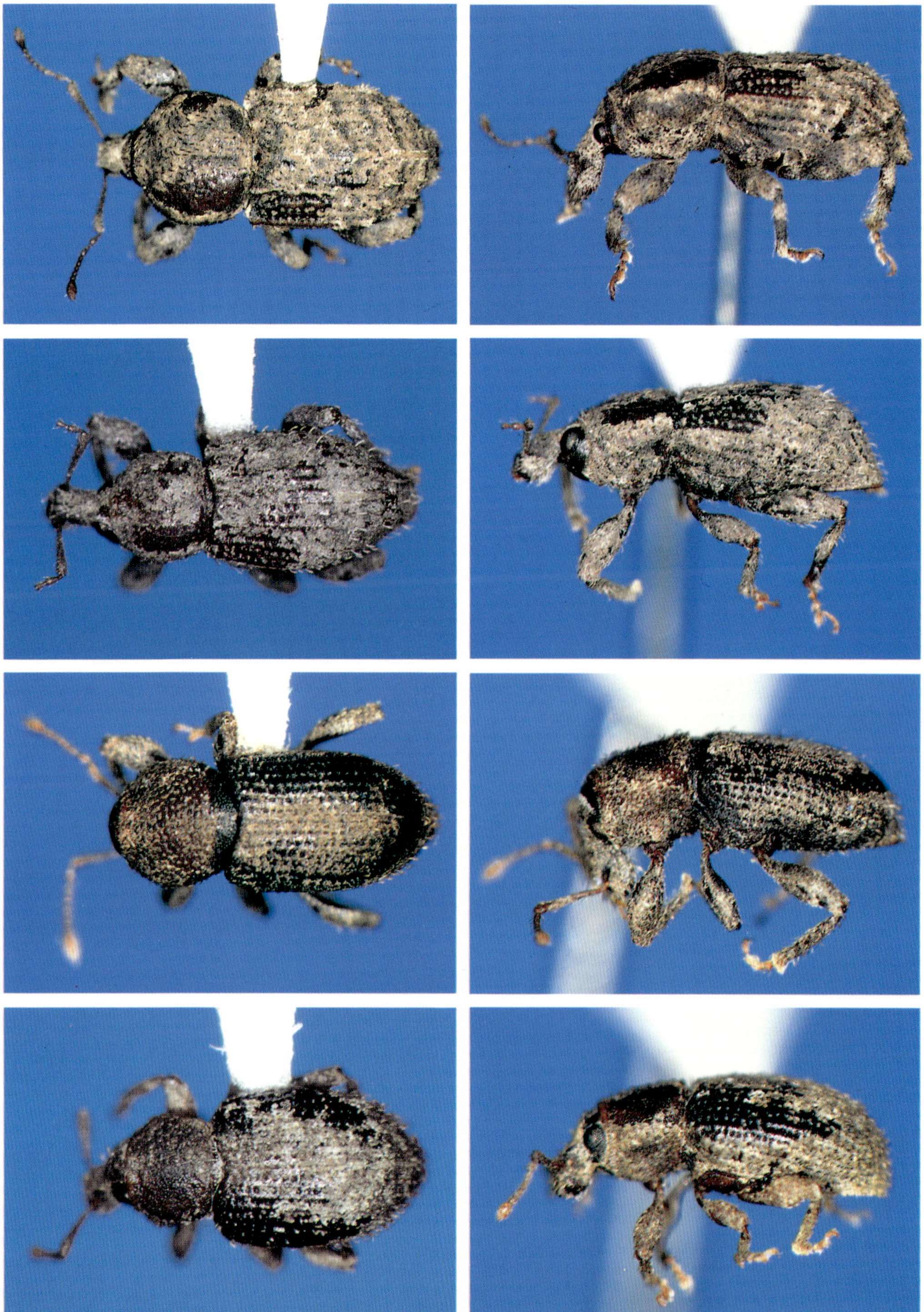

Plate 220 Adelognatha

1, 2 *Mandalotus* (*Mandalotus*) *squamosus* Lea
Lord Howe Island
Holotype male; 7.2 mm

3, 4 *Mandalotus* (*Mandalotus*) *sternocerus* Lea
Melbourne, Victoria
Holotype male; 7.2 mm

5, 6 *Mandalotus* (*Mandalotus*) *striatus* Lea
Leigh Creek, South Australia
Holotype male; 3.1 mm

7, 8 *Mandalotus* (*Mandalotus*) *subglaber* Lea
Forest Reefs, New South Wales
Holotype male; 4 mm

1	2
3	4
5	6
7	8

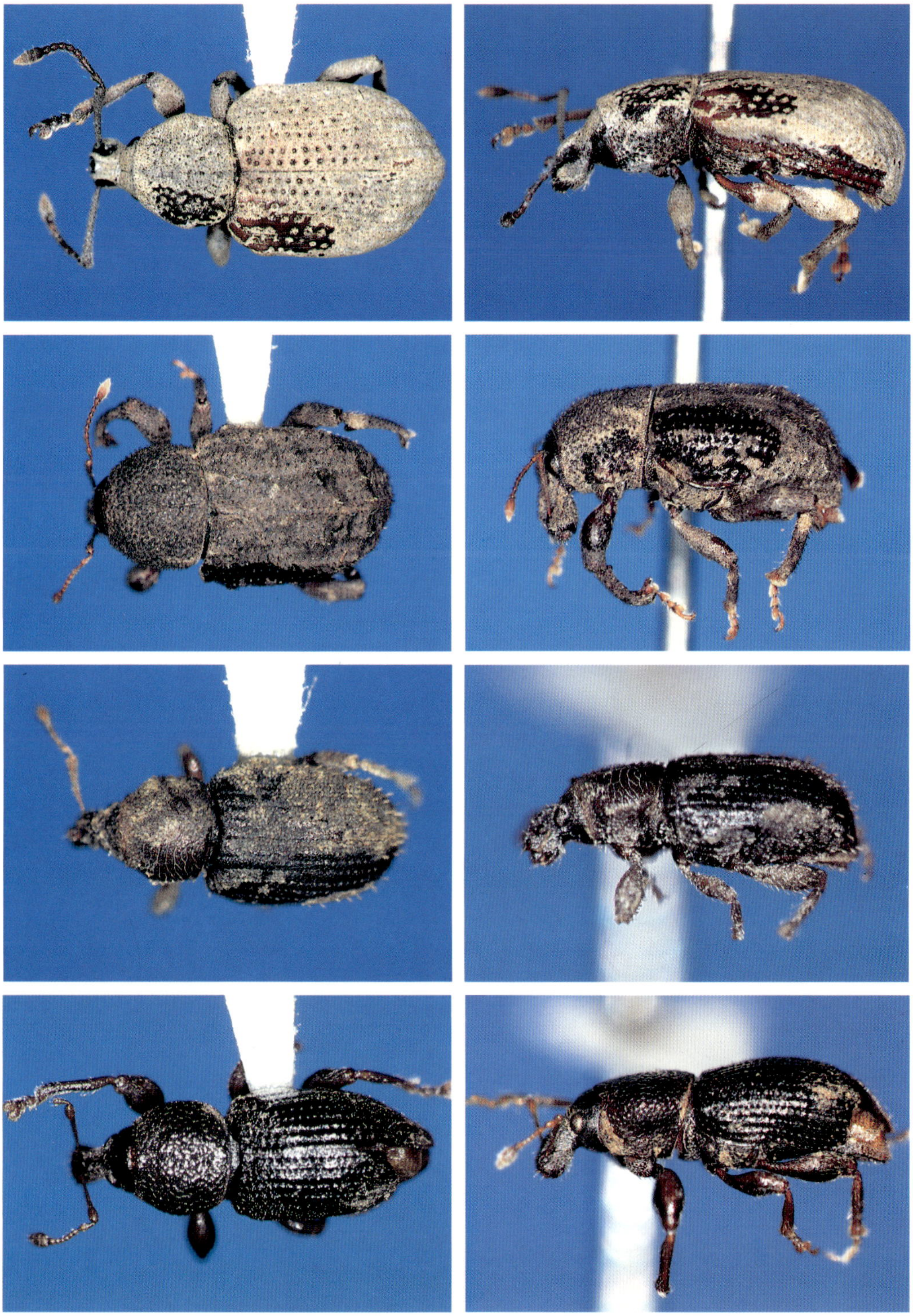

Plate 221 Adelognatha

1, 2 *Mandalotus* (*Mandalotus*) *suturalis* Lea
Armidale, New South Wales
Paratype male; 4 mm

3, 4 *Mandalotus* (*Mandalotus*) *taylori* Lea
Oberon, New South Wales
Holotype male; 7.5 mm

5, 6 *Mandalotus* (*Mandalotus*) *tenuicornis* Lea
Blackrocks, South Australia
Male; 2.8 mm

7, 8 *Mandalotus* (*Mandalotus*) *tibialis* Lea
Moss Vale, New South Wales
Male; 5.5 mm

1	2
3	4
5	6
7	8

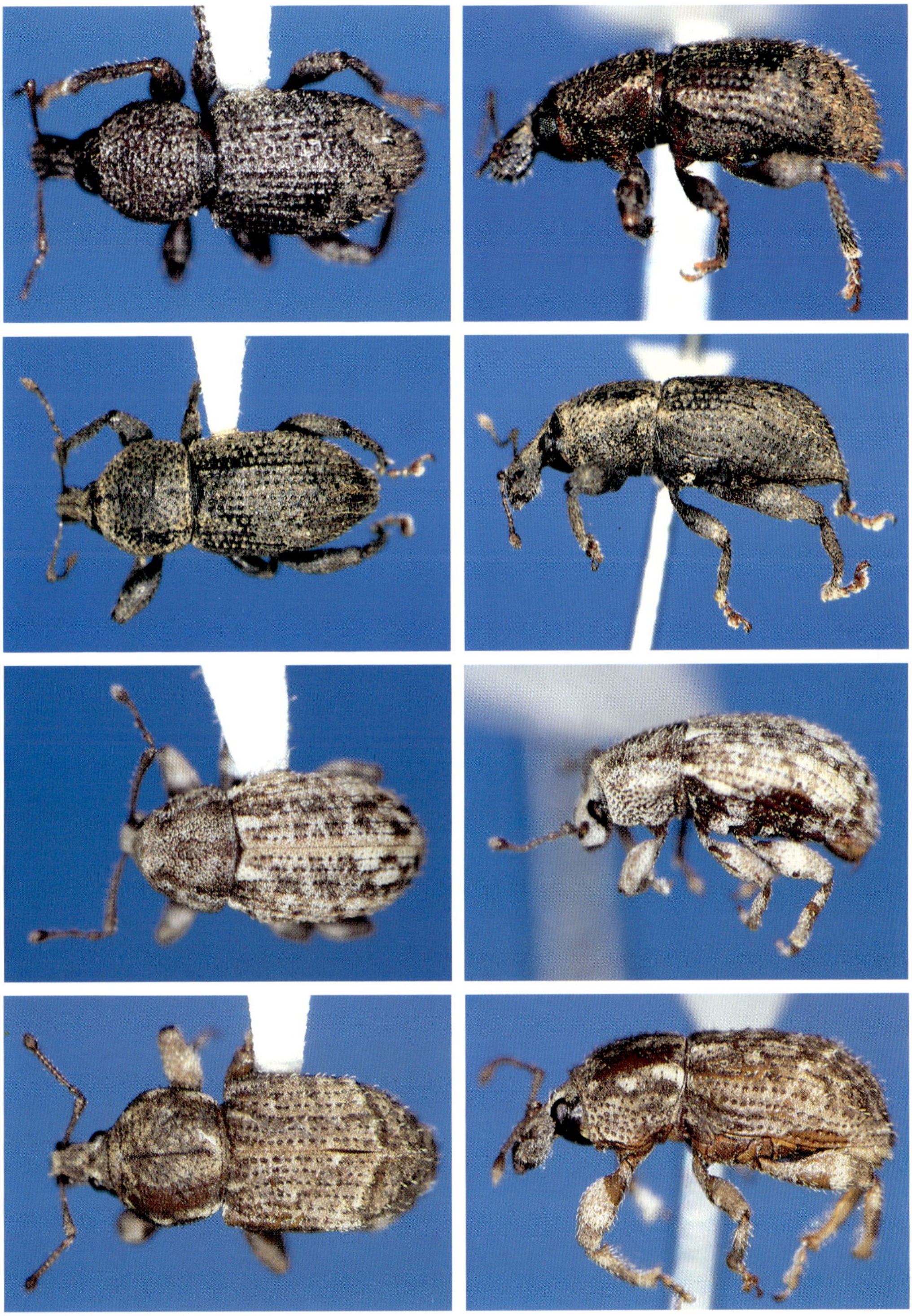

Plate 222 Adelognatha

1, 2 *Mandalotus* (*Mandalotus*) *transversus* Lea
Dorrigo, New South Wales
Holotype female; 5.6 mm

3, 4 *Mandalotus* (*Mandalotus*) *trisinuatus* Lea
Bright, Victoria
Holotype male; 3.5 mm

5, 6 *Mandalotus* (*Mandalotus*) *tuberculiventris* Lea
Sydney, New South Wales
Holotype male; 6 mm

7, 8 *Mandalotus* (*Mandalotus*) *uniformis* Lea
Mount Feathertop, Victoria
Holotype male; 4 mm

1	2
3	4
5	6
7	8

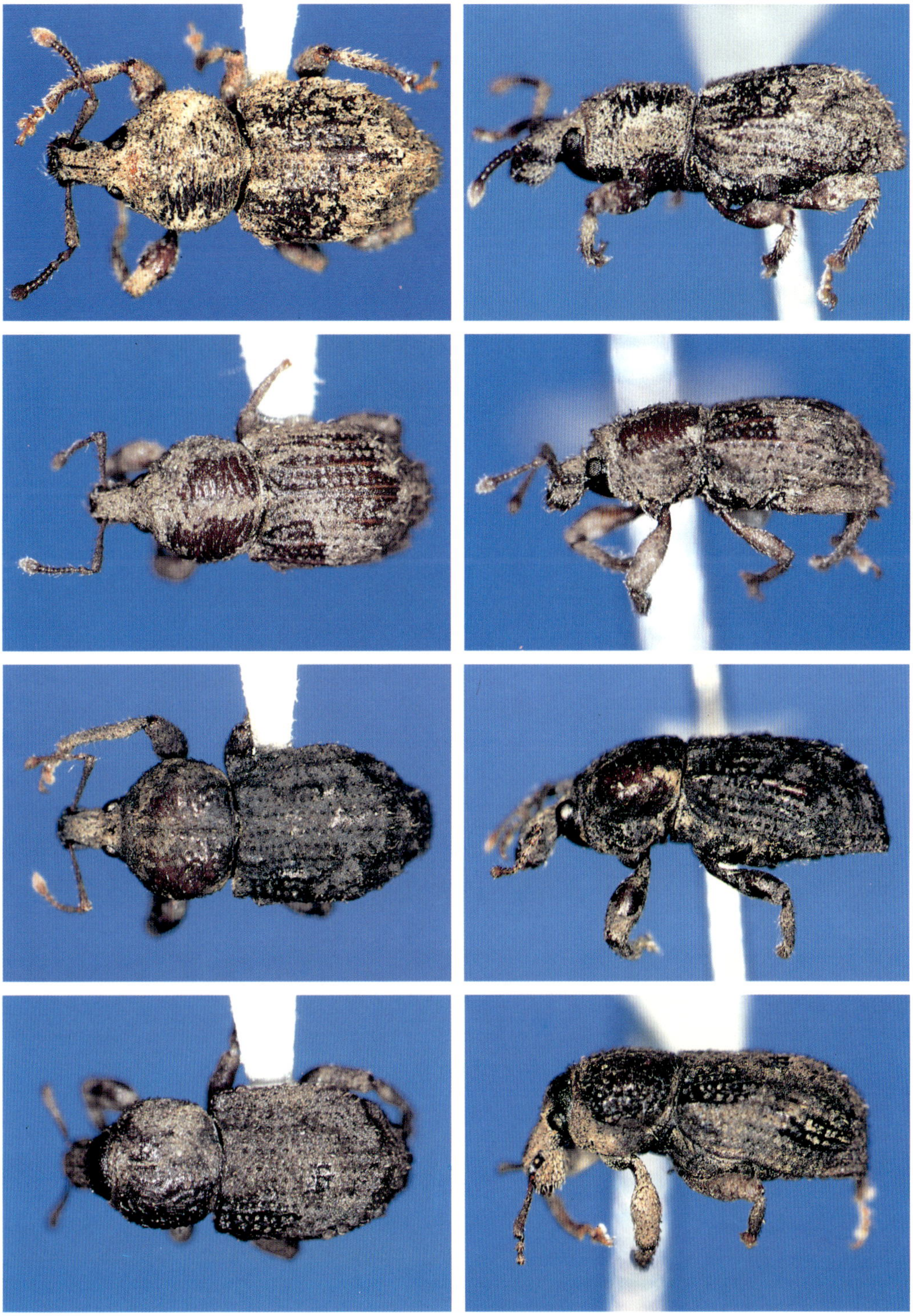

Plate 223 Adelognatha

1, 2 *Mandalotus* (*Mandalotus*) *valgus* Pascoe
Albion Park, New South Wales
Male; 7.7 mm

3, 4 *Mandalotus* (*Mandalotus*) *variabilis* Lea
Nubeena, Tasmania
Paratype male; 6.1 mm

5, 6 *Mandalotus* (*Mandalotus*) new species
Leadville, New South Wales
Male; 5.8 mm

7, 8 *Mandalotus* (*Bryodrassus*) *acutangulus* Lea
Blue Mountains, New South Wales
Male; 7 mm

1	2
3	4
5	6
7	8

Plate 224 Adelognatha

1, 2 *Mandalotus* (*Bryodrassus*) *amplicollis* Lea
Forest Reefs, New South Wales
Male; 6.4 mm

3, 4 *Mandalotus* (*Bryodrassus*) *bimaculatus* Lea
Tamborine Mountains, Queensland
Holotype male; 6.4 mm

5, 6 *Mandalotus* (*Bryodrassus*) *campylocnemis* Lea
Clarence River, New South Wales
Paratype male; 7.5 mm

7, 8 *Mandalotus* (*Bryodrassus*) *canalicornis* Lea
Armidale, New South Wales
Paratype male; 5 mm

1	2
3	4
5	6
7	8

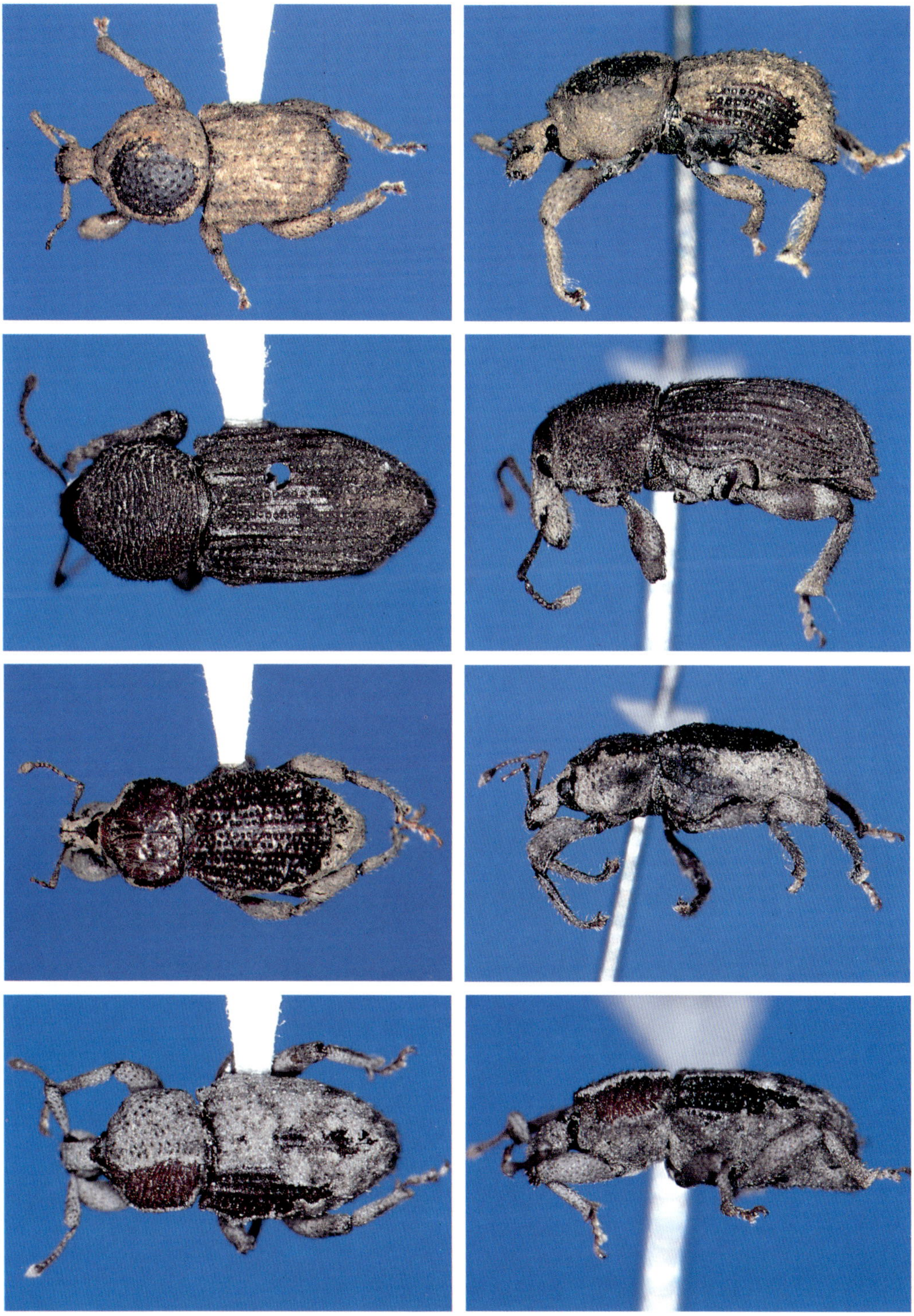

Plate 225 Adelognatha

1, 2 *Mandalotus* (*Bryodrassus*) *carinatipes* Lea
Mount Wilson, New South Wales
Male; 6.5 mm

3, 4 *Mandalotus* (*Bryodrassus*) *cinereus* Lea
"Bourke and Wilcania", New South Wales
"Darling River Flood, v – vi-1890"
Paratype female; 5.3 mm

5, 6 *Mandalotus* (*Bryodrassus*) *corrugicollis* Lea
Fletcher, Queensland
Male; 7 mm

7, 8 *Mandalotus* (*Bryodrassus*) *crawfordi* Blackburn
Nannum, South Australia
Male; 4.5 mm

1	2
3	4
5	6
7	8

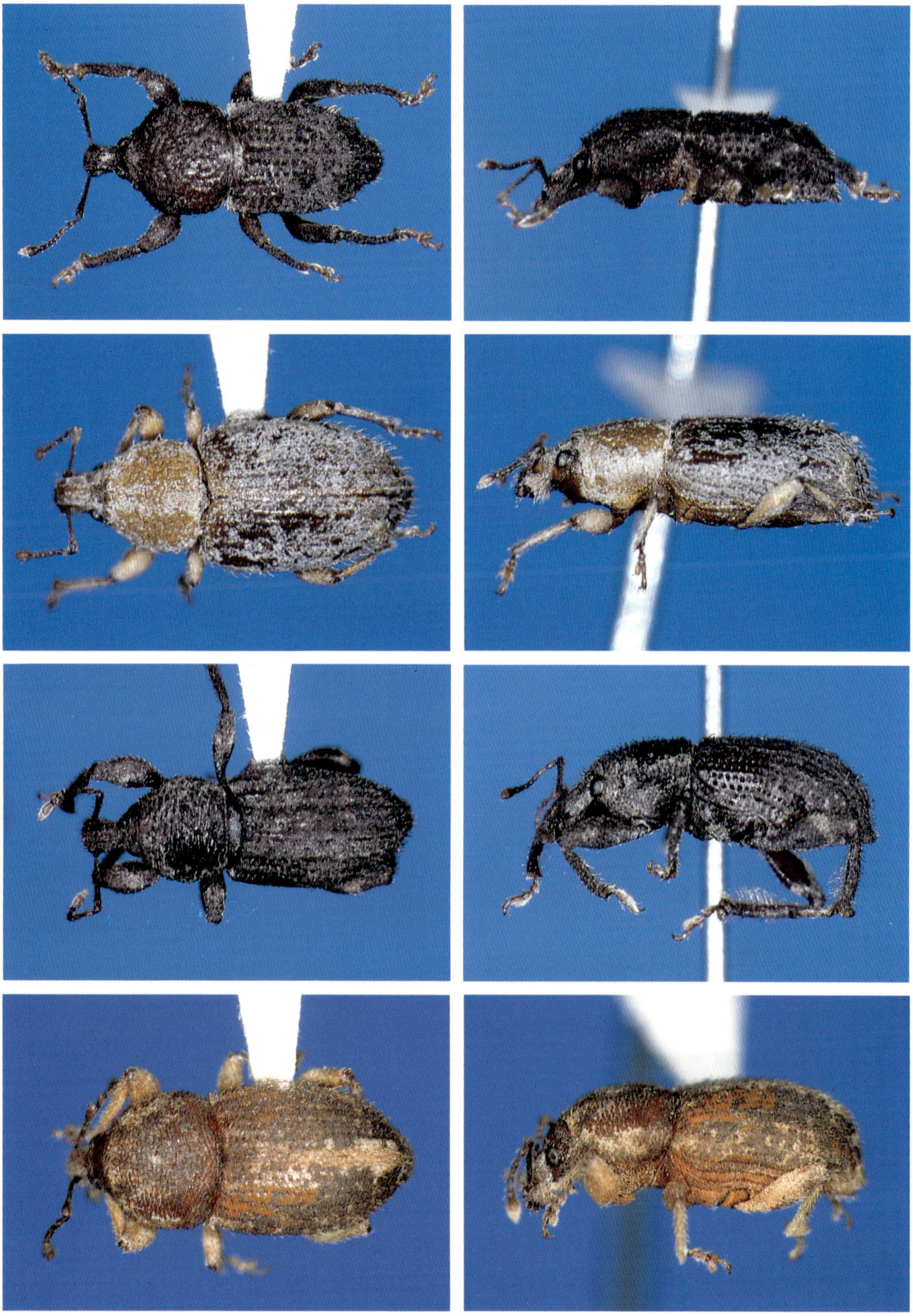

Plate 226 Adelognatha

1, 2 *Mandalotus* (*Bryodrassus*) *dentipes* Lea
Sydney, New South Wales
Holotype male; 6 mm

3, 4 *Mandalotus* (*Bryodrassus*) *dolens* Lea
Maryborough, Queensland
Holotype male; 5 mm

5, 6 *Mandalotus* (*Bryodrassus*) *exilis* Oke
Bendigo, Victoria
Holotype male; 2.8 mm

7, 8 *Mandalotus* (*Bryodrassus*) *femoralis* Lea
Nyreena, Queensland
Holotype male; 4.8 mm

1	2
3	4
5	6
7	8

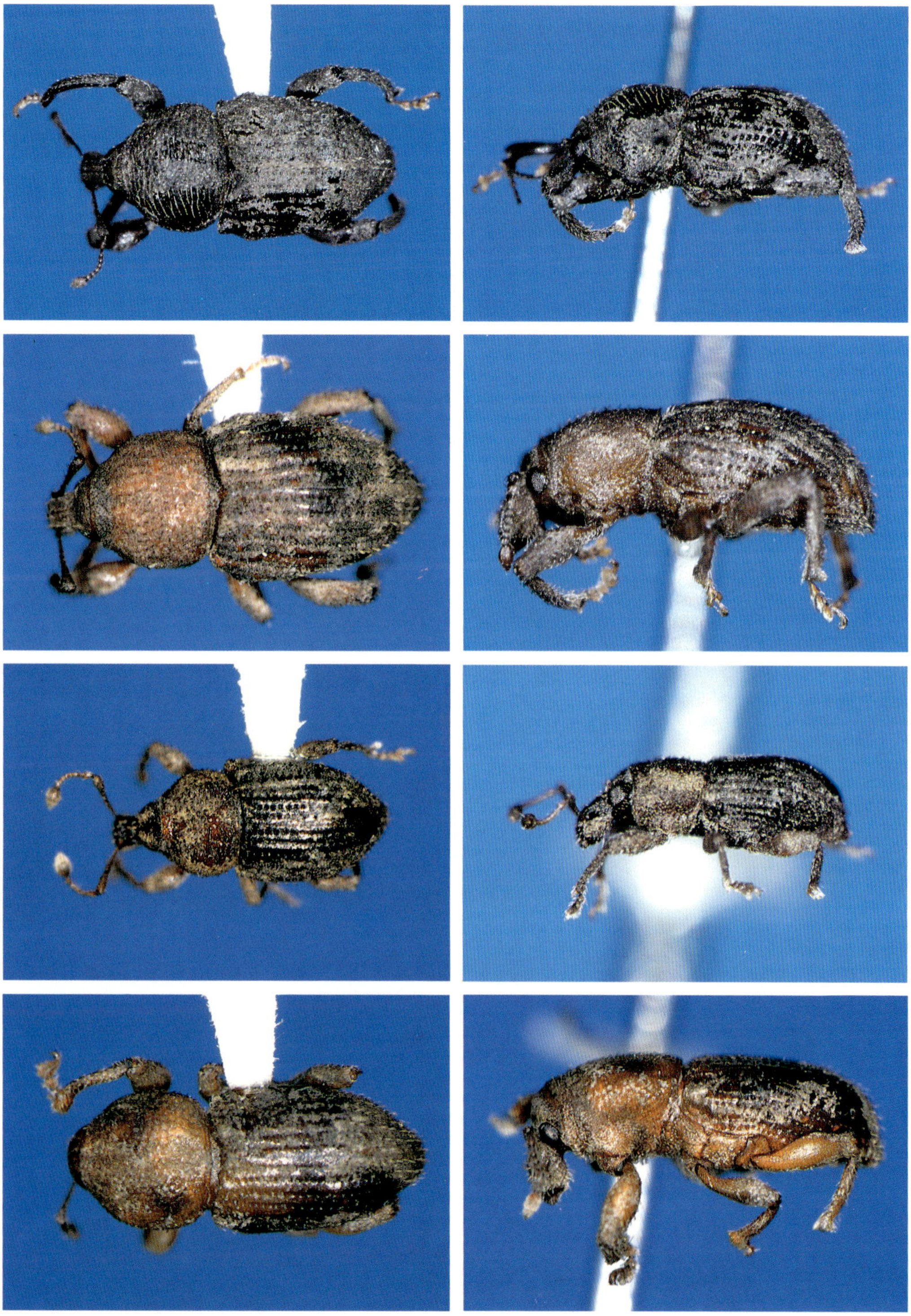

Plate 227 **Adelognatha**

1, 2 *Mandalotus* (*Bryodrassus*) *ferrugineus* Lea
Mount Tamborine, Queensland
Holotype male; 8.2 mm

3, 4 *Mandalotus* (*Bryodrassus*) *geminatus* Lea
Townsville, Queensland
Male, compared with holotype; 4.6 mm

5, 6 *Mandalotus* (*Bryodrassus*) *incisus* Lea
Blackheath, New South Wales
Male; 6.7 mm

7, 8 *Mandalotus* (*Bryodrassus*) *latus* Lea
Mount Wellington, Tasmania
Holotype male; 6.2 mm

1	2
3	4
5	6
7	8

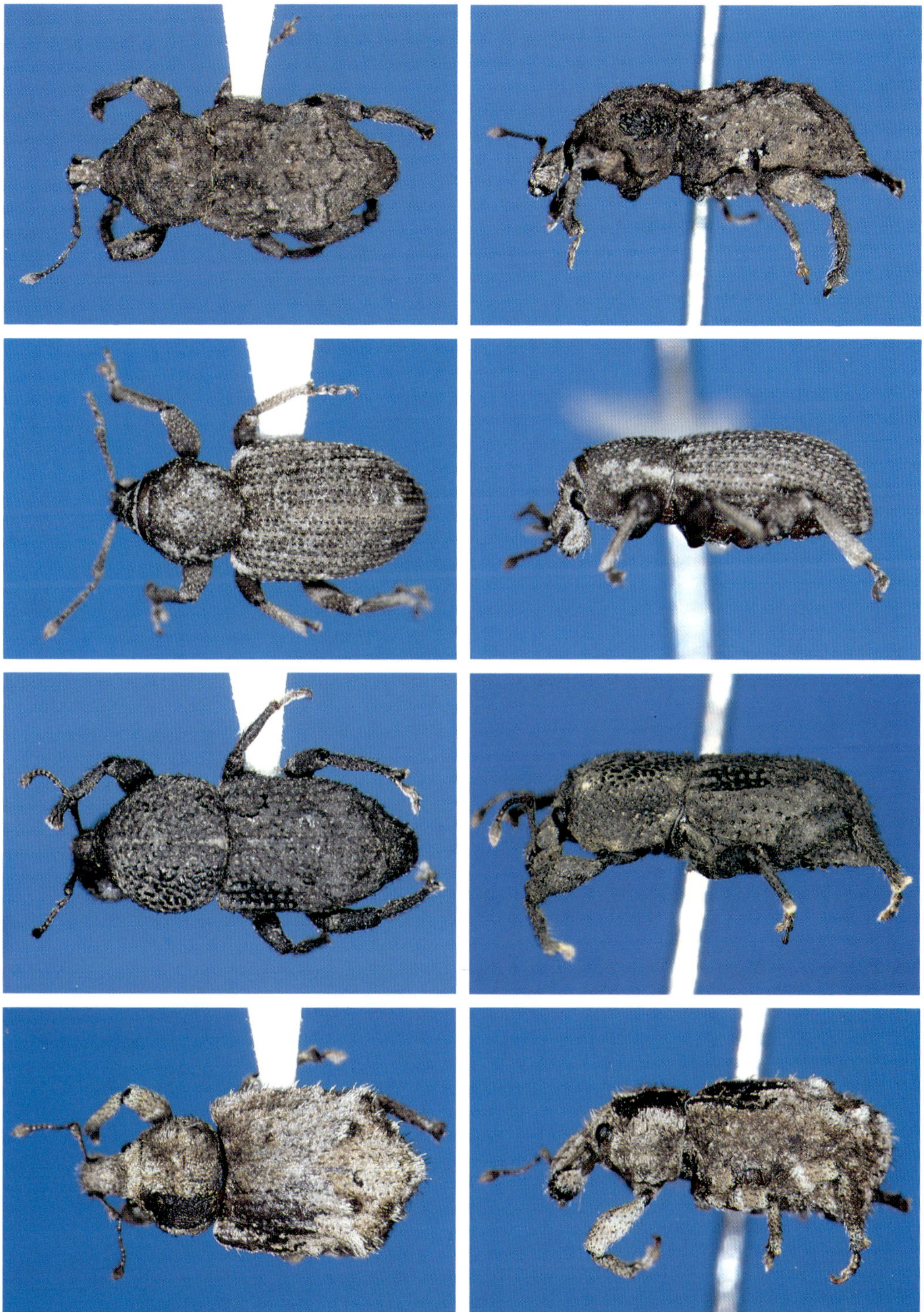

Plate 2 28 Adelognatha

1, 2 *Mandalotus* (*Bryodrassus*) *lucaris* Oke
Dorrigo, New South Wales
Holotype male; 2.7 mm

3, 4 *Mandalotus* (*Bryodrassus*) *melancholicus* Lea
Armidale, New South Wales
Holotype male; 6.3 mm

5, 6 *Mandalotus* (*Bryodrassus*) *miricollis* (Broun)
Owairaka, New Zealand
Female; 5.9 mm

7, 8 *Mandalotus* (*Bryodrassus*) *multicarinatus* Lea
Kulkyne, Victoria
Holotype male; 3.9 mm

1	2
3	4
5	6
7	8

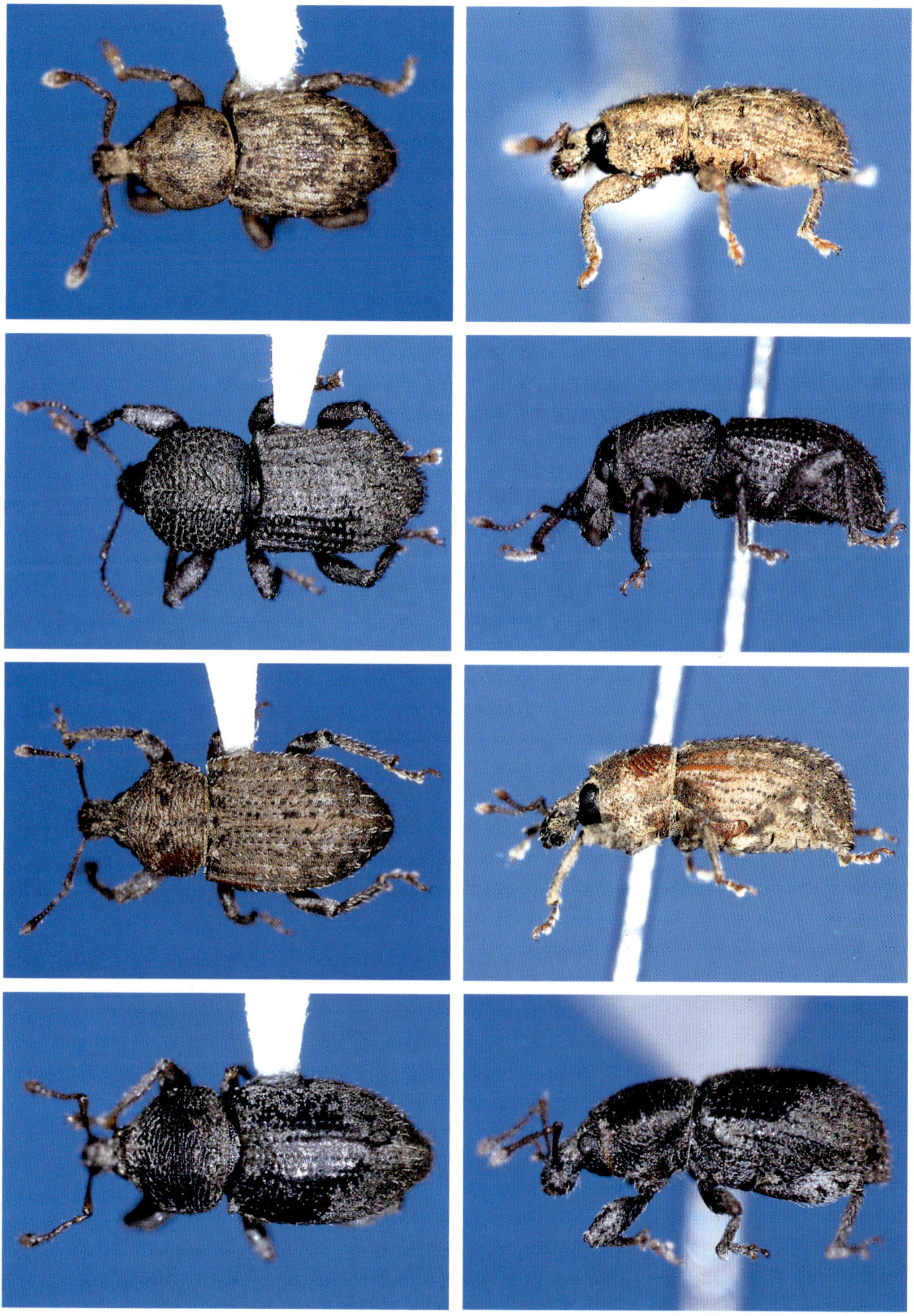

Plate 229 Adelognatha

1, 2 *Mandalotus* (*Bryodrassus*) *niger* Lea
Sydney, New South Wales
Male; 8 mm
Note the visible mesosternal process.

3, 4 *Mandalotus* (*Bryodrassus*) *oculivorus* Lea
Smoky Bay, South Australia
Holotype male; 2.4 mm

5, 6 *Mandalotus* (*Bryodrassus*) *prosternalis* Lea
Gosford, New South Wales
Holotype male; 2.9 mm
Note the visible prosternal process.

7, 8 *Mandalotus* (*Bryodrassus*) *pyrifer* Lea
Jindabyne, New South Wales
Holotype male; 7 mm

1	2
3	4
5	6
7	8

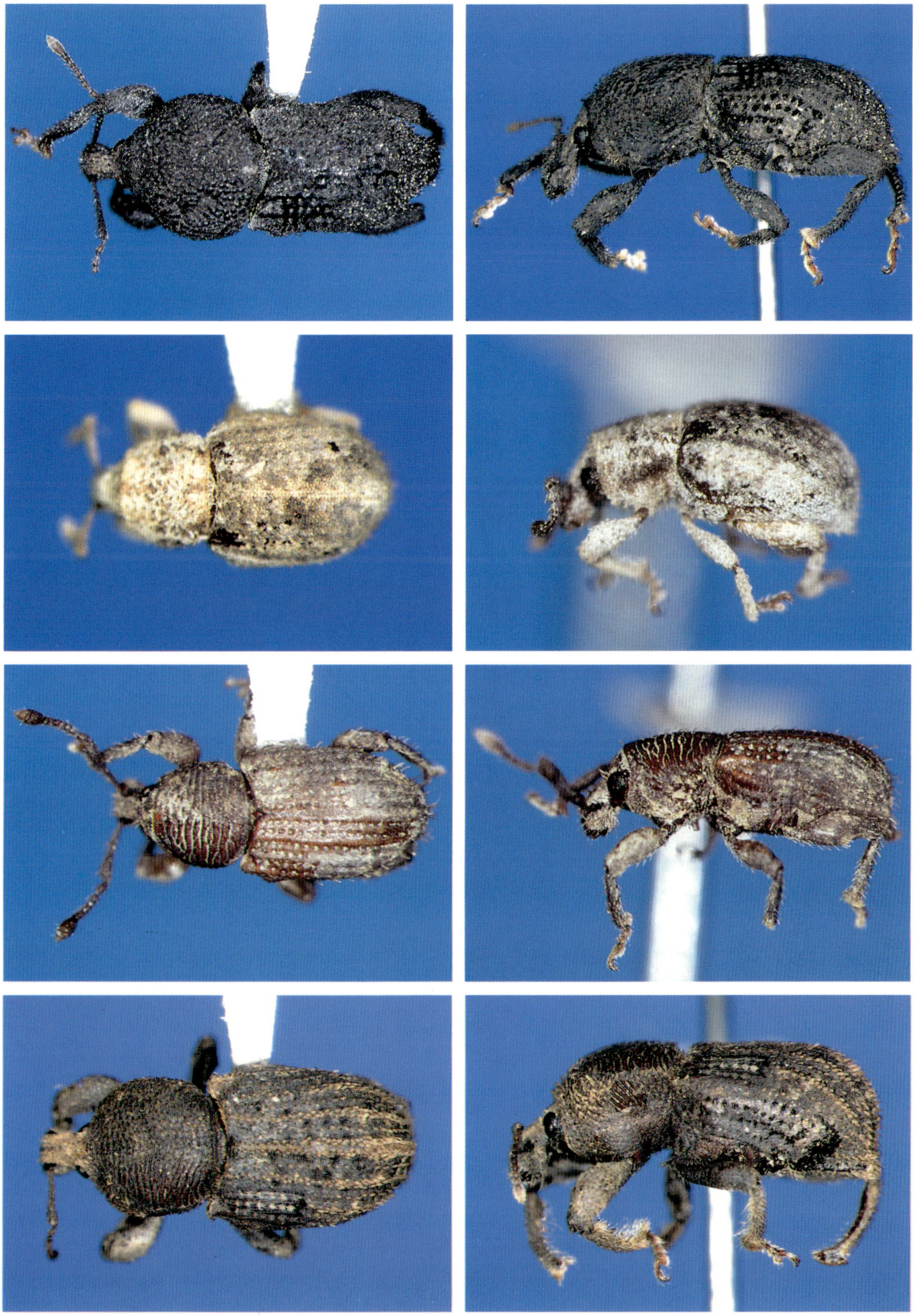

Plate 230 Adelognatha

1, 2 *Mandalotus* (*Bryodrassus*) *setosus* Lea
Windsor, New South Wales
Holotype male, abraded and damaged; 6.8 mm

3, 4 *Mandalotus* (*Bryodrassus*) *similis* Lea
Adelaide, South Australia
Male; 4.8 mm

5, 6 *Mandalotus* (*Bryodrassus*) subhumeralis Lea
Hynam, South Australia
Female; 3.5 mm

7, 8 *Mandalotus* (*Bryodrassus*) *villosipes* Lea
Barrington Tops, New South Wales
Paratype male; 4.6 mm

1	2
3	4
5	6
7	8

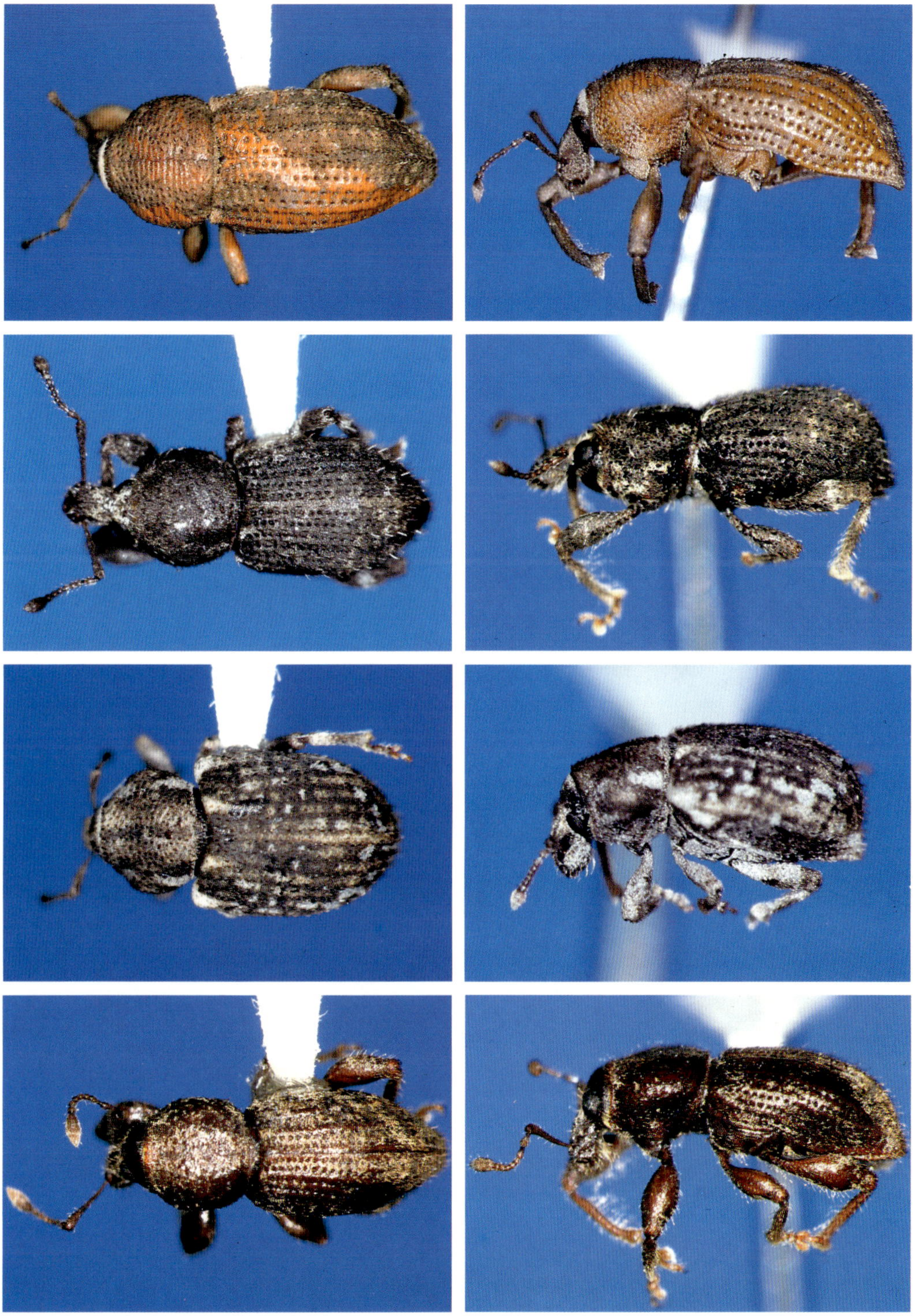

Plate 231 Adelognatha

1, 2 *Mandalotus* (*Paramandalotus*) *advenus* Blackburn
Mount Hotham, Victoria
Male; 3.9 mm
Note the visible metatibial tooth.

3, 4 *Mandalotus* (*Paramandalotus*) *alpinus* Lea
Victorian Alps
Paratype male; 4.4 mm

5, 6 *Mandalotus* (*Paramandalotus*) *bicarinatus* Lea
Hobart, Tasmania
Paratype male; 3.4 mm

7, 8 *Mandalotus* (*Paramandalotus*) *blackburni* Lea
Hobart, Tasmania
Holotype male; 5.7 mm

1	2
3	4
5	6
7	8

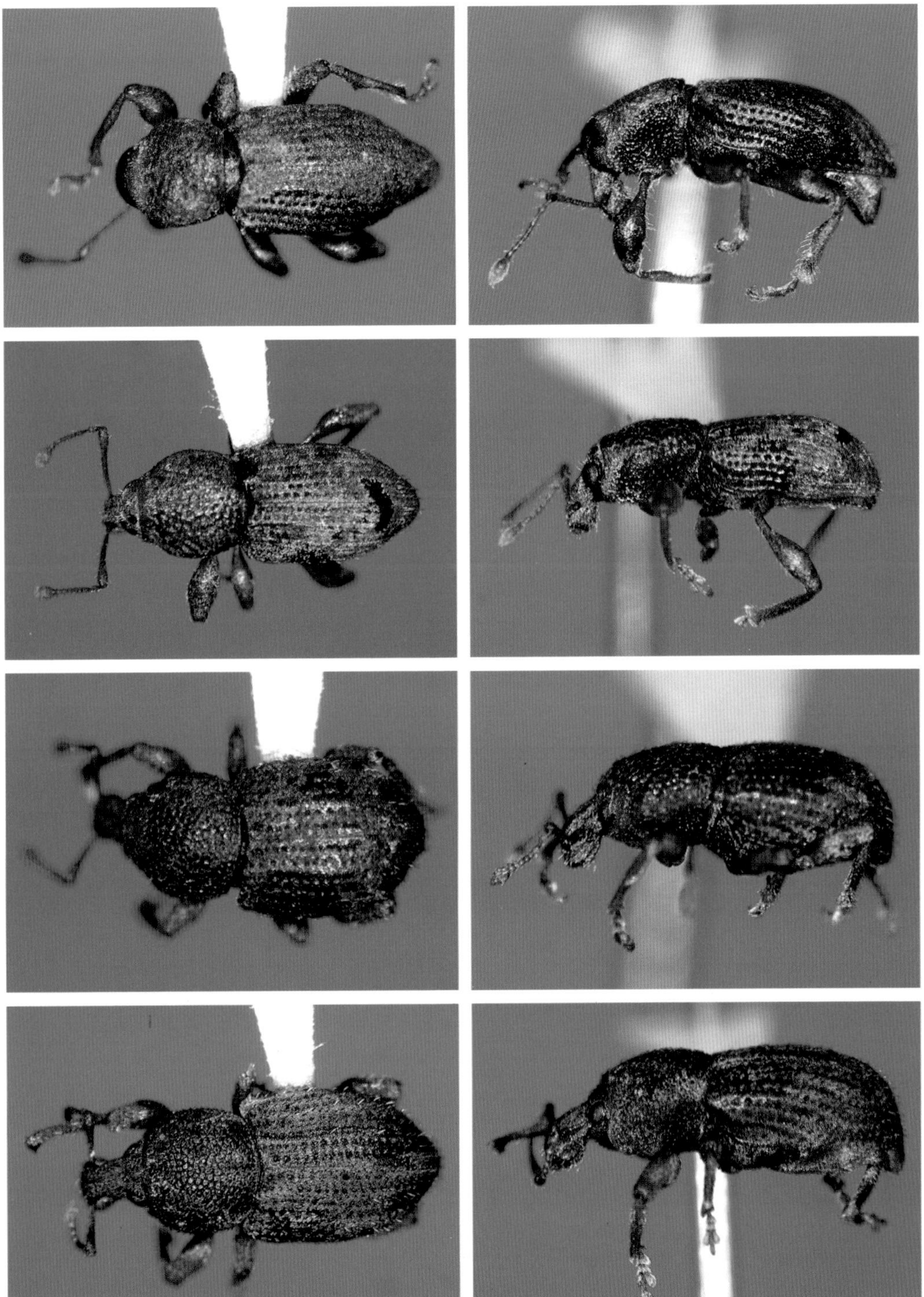

Plate 232 Adelognatha

1, 2 *Mandalotus* (*Paramandalotus*) *brevicarinatus* Lea
Launceston, Tasmania
Paratype male; 5.1 mm

3, 4 *Mandalotus* (*Paramandalotus*) *bryophagus* Lea
Hobart, Tasmania
Holotype male; 3.4 mm

5, 6 *Mandalotus* (*Paramandalotus*) *denticulatus* Lea
Belgrave, Victoria
Male, compared with holotype; 4 mm

7, 8 *Mandalotus* (*Paramandalotus*) *fulginosus* (Pascoe)
Frankford, Tasmania
Male; 6 mm
From the holotype of *carinativentris* Lea, a synonym.

1	2
3	4
5	6
7	8

Plate 233 Adelognatha

1, 2 *Mandalotus* (*Paramandalotus*) *granicollis* Lea
Mount Tomah, New South Wales
Holotype male; 3.6 mm

3, 4 *Mandalotus* (*Paramandalotus*) *granulatus* Lea
Mount Wellington, Tasmania
Holotype male; 7.7 mm

5, 6 *Mandalotus* (*Paramandalotus*) *gymnogaster* Lea
Lakes Entrance, Victoria
Paratype male; 3.5 mm

7, 8 *Mandalotus* (*Paramandalotus*) *hypulus* Lea
Tasmania
Holotype male; 5.4 mm

1	2
3	4
5	6
7	8

Plate 234 Adelognatha

1, 2 *Mandalotus* (*Paramandalotus*) *imitator* Lea
Swansea, Tasmania
Holotype male; 7 mm

3, 4 *Mandalotus* (*Paramandalotus*) *insignipes* Lea
Millgrave, Victoria
Holotype male; 3.8 mm
Note the remarkable metatibiae.

5, 6 *Mandalotus* (*Paramandalotus*) *inusitatus* Lea
Hobart, Tasmania
Holotype male; 3.2 mm

7, 8 *Mandalotus* (*Paramandalotus*) *litoralis* Lea
Hobart, Tasmania
Holotype male; 4.2 mm

1	2
3	4
5	6
7	8

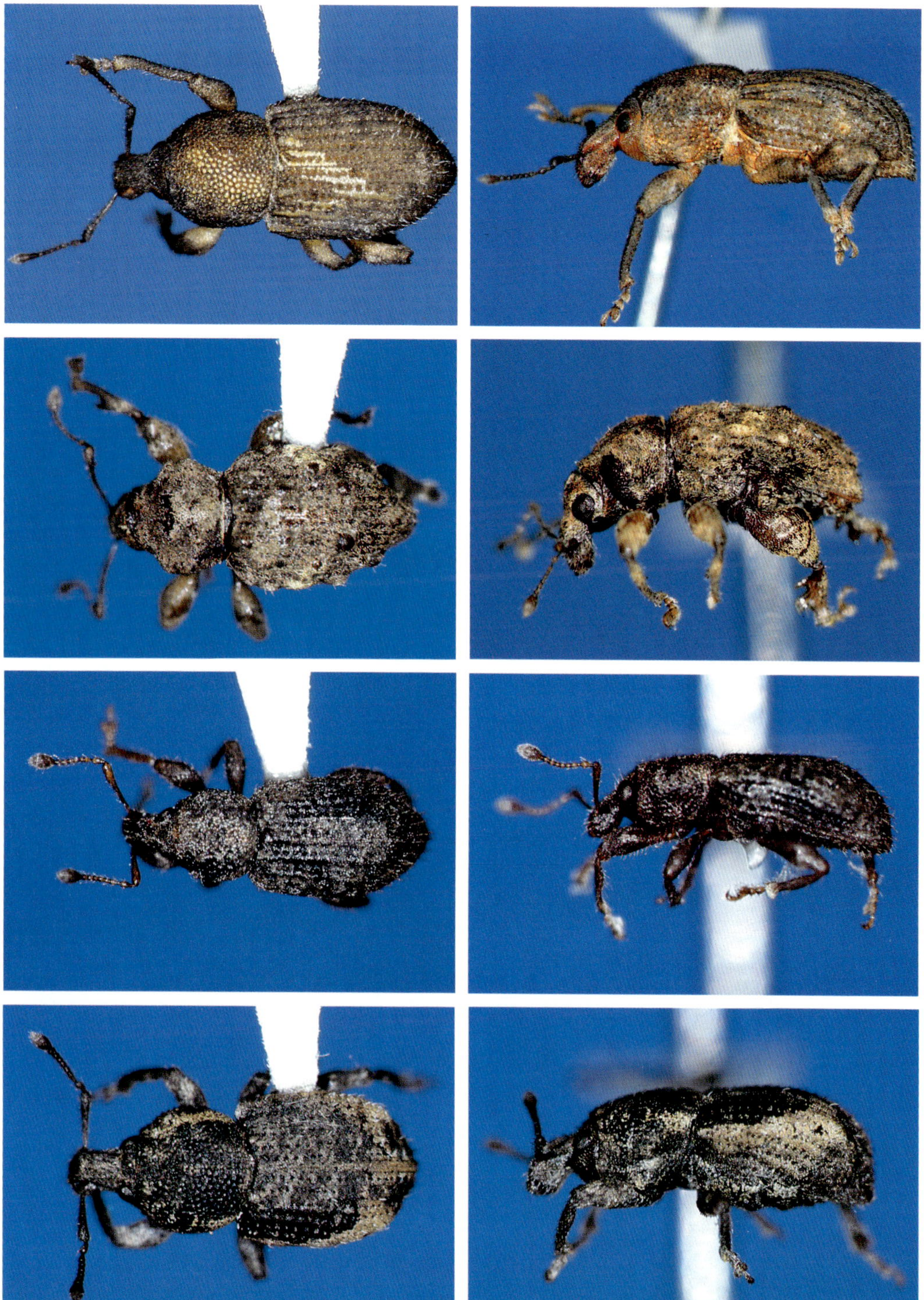

Plate 235 Adelognatha

1, 2 *Mandalotus* (*Paramandalotus*) *macrops* Lea
Healesville, Victoria
Holotype male; 3 mm

3, 4 *Mandalotus* (*Paramandalotus*) *maculatus* Lea
Bruny Island, Tasmania
Holotype female; 3.7 mm

5, 6 *Mandalotus* (*Paramandalotus*) *minutus* Lea
Hynam, South Australia
Male, compared with holotype; 3.1 mm

7, 8 *Mandalotus* (*Paramandalotus*) *murrayi* Lea
Pearson Island, South Australia
Paratype male; 3.5 mm

1	2
3	4
5	6
7	8

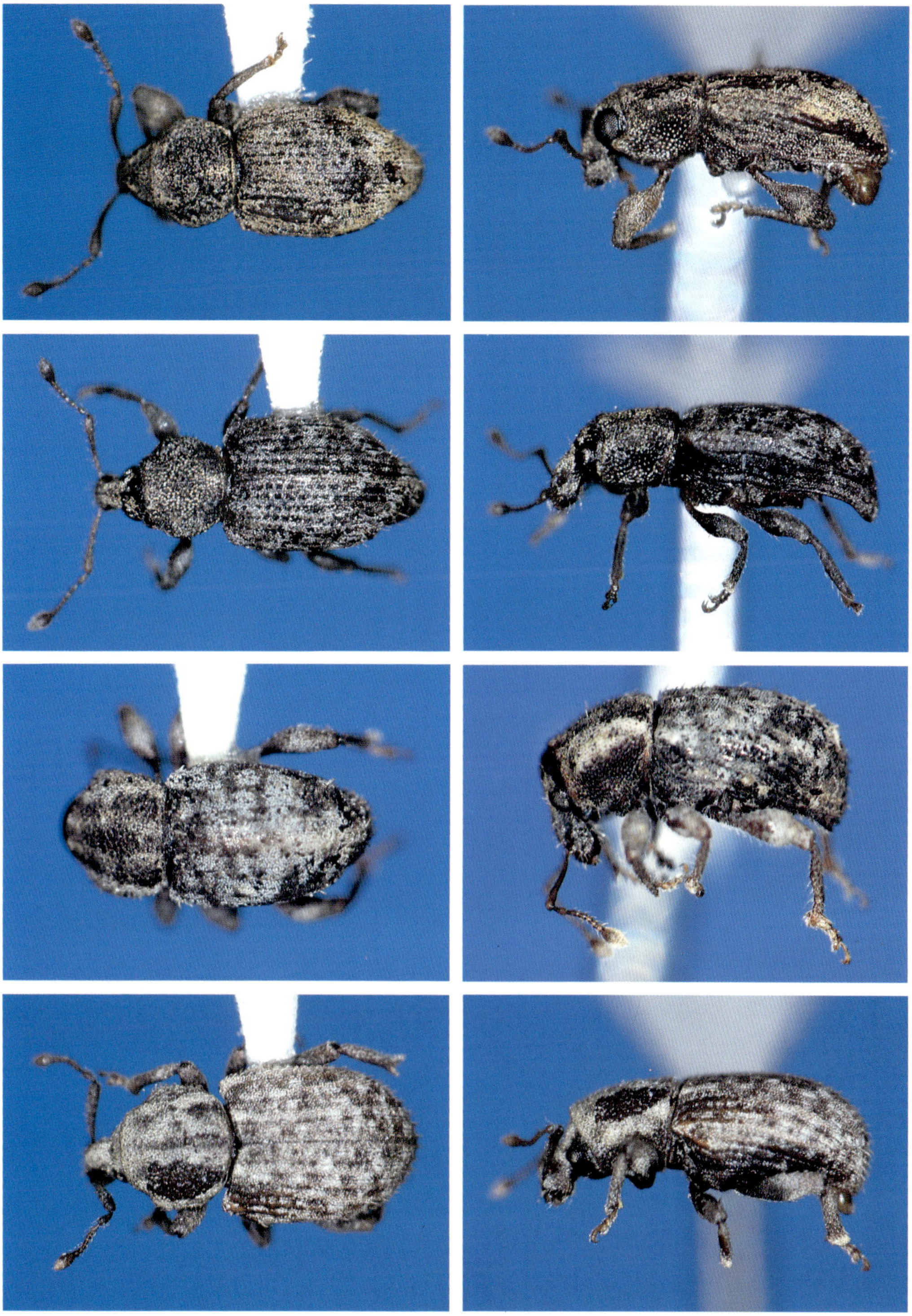

Plate 236 Adelognatha

1, 2 *Mandalotus* (*Paramandalotus*) *musivorus* Lea
Waratah, Tasmania
Male; 3 mm

3, 4 *Mandalotus* (*Paramandalotus*) *musivorus* Lea
Waratah, Tasmania
Female; 3.5 mm

5, 6 *Mandalotus* (*Paramandalotus*) *pentagonoderes* Lea
Warburton, Victoria
Holotype male; 7.2 mm

7, 8 *Mandalotus* (*Paramandalotus*) *punctiventris* Lea
Corny Point, South Australia
Male; 3.7 mm

1	2
3	4
5	6
7	8

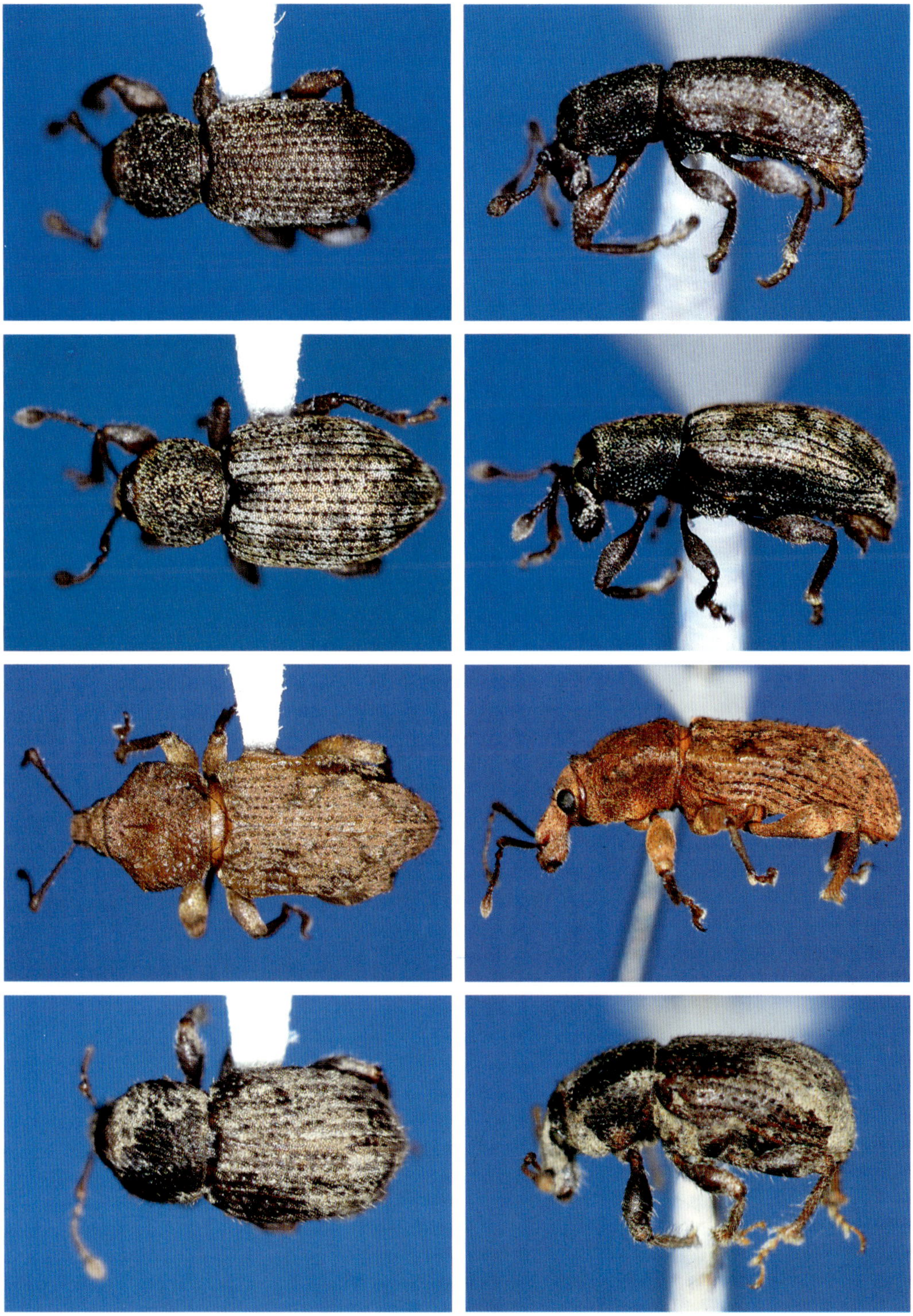

Plate 237 Adelognatha

1, 2 *Mandalotus* (*Paramandalotus*) *rufipes* Lea
Waratah, Tasmania
Holotype female; 3.2 mm

3, 4 *Mandalotus* (*Paramandalotus*) *sabulosus* Lea
Eden, New South Wales
Male; 5.4 mm

5, 6 *Mandalotus* (*Paramandalotus*) *severini* Lea
Tasmania
Holotype male; 8 mm

7, 8 *Mandalotus* (*Paramandalotus*) *sterilis* Erichson
Warrnambool, Victoria
Male; 7.2 mm

1	2
3	4
5	6
7	8

Plate 238 Adelognatha

1, 2 *Mandalotus* (*Paramandalotus*) *sydneyensis* Lea
Sydney, New South Wales
Male; 5 mm

3, 4 *Mandalotus* (*Paramandalotus*) *tuberipennis* Lea
Warburton, Victoria, 1,200 m
Paratype male; 4.7 mm

5, 6 *Mandalotus* (*Paramandalotus*) *vacillans* Lea
Tasmania
Paratype male; 4.1 mm

7, 8 *Mandalotus* (*Paramandalotus*) *vigilans* Lea
Millgrave, Victoria
Holotype male; 3.2 mm

1	2
3	4
5	6
7	8

Plate 239 Adelognatha

1, 2 *Matesia maculata* Lea
King Sound, Western Australia
Female; 7 mm

3, 4 *Merimnetes australis* (Boheman)
Hobart, Tasmania
Female; 6.7 mm
From a paratype of *Merimnetes fagi* Lea, a synonym.

5, 6 *Merimnetes celmisiae* Lea
Schlink Pass, Kosciusko National Park, New South Wales
Male, compared with holotype; 3.7 mm

7, 8 *Merimnetes decipiens* (Lea)
Mount Wellington, Tasmania
Female; 3.7 mm

1	2
3	4
5	6
7	8

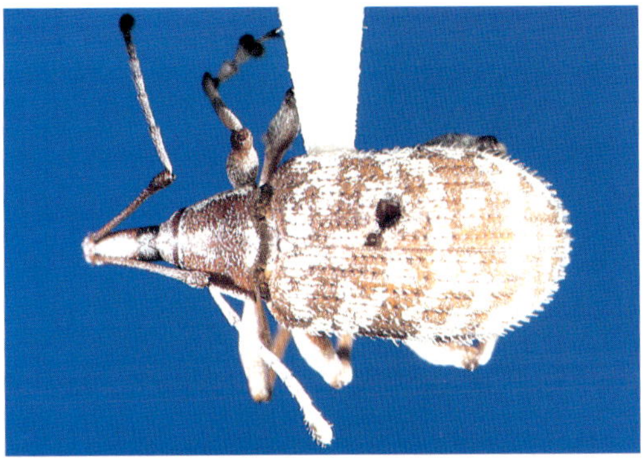

Plate 240 Adelognatha

1, 2 *Merimnetes montanus* Lea
Summit of Mount Wellington, Tasmania
Paratype male; 7 mm

3, 4 *Merimnetes montanus* Lea
Summit of Mount Wellington, Tasmania
Paratype female; 7.3 mm

5, 6 *Merimnetes simplicipennis* Lea
Mount Wellington, Tasmania
Paratype female; 8.5 mm

7, 8 *Merimnetes viridis* Lea
Schlink Pass, Kosciusko National Park, New South Wales
Female compared with holotype; 5.4 mm

1	2
3	4
5	6
7	8

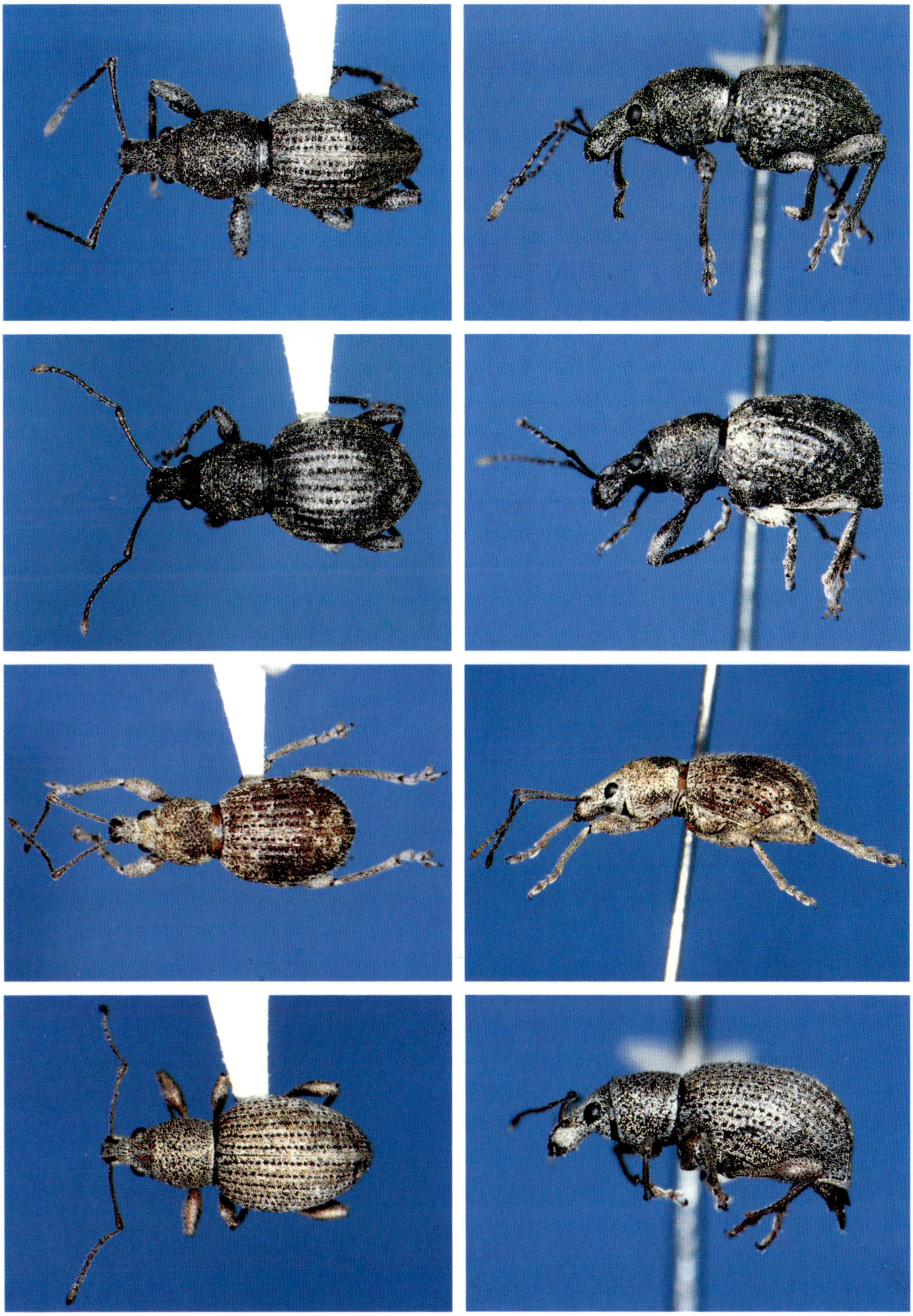

Plate 241 Adelognatha

1, 2 *Hackeria? aberrans* (Lea)
Daly River, Northern Territory
Holotype male; 6.5 mm

3, 4 *Myllocerus abstersus* (Pascoe)
Cue, Western Australia
Male; 7.4 mm

5, 6 *Myllocerus abstersus* (Pascoe)
Cue, Western Australia
Female; 9.1 mm

7, 8 *Myllocerus acutidens* Lea
Sherlock River, Western Australia
Holotype female; 9 mm

Note: Many species of *Myllocerus* have two or more colour forms (Plate 243:1–4, for example), and care must be taken not to be confused by varieties which differ in colour or pattern from the forms illustrated. Moreover, the shapes of the sexes of some species are significantly different (Plate 253:5,7, for example).

1	2
3	4
5	6
7	8

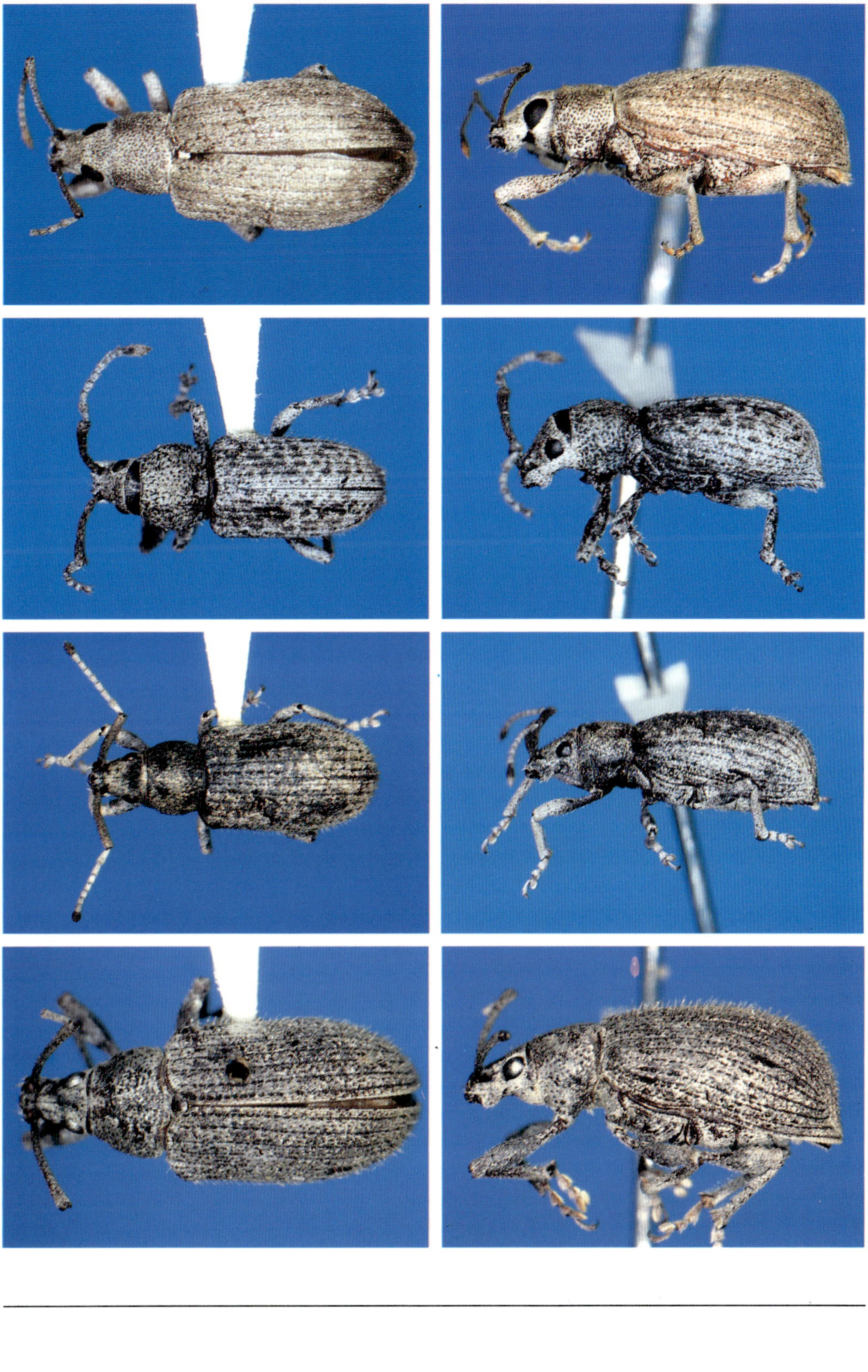

Plate 242 Adelognatha

1, 2 *Myllocerus aeruginosus* Lea
Groote Eylandt, Northern Territory
Paratype male; 4.6 mm

3, 4 *Myllocerus amblyrhinus* Lea
Roebuck Bay, Western Australia
Holotype female; 5.5 mm

5, 6 *Myllocerus angustibasis* Lea
Fortescue River, Western Australia
Holotype female; 4.6 mm

7, 8 *Myllocerus anoplus* Lea
Dalby, Queensland
Holotype female; 3.5 mm

1	2
3	4
5	6
7	8

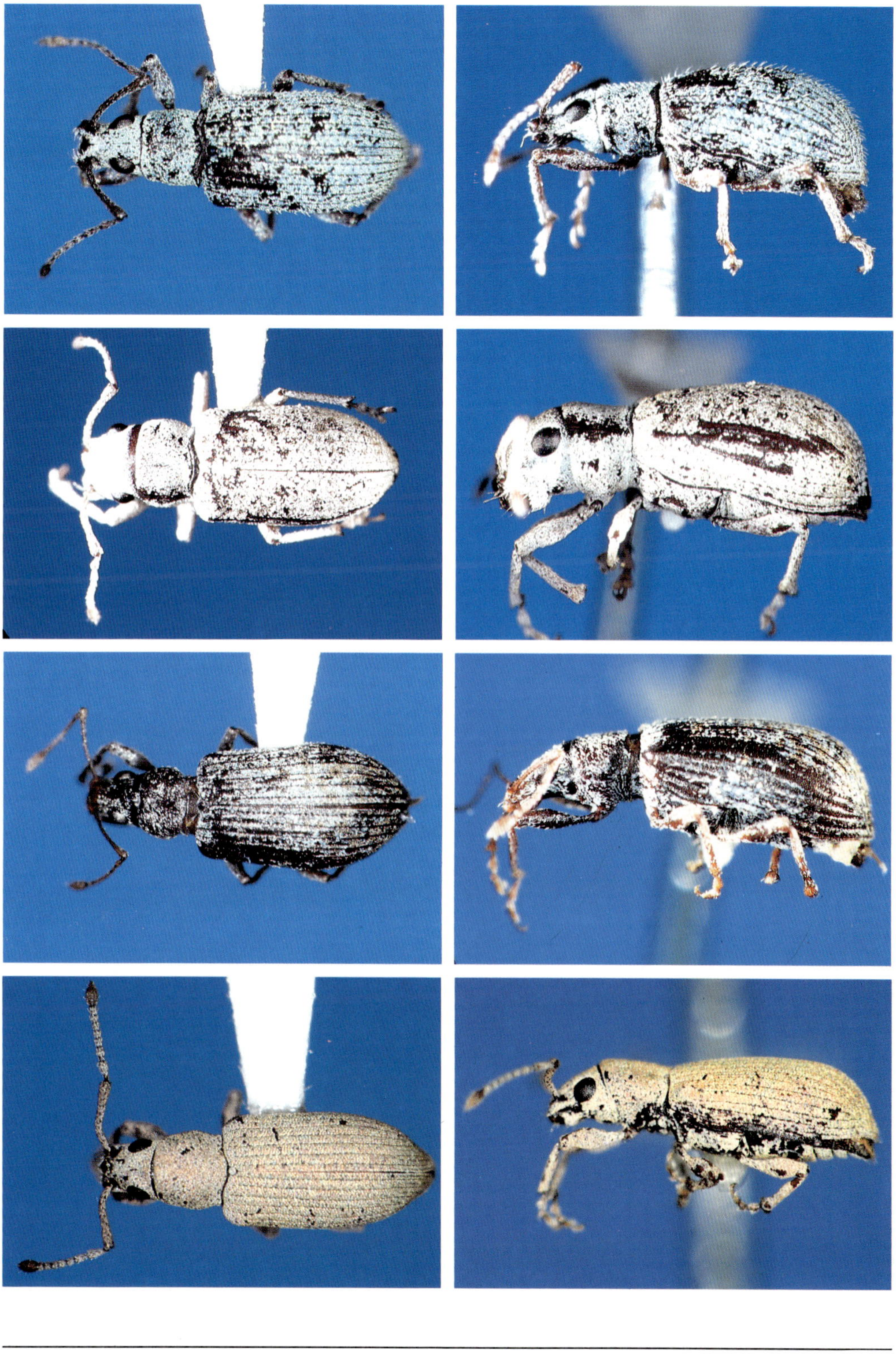

Plate 243 Adelognatha

1, 2 *Myllocerus aphthosus* Pascoe, colour form
Cairns, Queensland
Female; 7.7 mm

3, 4 *Myllocerus aphthosus* Pascoe, colour form
Stewart River, Queensland
Male; 7. 8 mm

5, 6 *Myllocerus armipectus* Lea
King River, Northern Territory
Female; 6.1 mm

7, 8 *Myllocerus armipes* Lea
Endeavour River, Queensland
Holotype male; 6.1 mm

1	2
3	4
5	6
7	8

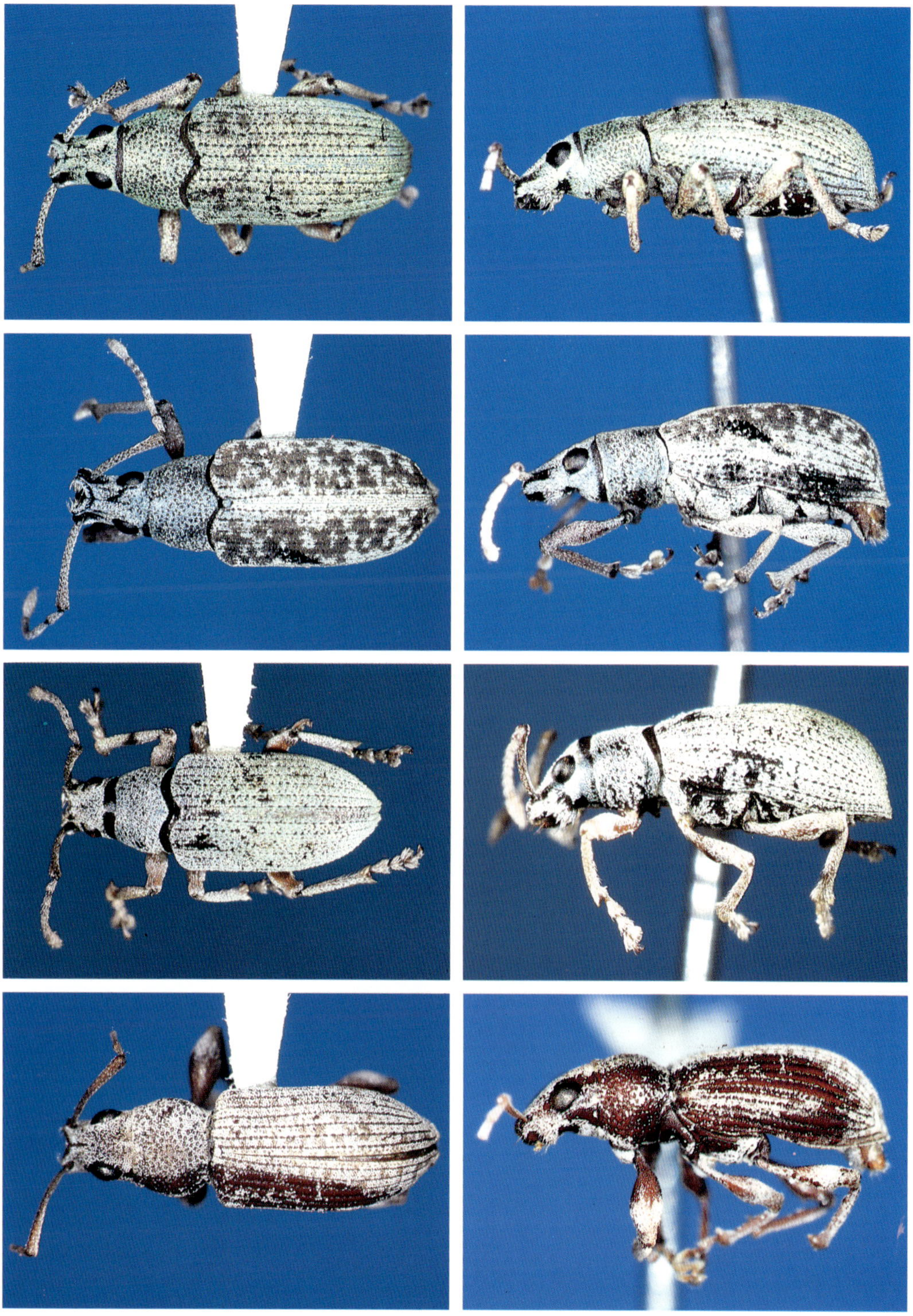

Plate 244 Adelognatha

1, 2 *Myllocerus ashi* Lea
Ooldea, South Australia
Female; 6.6 mm

3, 4 *Myllocerus aurifex* Pascoe
Tennant Creek, Northern Territory
Male; 6.8 mm

5, 6 *Myllocerus baeodontomerus* Lea
Cunnamulla, Queensland
Holotype male; 5.8 mm

7, 8 *Myllocerus basicollis* Lea
Charters Towers, Queensland
Holotype male; 6.2 mm

1	2
3	4
5	6
7	8

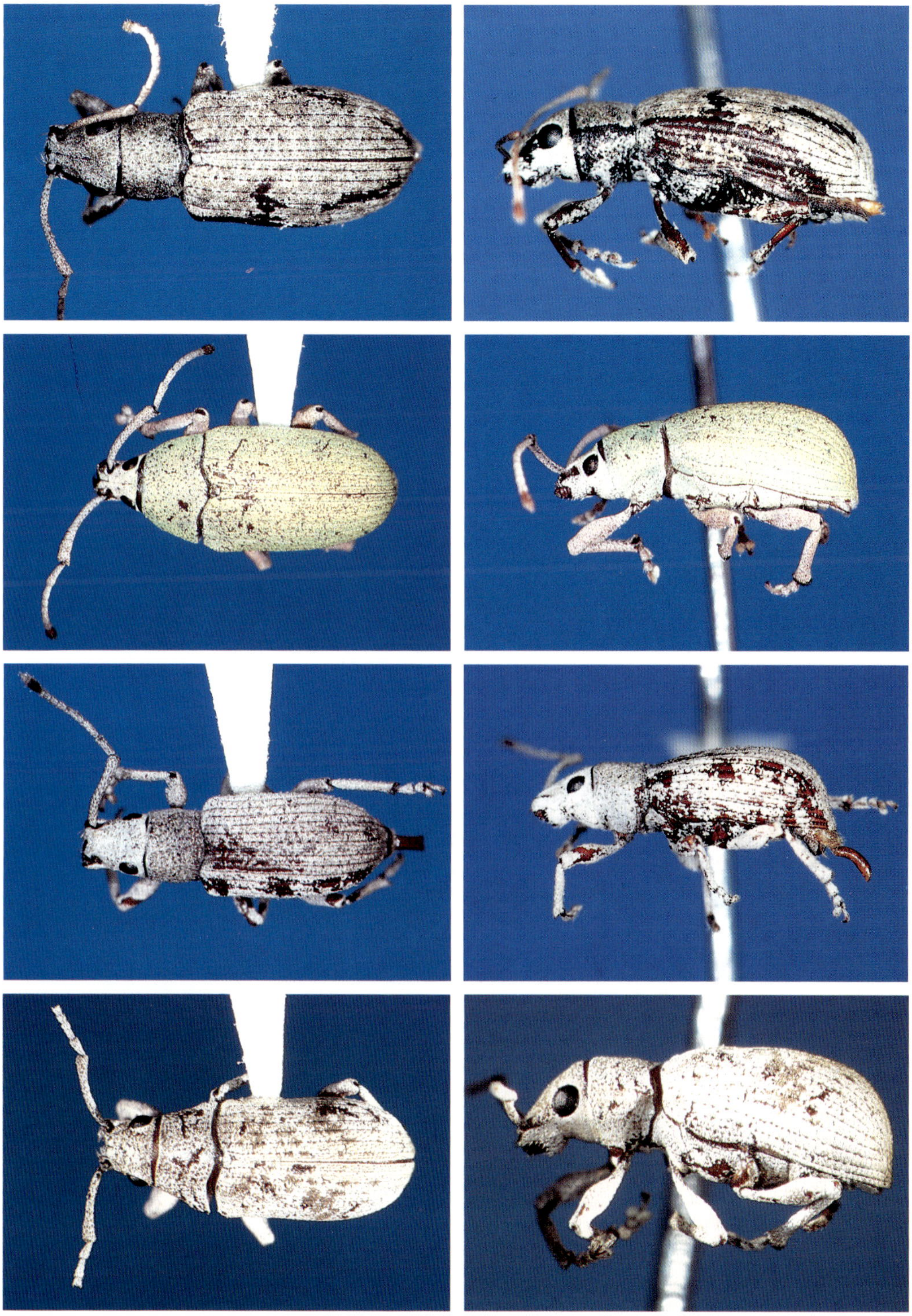

Plate 245 Adelognatha

1, 2 *Myllocerus bifasciatipennis* Lea
Northern Territory
Paratype male; 4.7 mm

3, 4 *Myllocerus bilineater* Lea
Port Darwin, Northern Territory
Paratype female; 6 mm

5, 6 *Myllocerus canalicornis* Lea, grey form
Northwest Western Australia
Holotype female; 5.2 mm

7, 8 *Myllocerus carinatus* Lea
King Sound, Western Australia
Female; 7.7 mm

1	2
3	4
5	6
7	8

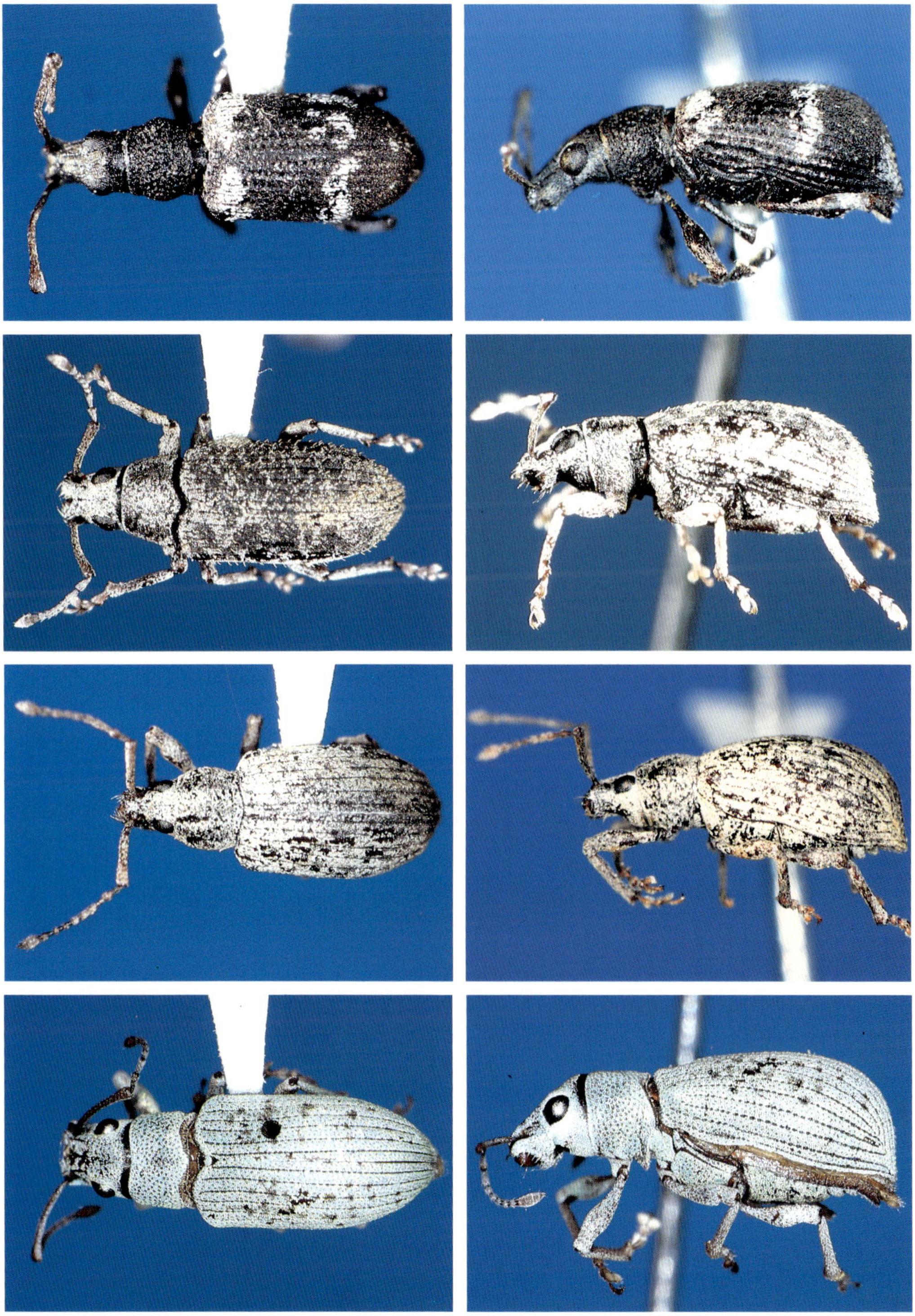

Plate 246 Adelognatha

1, 2 *Myllocerus castor* Lea
Maryborough, Queensland
Male; 6.2 mm

3, 4 *Myllocerus chaunoderes* Lea
Groote Eylandt, Northern Territory
Paratype female; 6.1 mm

5, 6 *Myllocerus cineracens* Pascoe
Charters Towers to Hughenden, Queensland
Female; 6.8 mm

7, 8 *Myllocerus confinis* Lea
Hergott Springs, South Australia
Paratype female; 7.8 mm

1	2
3	4
5	6
7	8

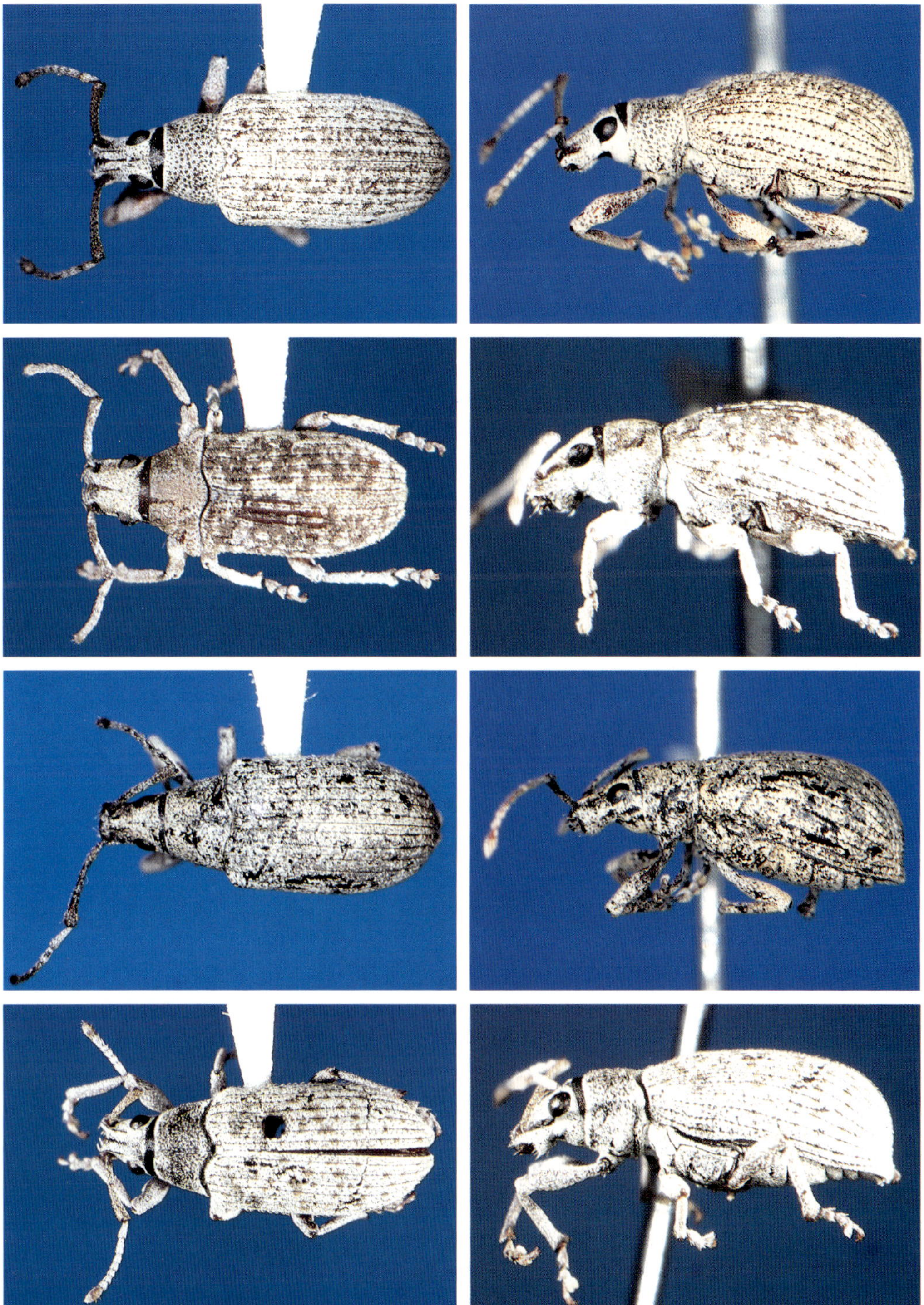

Plate 247 Adelognatha

1, 2 *Myllocerus constricticollis* Lea
Tennant Creek, Northern Territory
Holotype female; 4.9 mm

3, 4 *Myllocerus cyrtops* Lea
Northwest Western Australia
Paratype male; 6.5 mm

5, 6 *Myllocerus darwini* Lea
Darwin, Northern Territory
Paratype male; 6.2 mm

7, 8 *Myllocerus decipiens* Lea
Groote Eylandt, Northern Territory
Holotype male; 4.7 mm

1	2
3	4
5	6
7	8

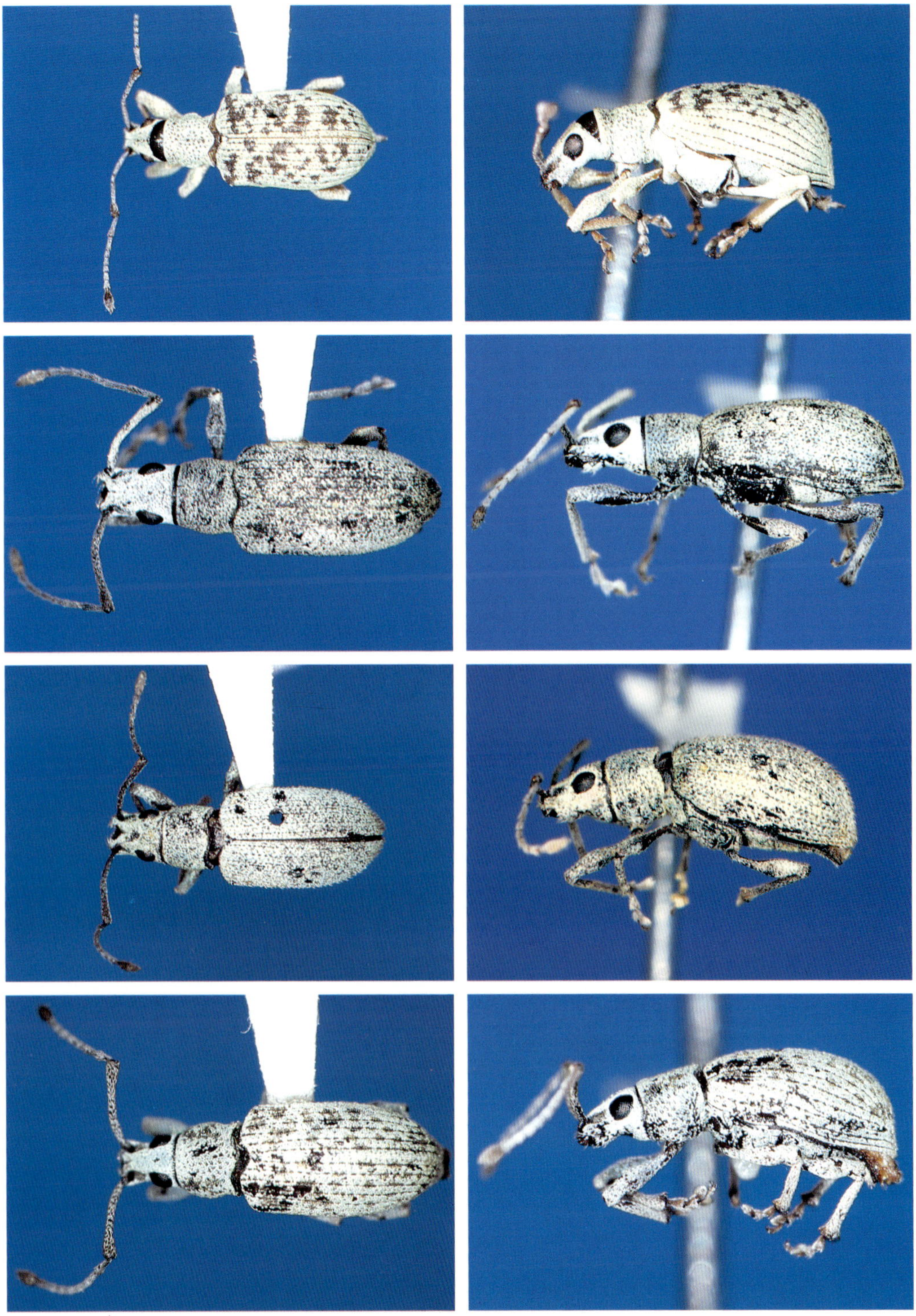

Plate 248 Adelognatha

1, 2 *Myllocerus decretus* Pascoe
Champion Bay, Western Australia
Holotype female; 8 mm

3, 4 *Myllocerus doddi* Lea
Fortescue River, Western Australia
Holotype male; 7.2 mm

5, 6 *Myllocerus duplicatus* Pascoe
Somerset, Queensland
Holotype female; 5.2 mm

7, 8 *Myllocerus echinatus* Lea
Cairns District, Queensland
Male; 4.4 mm

1	2
3	4
5	6
7	8

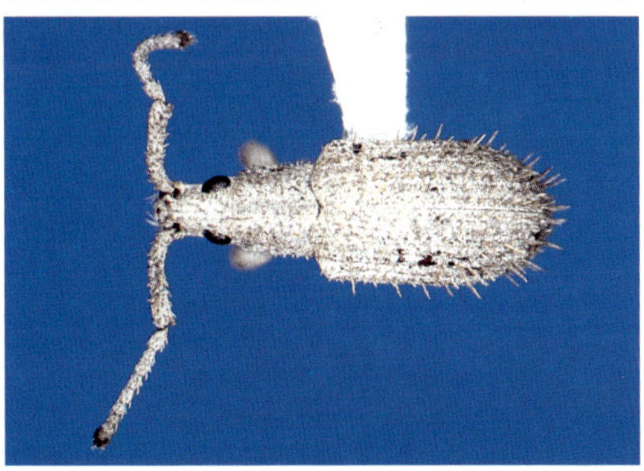

Plate 249 Adelognatha

1, 2 *Myllocerus elegans* Lea
Roebourne, Western Australia
Holotype male; 6.4 mm

3, 4 *Myllocerus exilis* Lea
Fortescue River, Hammersley Range, Western Australia
Male; 4.5 mm

5, 6 *Myllocerus foveiceps* Lea
Darwin, Northern Territory
Female, compared with type; 5.4 mm

7, 8 *Myllocerus foveifrons* Lea
Cunnamulla, Queensland
Paratype female; 7.1 mm

1	2
3	4
5	6
7	8

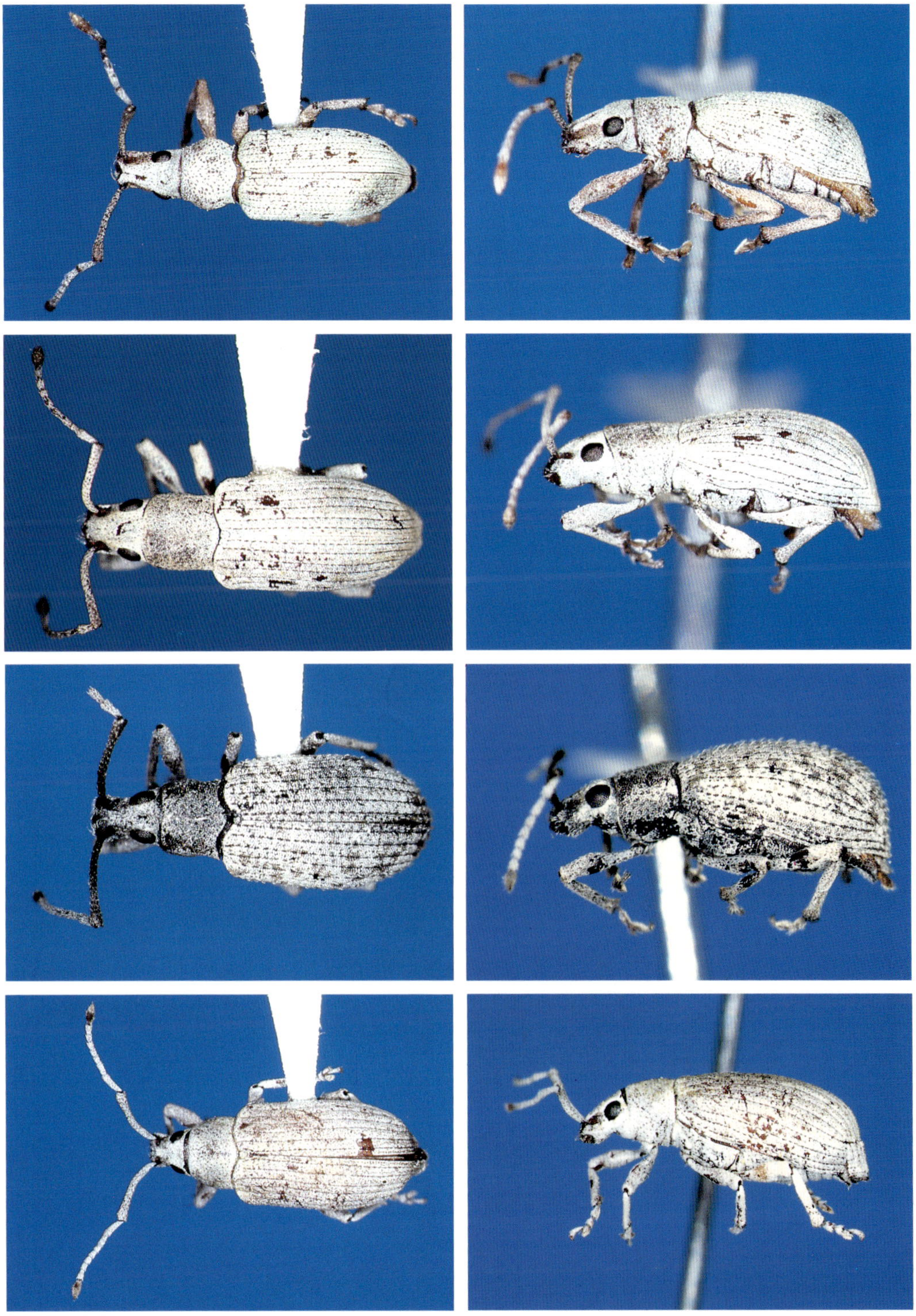

Plate 250 Adelognatha

1, 2 *Myllocerus fugitivus* Lea
Tennant Creek, Northern Territory
Holotype female; 5.7 mm

3, 4 *Myllocerus fuscomaculatus* Lea
Cairns, Queensland
Holotype female, damaged and abraded; 5.1 mm

5, 6 *Myllocerus glaucinus* Pascoe
Champion Bay, Western Australia
Holotype male; 6.7 mm

7, 8 *Myllocerus bicolor* (Blackburn)
South Australia
Paratype female; 6.1 mm

1	2
3	4
5	6
7	8

Plate 251 Adelognatha

1, 2 *Myllocerus herbaceus* Pascoe
Champion Bay, Western Australia
Female, compared with holotype; 6 mm

3, 4 *Myllocerus herbivorus* Lea
Connection Island, Groote Eylandt, Northern Territory
Paratype female; 4.9 mm

5, 6 *Myllocerus hilli* Lea
Darwin, Northern Territory
Paratype female; 6.1 mm

7, 8 *Myllocerus incisicollis* (Lea)
Denham, Western Australia
Holotype male; 4.3 mm

1	2
3	4
5	6
7	8

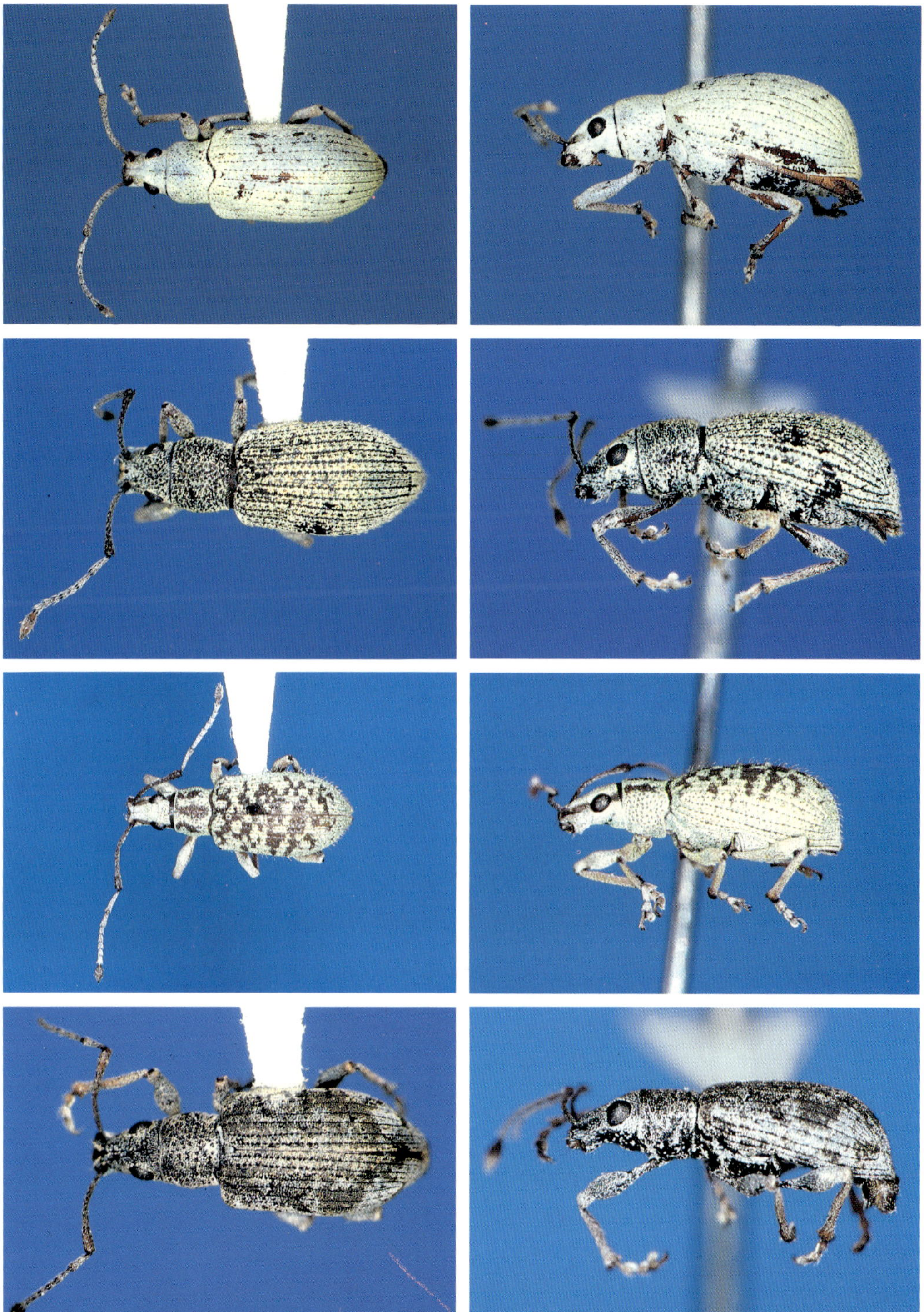

Plate 25 2 Adelognatha

1, 2 *Myllocerus inermipes* Lea
Daly River, Northern Territory
Holotype female; 4.5 mm

3, 4 *Myllocerus intercoxalis* Lea
Tennant Creek, Northern Territory
Holotype female; 8.5 mm

5, 6 *Myllocerus latibasis* Lea
Tennant Creek, Northern Territory
Paratype female; 8.1 mm

7, 8 *Myllocerus laticollis* Lea
Bathurst Head, Queensland
Male, teneral; 6.7 mm; note mandibular cusp.

1	2
3	4
5	6
7	8

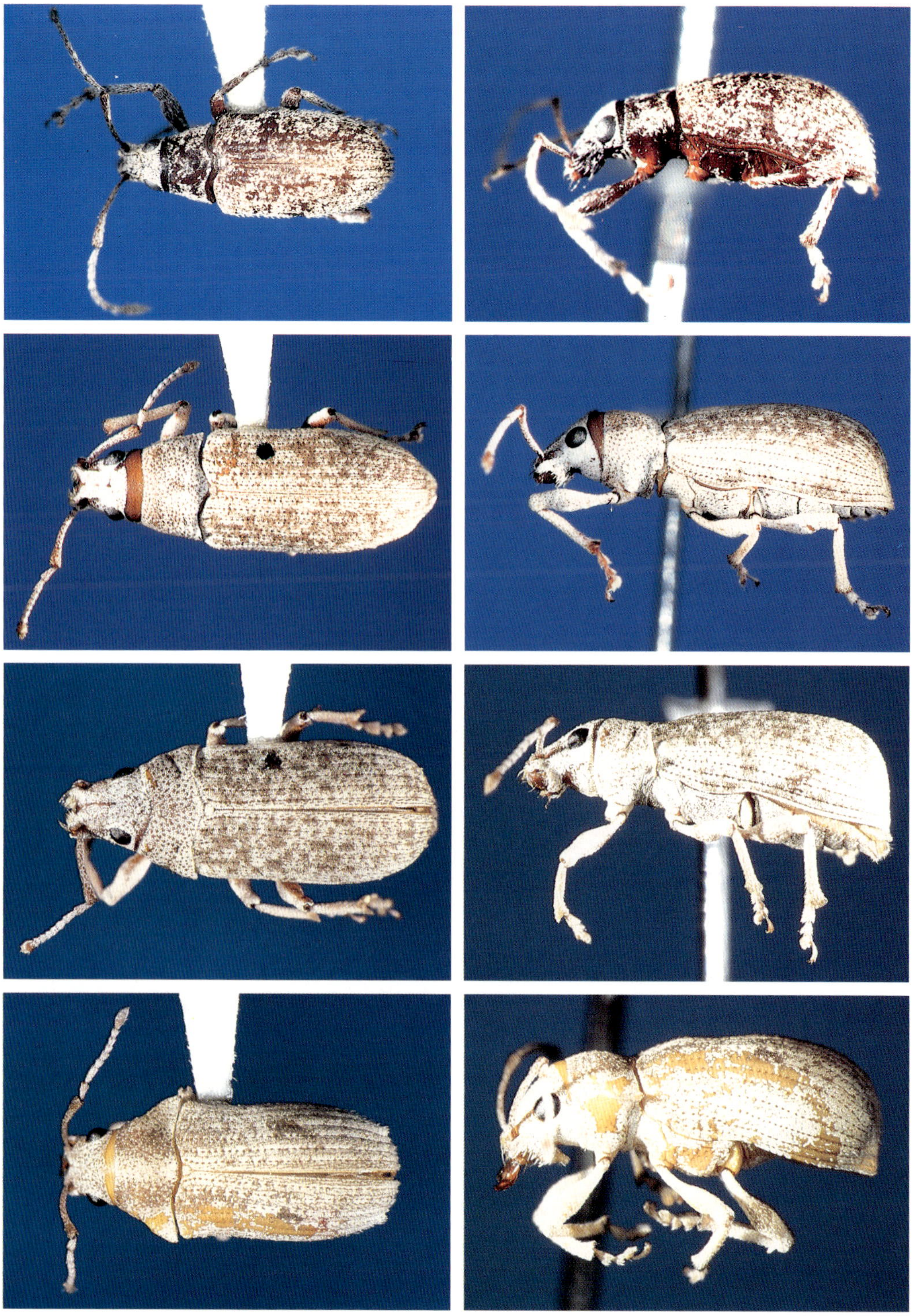

Plate 253 Adelognatha

1, 2 *Myllocerus longicornis* Pascoe
Champion Bay, Western Australia
Holotype female; 5.4 mm
Abraded; most squamae lost.

3, 4 *Myllocerus longus* Lea
Broome, Western Australia
Female; 7.1 mm

5, 6 *Myllocerus maculatus* (Blackburn)
Everard Range, Western Australia
Paratype male; 4.5 mm

7, 8 *Myllocerus maculatus* (Blackburn)
Everard Range, Western Australia
Paratype female; 5.1 mm

1	2
3	4
5	6
7	8

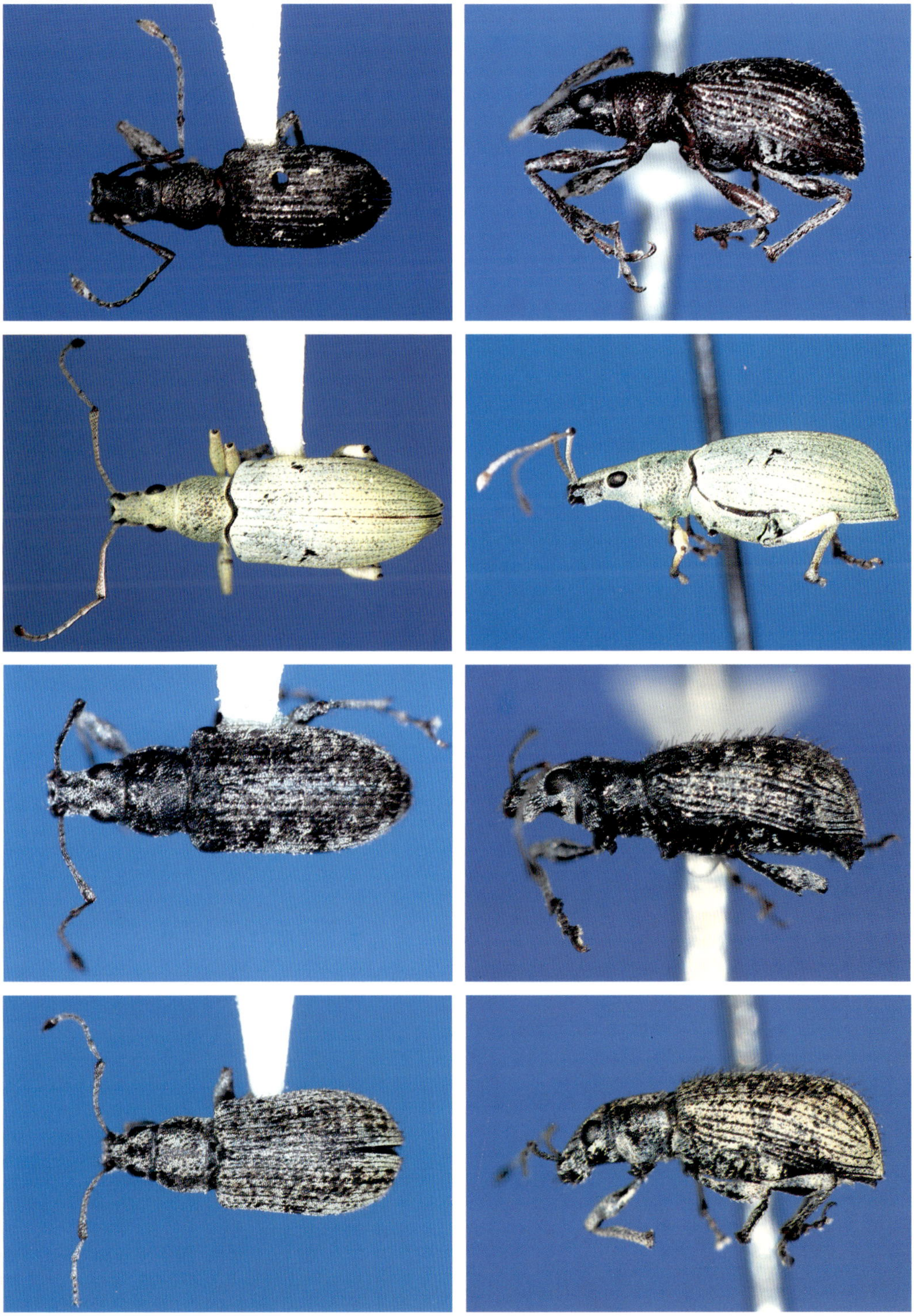

Plate 254 Adelognatha

1, 2 *Myllocerus mastersi* Lea
Northwest Western Australia
Paratype female; 4.6 mm

3, 4 *Myllocerus melvillensis* Lea
Melville Island, Northern Territory
Holotype male?; 4.6 mm

5, 6 *Myllocerus minusculus* Lea
Gayndah, Queensland
Holotype male; 3.6 mm

7, 8 *Myllocerus mirabilis* Lea
Northwest Western Australia
Holotype male; 5.7 mm

1	2
3	4
5	6
7	8

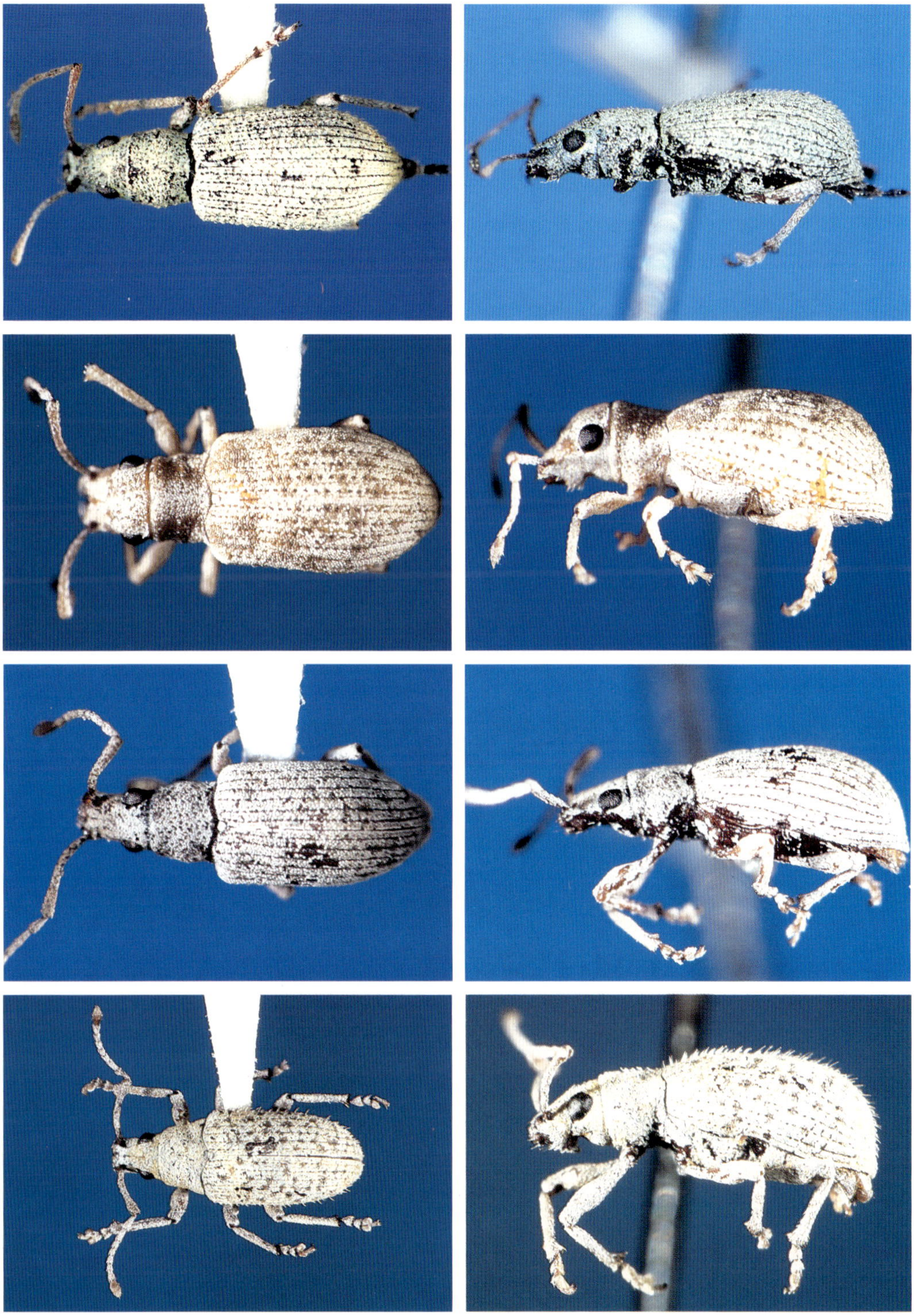

Plate 255 Adelognatha

1, 2 *Myllocerus multimaculatus* Lea
Cunnamulla, Queensland
Female; 6.7 mm

3, 4 *Myllocerus nasutus* Pascoe
Derby, Western Australia
Female; 7 mm; determined by Lea

5, 6 *Myllocerus niveus* Lea
Somerset, Queensland (a D'Albertis specimen)
Male; 5.9 mm

7, 8 *Myllocerus obliquifasciatus* Lea
Ankertel, Western Australia
Paratype male; 5.8 mm

1	2
3	4
5	6
7	8

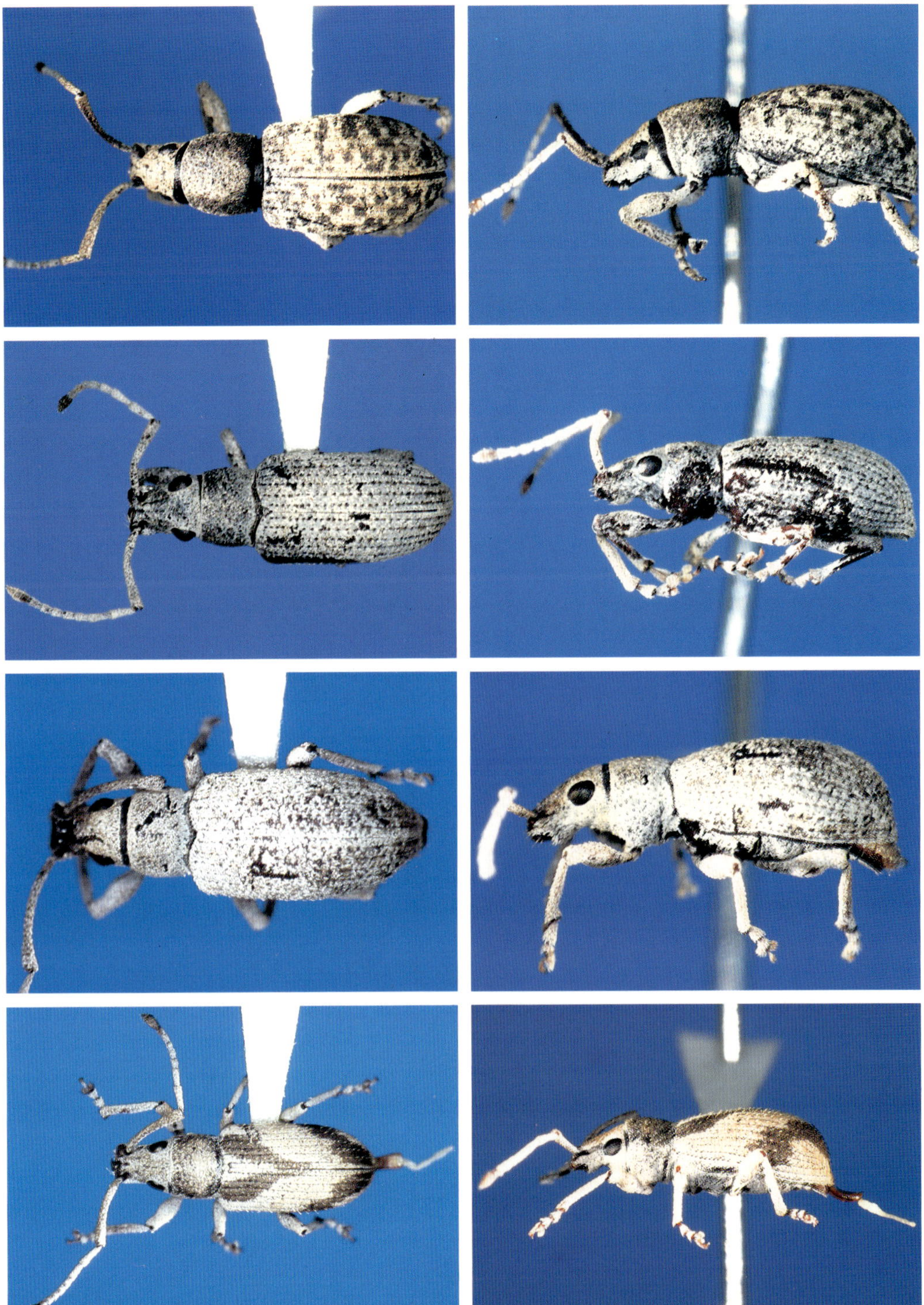

Plate 256 Adelognatha

1, 2 *Myllocerus obscurus* Lea
Darwin, Northern Territory
Female; 6.1 mm

3, 4 *Myllocerus perarmatus* Lea
Magnetic Island, Queensland
Paratype; 7.3 mm

5, 6 *Myllocerus pictus* Lea
Groote Eylandt, Northern Territory
Paratype female; 5.4 mm

7, 8 *Myllocerus pollux* Lea
Endeavour River, Queensland
Holotype male; 7.2 mm

1	2
3	4
5	6
7	8

Plate 257 Adelognatha

1, 2 *Myllocerus prosternalis* Lea
North Queensland
Holotype female; 8.5 mm

3, 4 *Myllocerus pteroderes* Lea
Hermannsburg, Central Australia
Holotype; 8.5 mm

5, 6 *Myllocerus quadricolor* Lea
49 km east of Darwin, Northern Territory
Male; 3.8 mm

7, 8 *Myllocerus rugicollis* Lea
Mount Tamborine, Queensland
Female; 5.7 mm

1	2
3	4
5	6
7	8

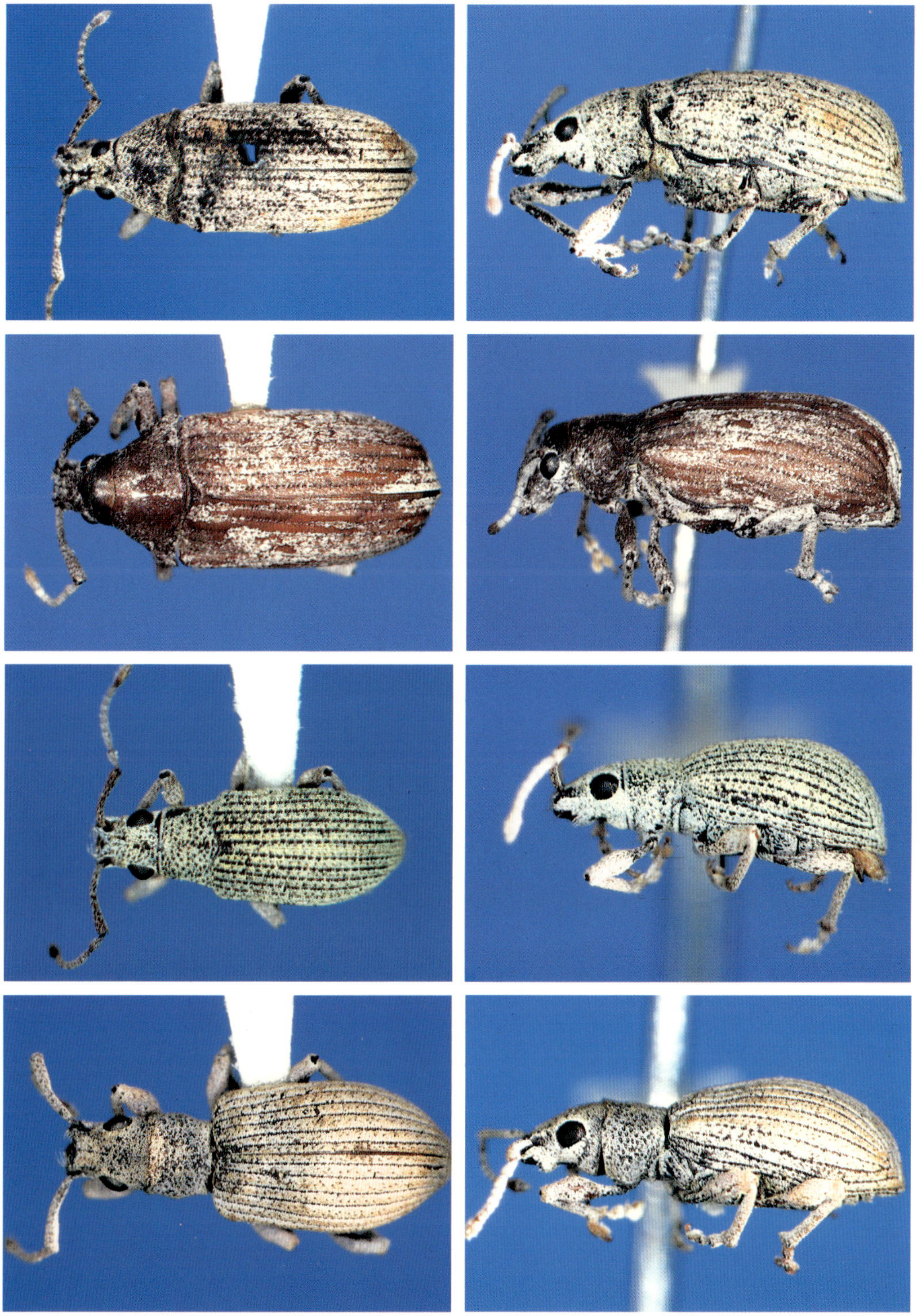

Plate 258 Adelognatha

1, 2 *Myllocerus rusticus* Lea
Western Australia
Female, compared with holotype; 8.7 mm

3, 4 *Myllocerus scitulus* Lea
Dalby, Queensland
Holotype male; 4.2 mm

5, 6 *Myllocerus setistriatus* Lea
Coen River, Queensland
Holotype female; 6.1 mm

7, 8 *Myllocerus sordidus* Lea
Geraldton, Western Australia
Paratype female; 6.3 mm

1	2
3	4
5	6
7	8

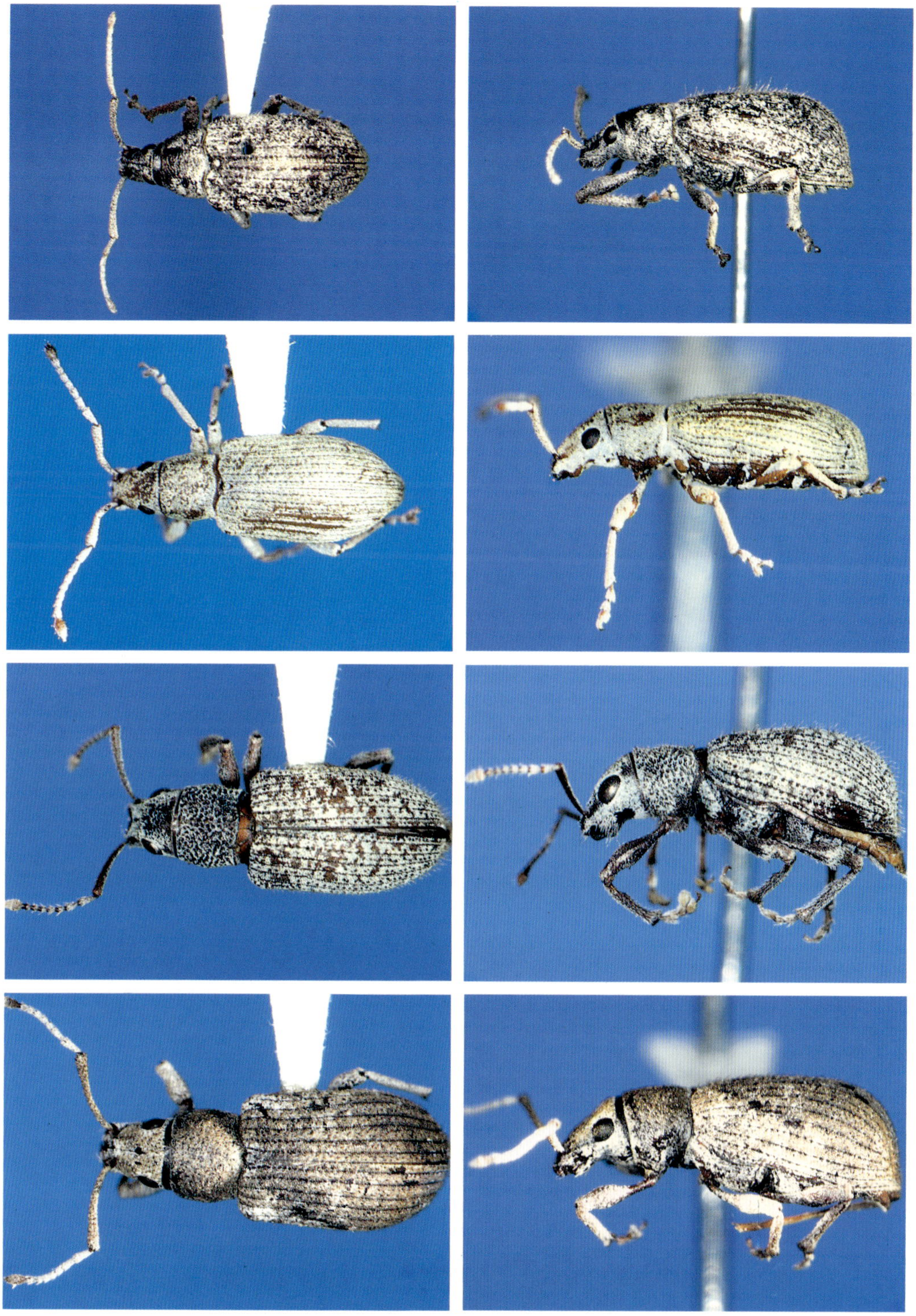

Plate 259 Adelognatha

1, 2 *Myllocerus speciosus* Blackburn
Queensland
Paratype male; 5 mm

3, 4 *Myllocerus squamicornis* Lea
Cleve, South australia
Holotype female; 5.4 mm

5, 6 *Myllocerus subrostralis* Lea
Northwest Western Australia
Holotype female; 7.8 mm

7, 8 *Myllocerus sulcicornis* Lea
Dalby, Queensland
Holotype female; 5 mm

1	2
3	4
5	6
7	8

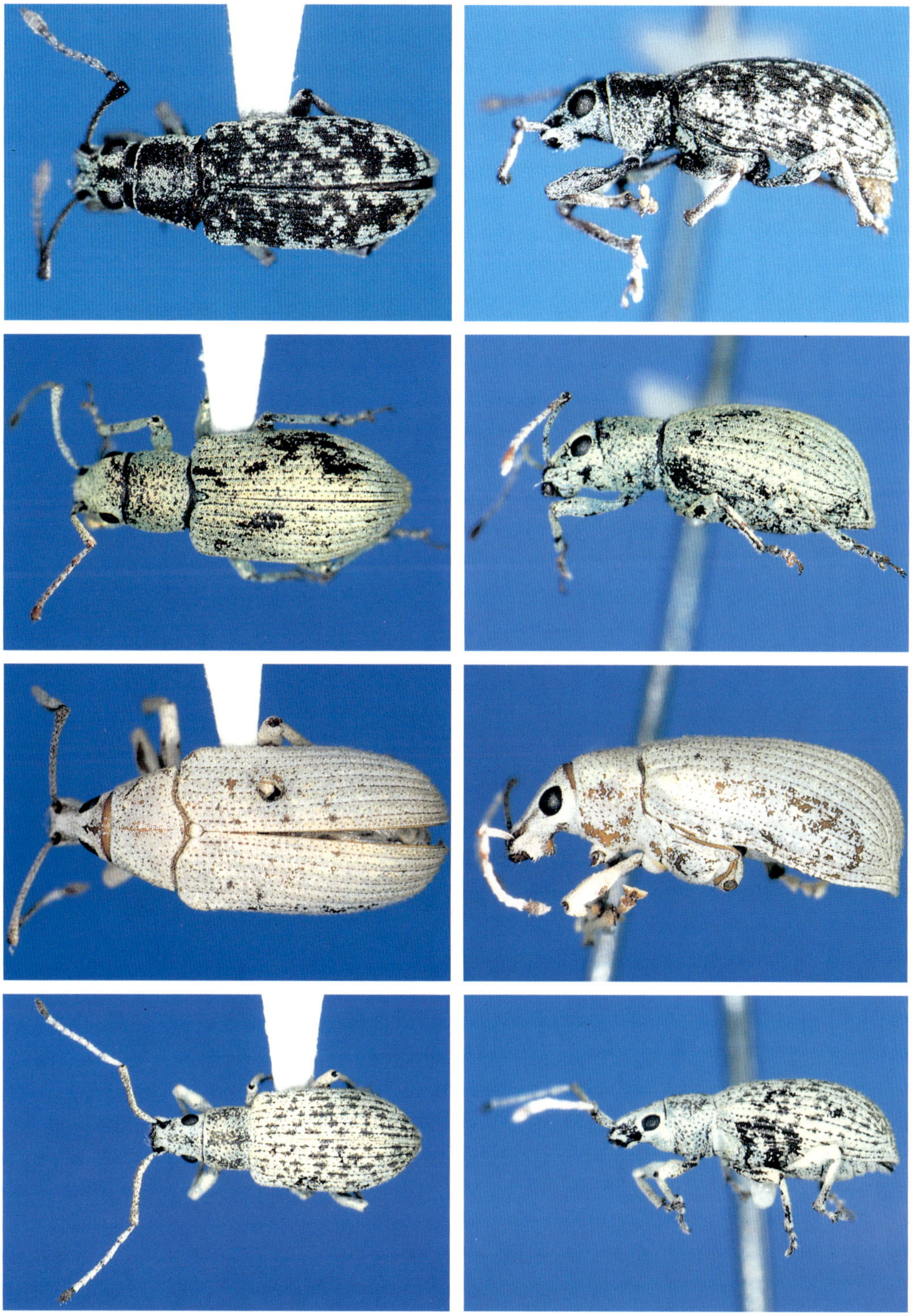

Plate 260 Adelognatha

1, 2 *Myllocerus suturalis* Lea
Gayndah, Queensland
Holotype female; 4.8 mm

3, 4 *Myllocerus tatei* Blackburn
Idracowra, Northern Territory
Syntype female; 7.7 mm

5, 6 *Myllocerus taylori* Lea
Northern Queensland
Holotype female; 7.2 mm

7, 8 *Myllocerus tibialis* Lea
Coen River, Queensland
Holotype female; 8 mm

1	**2**
3	**4**
5	**6**
7	**8**

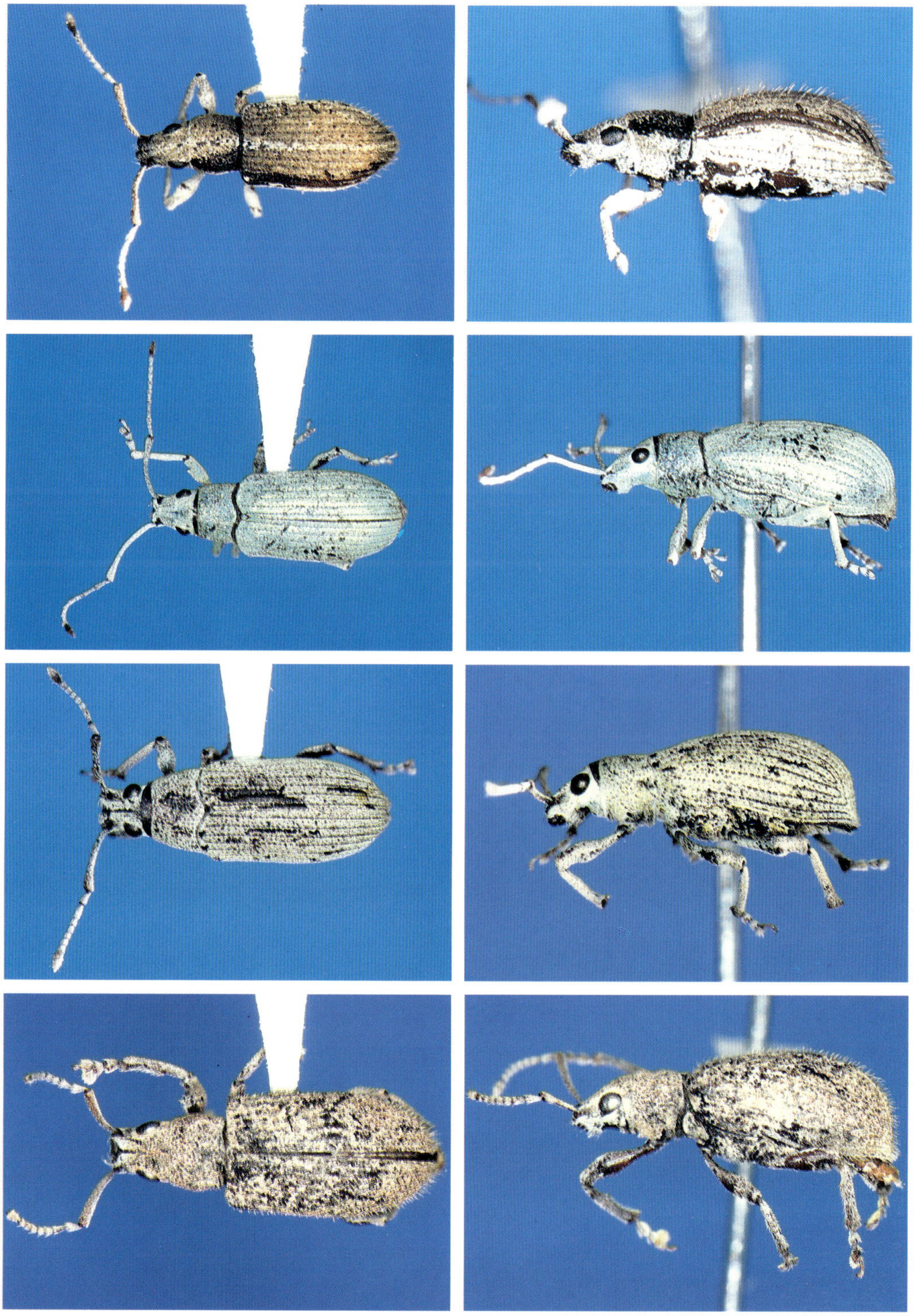

Plate 261 Adelognatha

1, 2 *Myllocerus torridus* Blackburn
Northern Territory
Paratype male; 6.5 mm

3, 4 *Myllocerus trepidus* Pascoe
Somerset, Queensland
Female, compared with holotype; 4.8 mm

5, 6 *Myllocerus tricarinirostris* Lea
Kingoonya, South Australia
Holotype male; 5.8 mm

7, 8 *Myllocerus trilineatus* Lea
Eidsvold, Queensland
Female; 6.6 mm

1	2
3	4
5	6
7	8

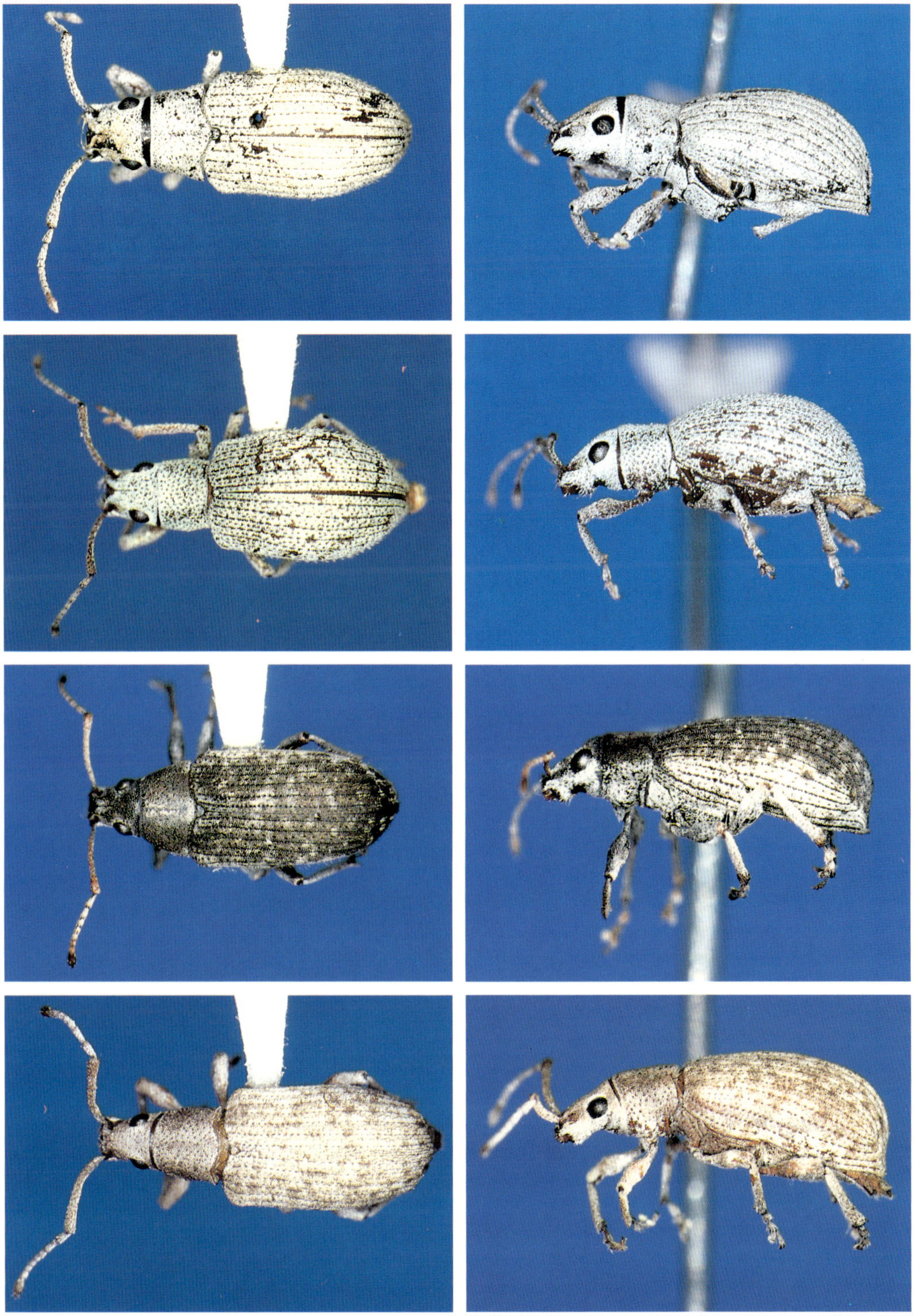

Plate 262 Adelognatha

1, 2 *Myllocerus tristriatus* Lea
Bathurst Island, Northern Territory
Male; 5.6 mm

3, 4 *Myllocerus tristis* Lea
Clermont, Queensland
Paratype male; 5.9 mm

5, 6 *Myllocerus usitatus* Lea
Whitton, New South Wales
Paratype female; 5 mm

7, 8 *Myllocerus varius* Lea
Cue, Western Australia
Holotype female; 5.7 mm

1	2
3	4
5	6
7	8

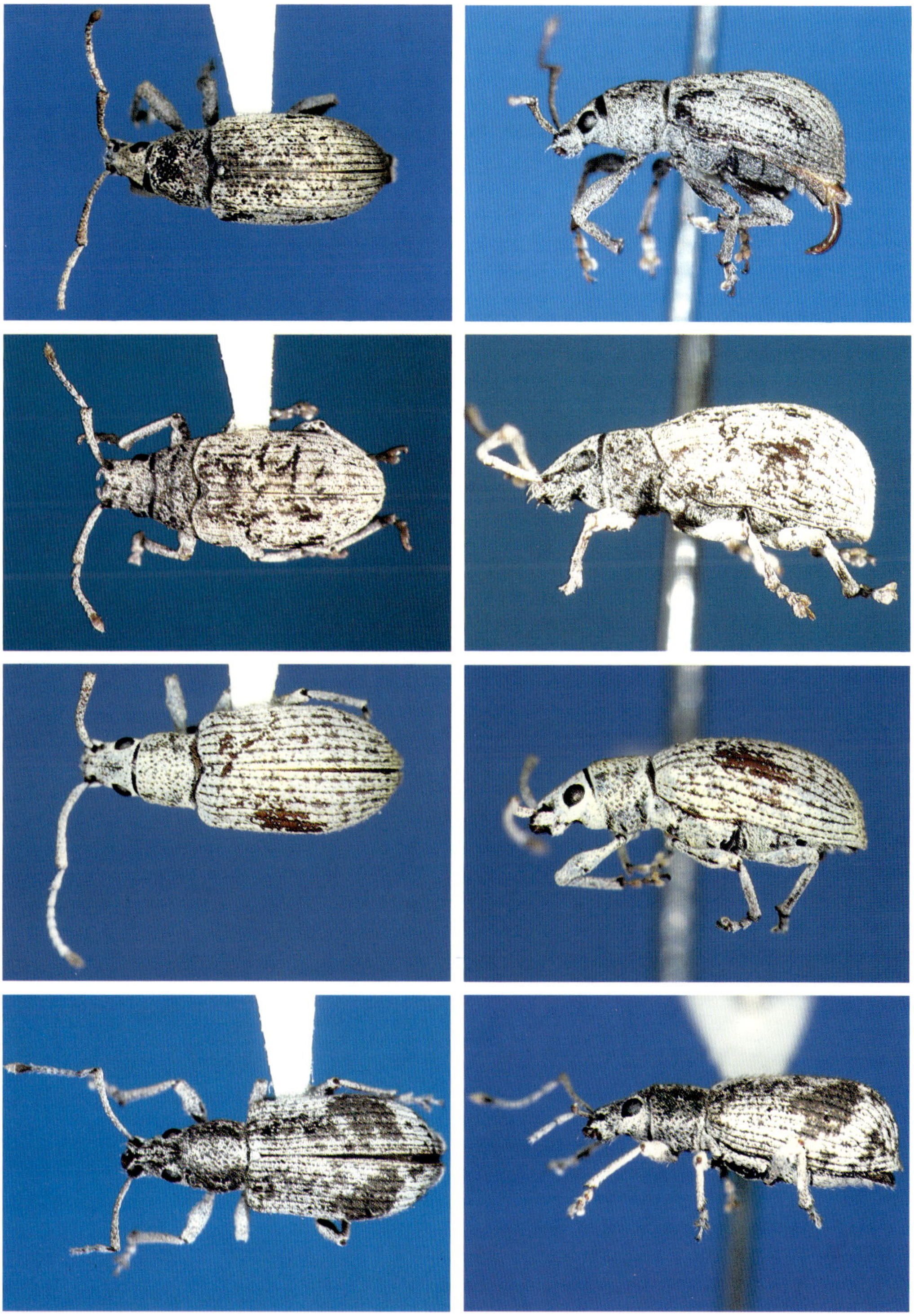

Plate 263 Adelognatha

1, 2 *Myllocerus villosipennis* Lea
Ooldea, South Australia
Paratype female; 4.8 mm

3, 4 *Myllocerus viridimicans* Lea
Ooldea, South Australia
Holotype female; 5.1 mm

5, 6 *Myllocerus viridipictus* Lea
Cape York, Queensland
Holotype male; 4.9 mm

7, 8 *Myllocerus zopherus* Lea
Port Augusta, South Australia
Holotype male; 4.6 mm

1	2
3	4
5	6
7	8

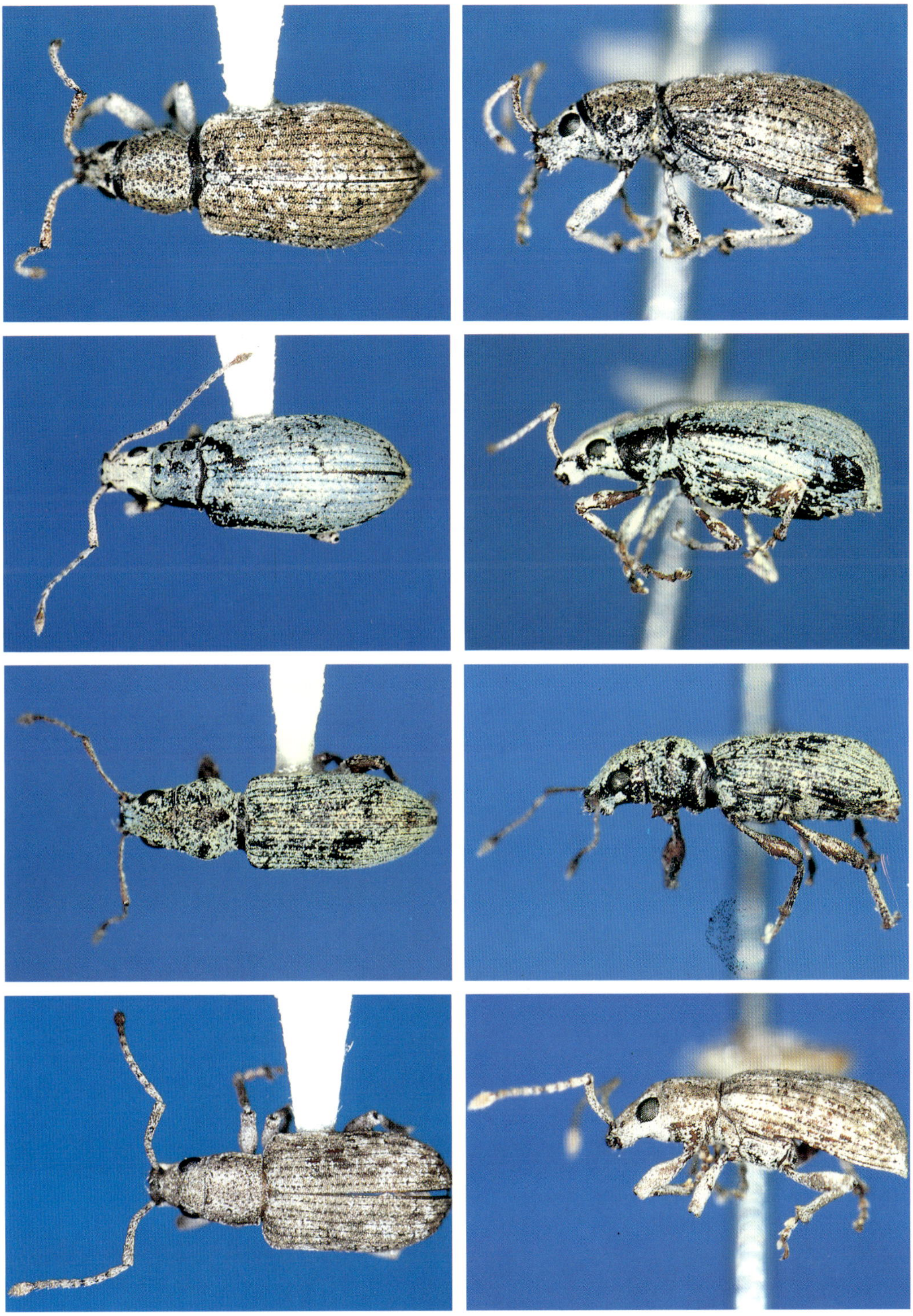

Plate 264 Adelognatha

1, 2 *Myllocerus* new species
Cape Range, Western Australia
Male; 6 mm

3, 4 *Neomerimnetes flindersiae* Marshall
Mount Tamborine, Queensland
Male; 3.8 mm

5, 6 *Neomerimnetes inflatus* Lea
Gympie, Queensland
Paratype male; 3.6 mm

7, 8 *Neomerimnetes sobrinus* Lea
Bundaberg, Queensland
Female; 5.3 mm

1	2
3	4
5	6
7	8

Plate 265 Adelognatha

1, 2 *Ochrometa amoena* Pascoe
Mullewa, Western Australia
Male; 10 mm

3, 4 *Ocynoma cordipennis* Pascoe
Western Australia
Holotype female; 4.8 mm

5, 6 *Oophthalmus pistor* (Germar)
Gawler, South Australia
Male; 10 mm

7, 8 *Oophthalmus rostralis* (Blackburn)
Moonta or Adelaide, South Australia
Paratype male; 6.2 mm

1	2
3	4
5	6
7	8

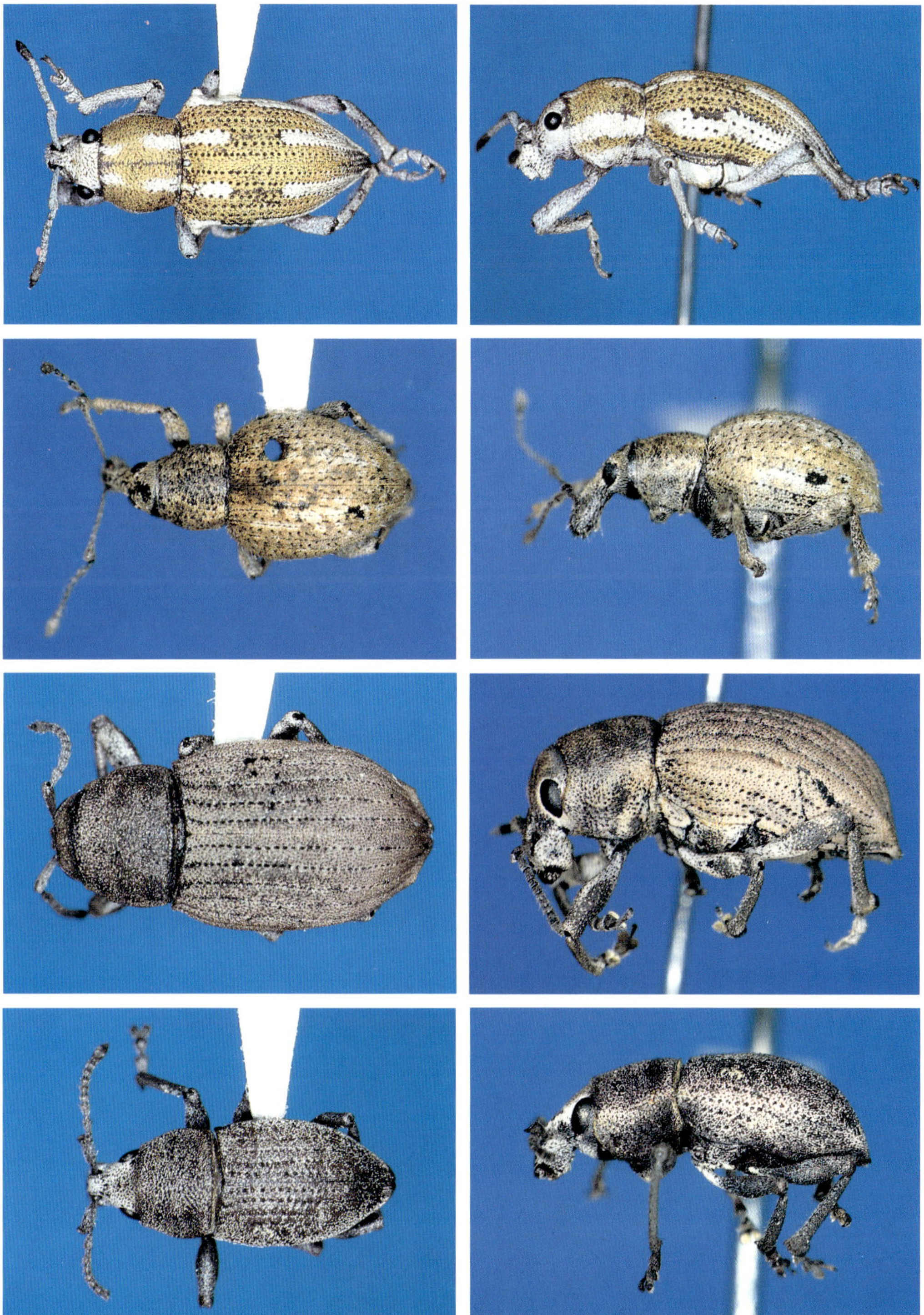

Plate 266 Adelognatha

1, 2 *Oopthalmus viridis* (Oke)
Hatton Lakes, Victoria
Male; 6.5 mm

3, 4 *Oribius albivarius* (Lea)
Coen River, Queensland
Holotype female; 5.2 mm

5, 6 *Oribius equinus* (Lea)
"Australien" (Queensland)
Paratype male; 3.8 mm

7, 8 *Oribius gestroi* (Pascoe)
Somerset, Queensland (a D'Albertis example)
Female; 6.7 mm

1	2
3	4
5	6
7	8

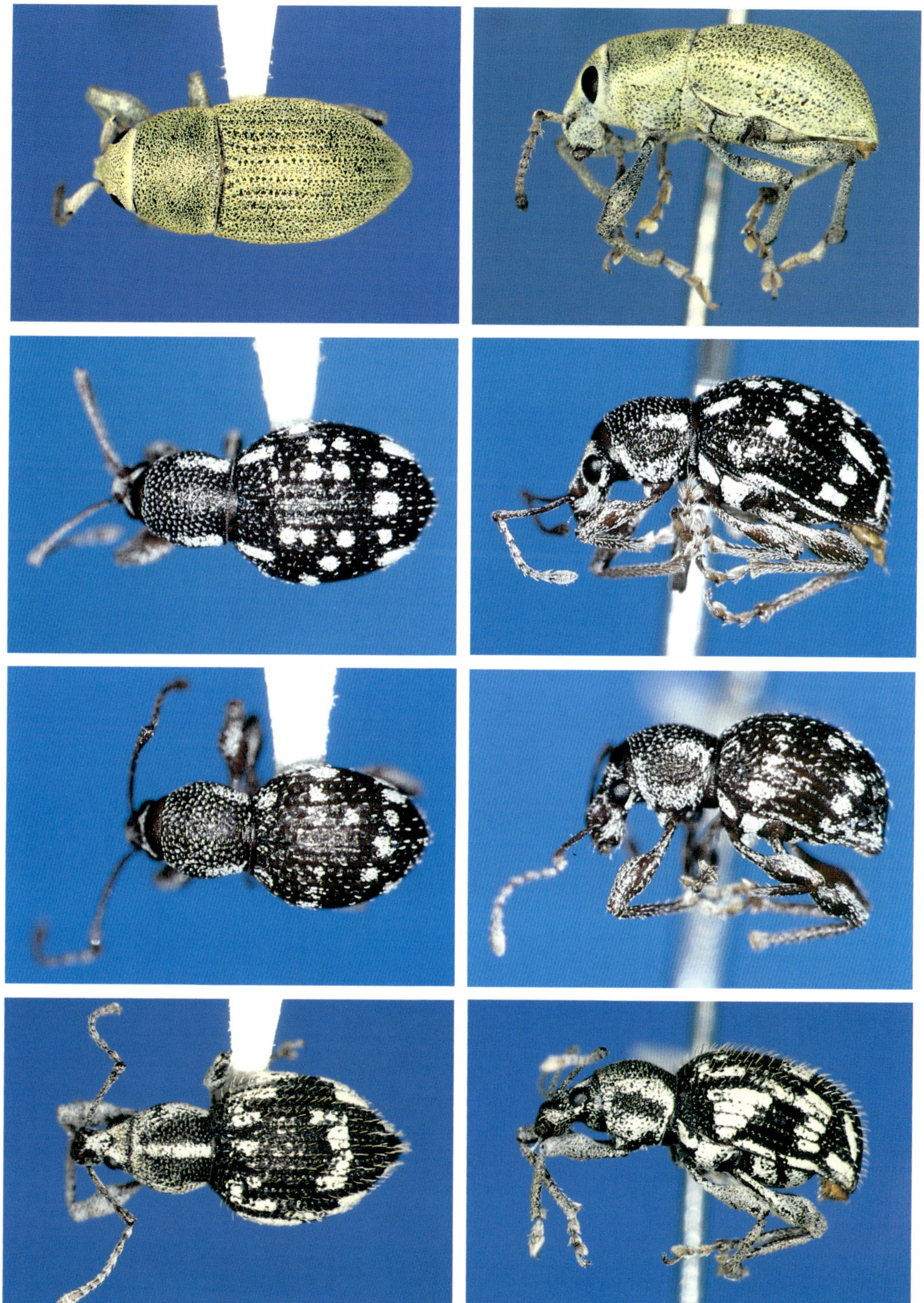

Plate 267 Adelognatha

1, 2 *Oribius jansoni* (Pascoe)
Bloomfield River, Queensland
Male; 5.3 mm

3, 4 *Oribius tessellatus* (Blanchard)
Darnley Island, Torres Strait, Queensland
Male; 8 mm

5, 6 *Oribius trivittatus* (Lea)
Darnley Island, Torres Strait, Queensland
Holotype male; 5 mm

7, 8 *Pantorytes stanleyanus* (White)
Edge Hill, Queensland
Male; 12 mm
The pale elytral maculae are green.

1	2
3	4
5	6
7	8

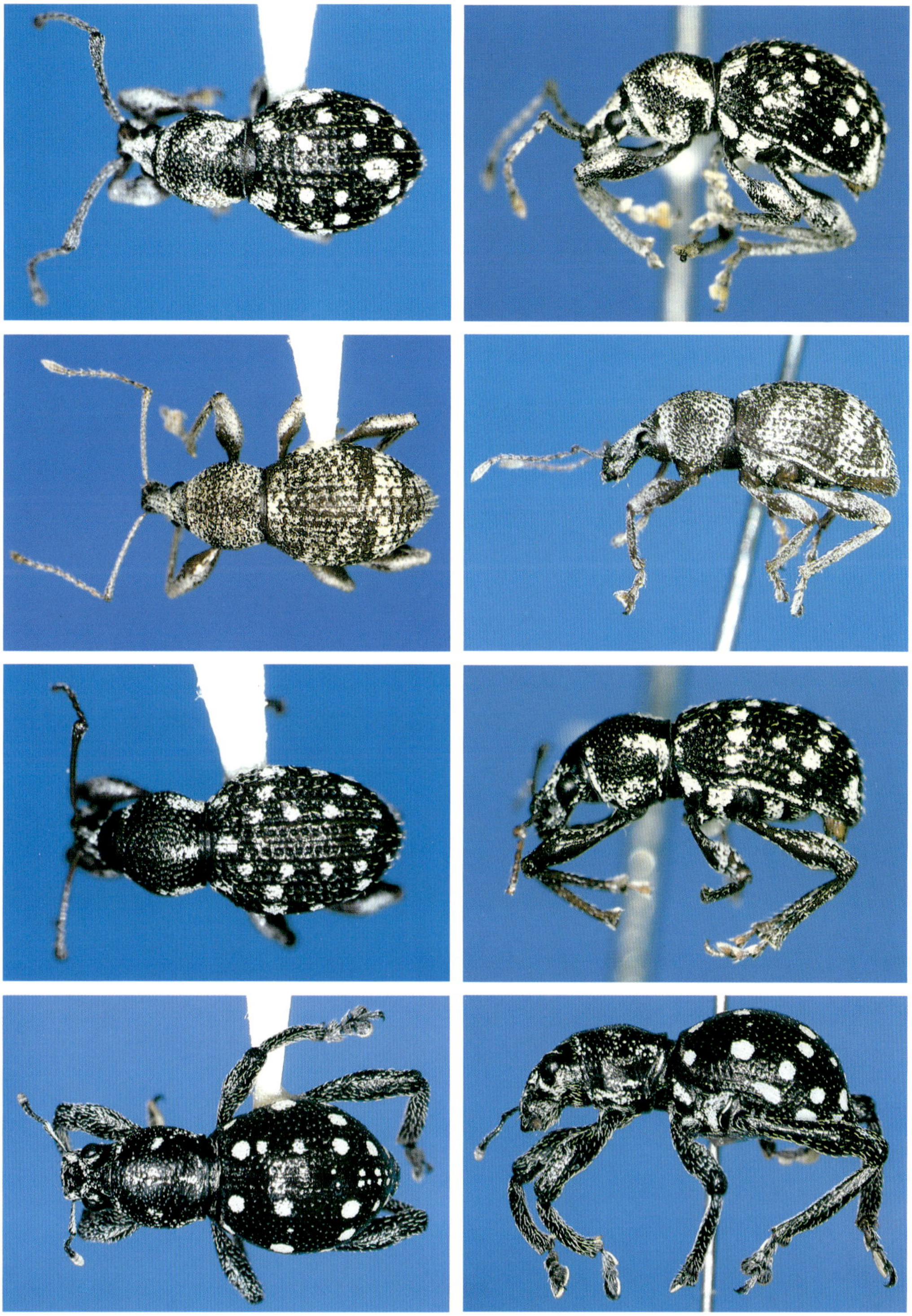

Plate 268 Adelognatha

1, 2 *Pascoellus fissiceps* (Lea)
Geraldton, Western Australia
Holotype female; 4.7 mm

2, 3 *Pascoellus militaris* (Lea)
Swan River, Western Australia
Holotype male; 5.3 mm

5, 6 *Pascoellus squalidus* (Blackburn)
Leigh Creek, South Australia
Paratype male; 4.3 mm
The specimen had been attacked by mold.

7, 8 *Pascoellus vittaticeps* (Blackburn)
Western Australia
Paratype female; 3.6 mm

1	2
3	4
5	6
7	8

Plate 269 Adelognatha

1, 2 *Peripagis rufipes* Pascoe
New South Wales
Female; 13 mm

3, 4 *Perperus angustibasis* Lea
Bunya Mountains, Queensland
Male; 6.7 mm

5, 6 *Perperus conloni* Lea
King Island, Bass Strait
Paratype male; 8 mm

7, 8 *Perperus convexipennis* Blackburn
Maitland, New South Wales
Paratype female; 8.5 mm

1	2
3	4
5	6
7	8

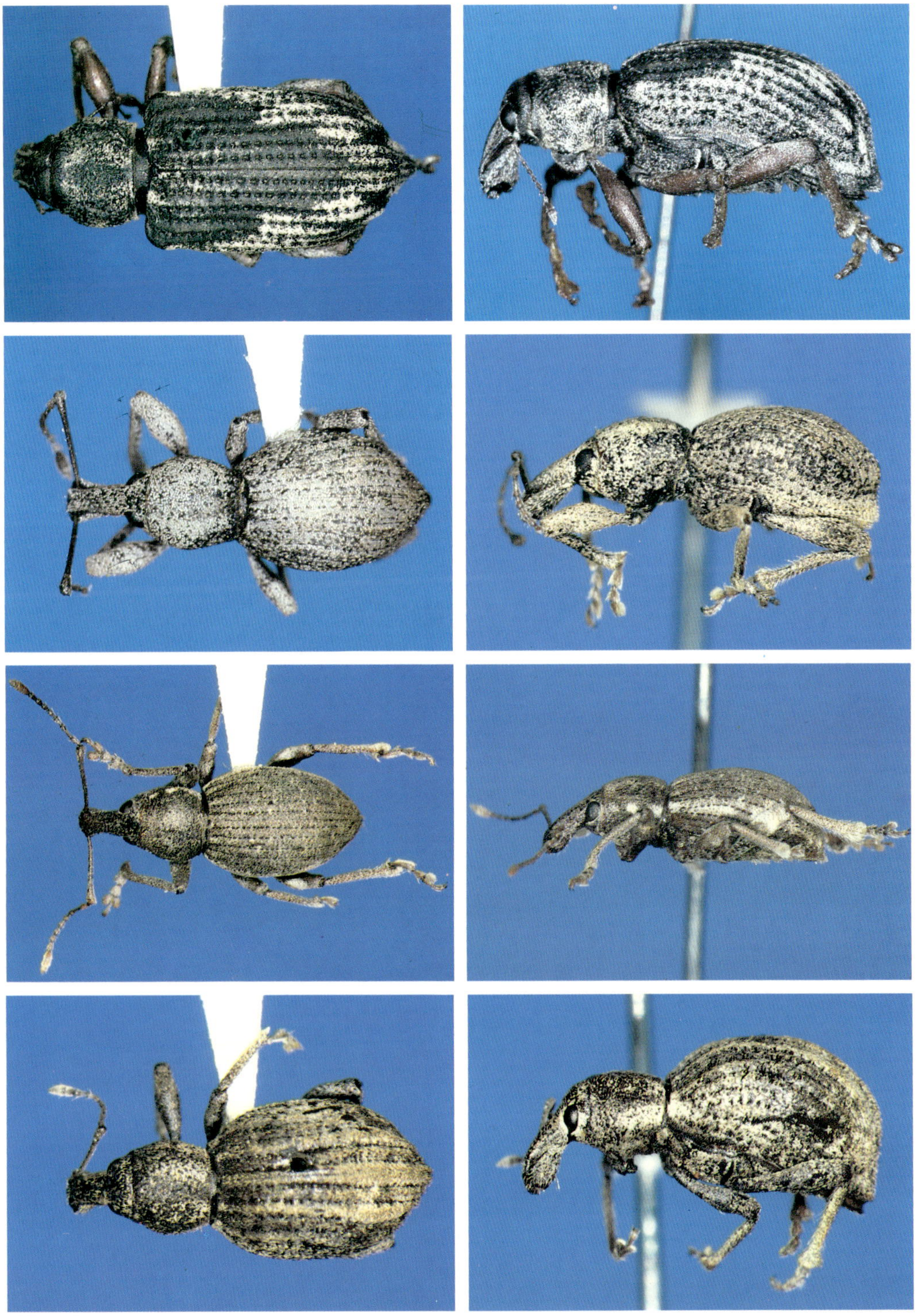

Plate 270 Adelognatha

1, 2 *Perperus innocuus* Boheman
New South Wales
Female; 8 mm
From the holotype of the synonym *delens* Blackburn.

3, 4 *Perperus insularis* Boheman
North Sydney, New South Wales
Male, compared with type; 7 mm

5,6 *Perperus lateralis* (Boisduval)
Moruya, New South Wales
Female; 12 mm

7, 8 *Perperus obscurus* Boheman
New South Wales ?
Holotype female; 6.5 mm

1	2
3	4
5	6
7	8

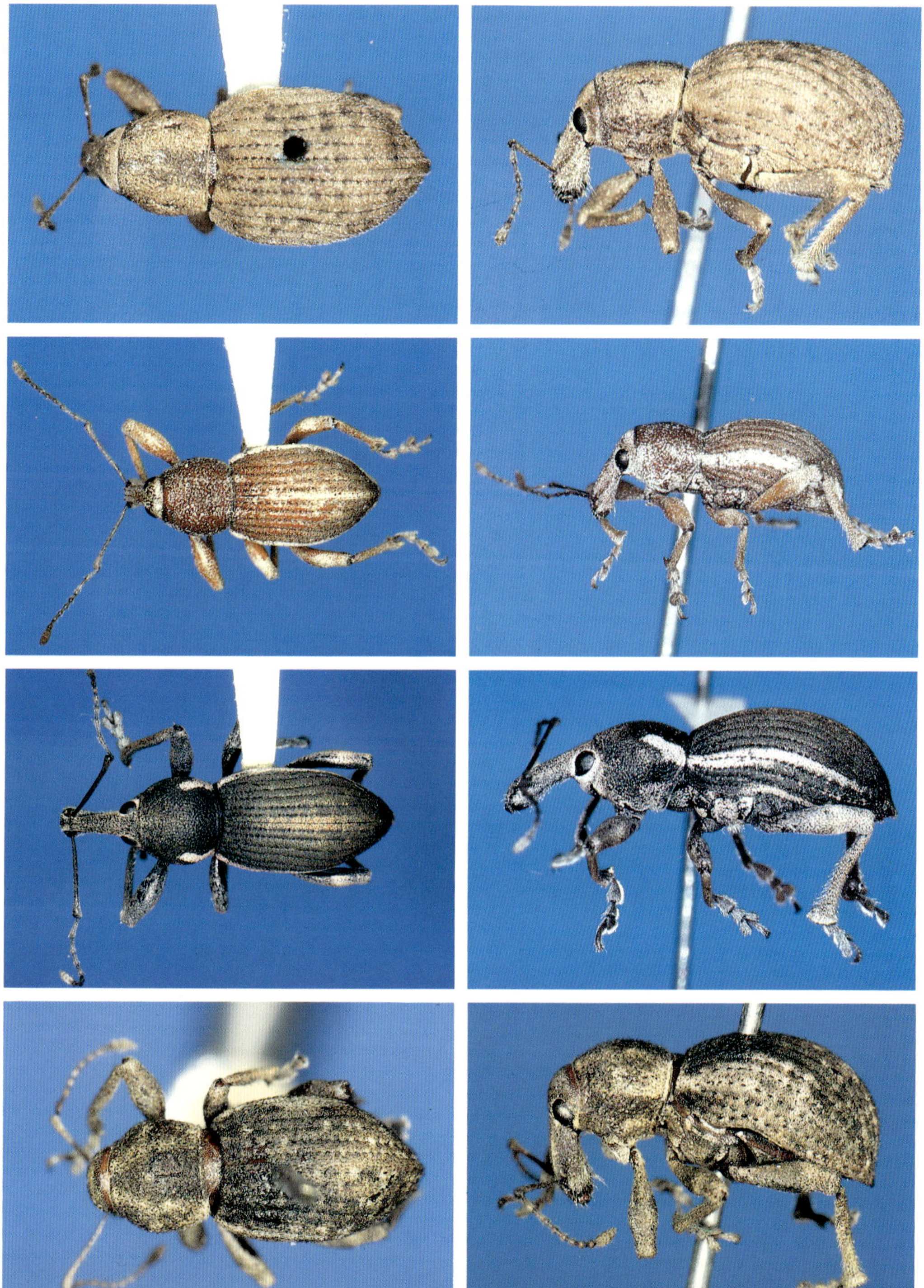

Plate 271 Adelognatha

1, 2 *Perperus sinuatus* (Blackburn)
Alps, Victoria
Paratype female; 9 mm

3, 4 *Perperus tuberculatus* Lea
South Gippsland, Victoria
Male; 7 mm

5, 6 *Perperus urticarum* Pascoe
Gayndah, Queensland
Topotype female; 7.5 mm

7, 8 *Perperus variegatus* Pascoe
Rockhampton, Queensland
Paratype male; 3.4 mm

1	2
3	4
5	6
7	8

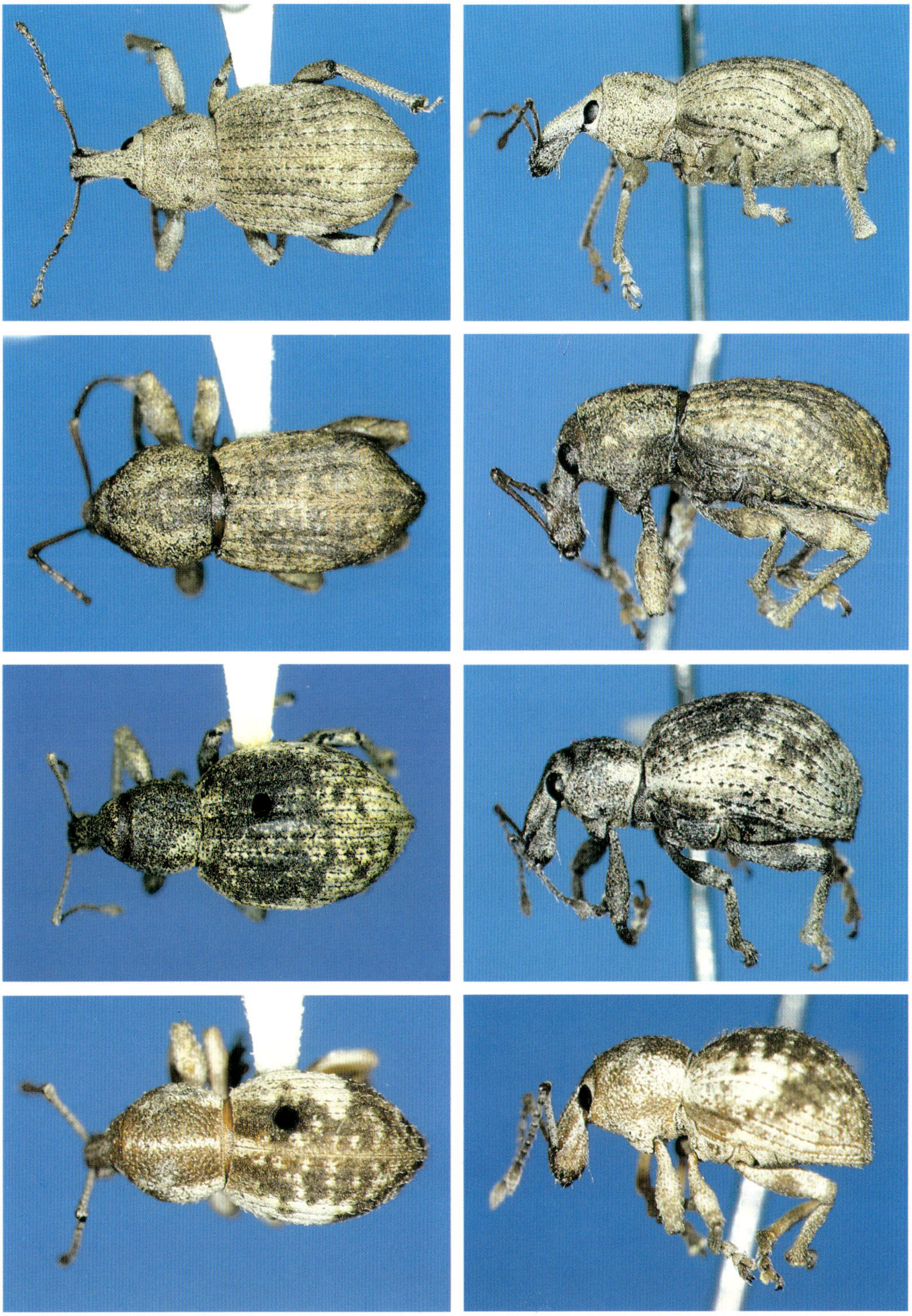

Plate 272 Adelognatha

1, 2 *Polyphrades apicalis* Lea
Queensland
Holotype female; 4.8 mm

3, 4 *Polyphrades basalis* Lea
Darwin, Northern Territory
Holotype male; 6.3 mm

5, 6 *Polyphrades biplagiatus* Pascoe
Western Australia
Female; 4.5 mm

7, 8 *Polyphrades cordatus* Lea
Forest Reefs, New South Wales
Holotype female; 2.7 mm

Note: *Polyphrades*, as constituted in literature, is compound. Pending publication of my revision of the genus in the forthcoming volume on the Adelognatha in this series, I leave most of names here and on Plates 273 – 282, 299 temporarily in *Polyphrades* as they now rest in literature.

1	2
3	4
5	6
7	8

Plate 273 Adelognatha

1, 2 *Polyphrades cordipennis* Lea
"Charters Towers to Concurry", Queensland
Topotype female; 9.3 mm

3, 4 *Polyphrades duriusculus* Lea
Cairns, Queensland
Holotype male; 5.2 mm

5, 6 *Polyphrades emblematicus* Lea
Northwest Western Australia
Holotype female; 5.2 mm

7, 8 *Polyphrades exoletus* Lea
Swan River, Western Australia
Paratype male; 5.9 mm

1	2
3	4
5	6
7	8

Plate 274 Adelognatha

1, 2 *Polyphrades extenuatus* Lea
Swan River
Holotype male; 4.3 mm

3, 4 *Polyphrades farinosus* Lea
Finke River, Northern Territory
Holotype female ?; 16.5 mm

5, 6 *Polyphrades fortis* Blackburn
Victoria
Female; 16.5 mm

7, 8 *Polyphrades gibbipennis* Lea
Finke River, Northern Territory
Paratype male; 10 mm

1	2
3	4
5	6
7	8

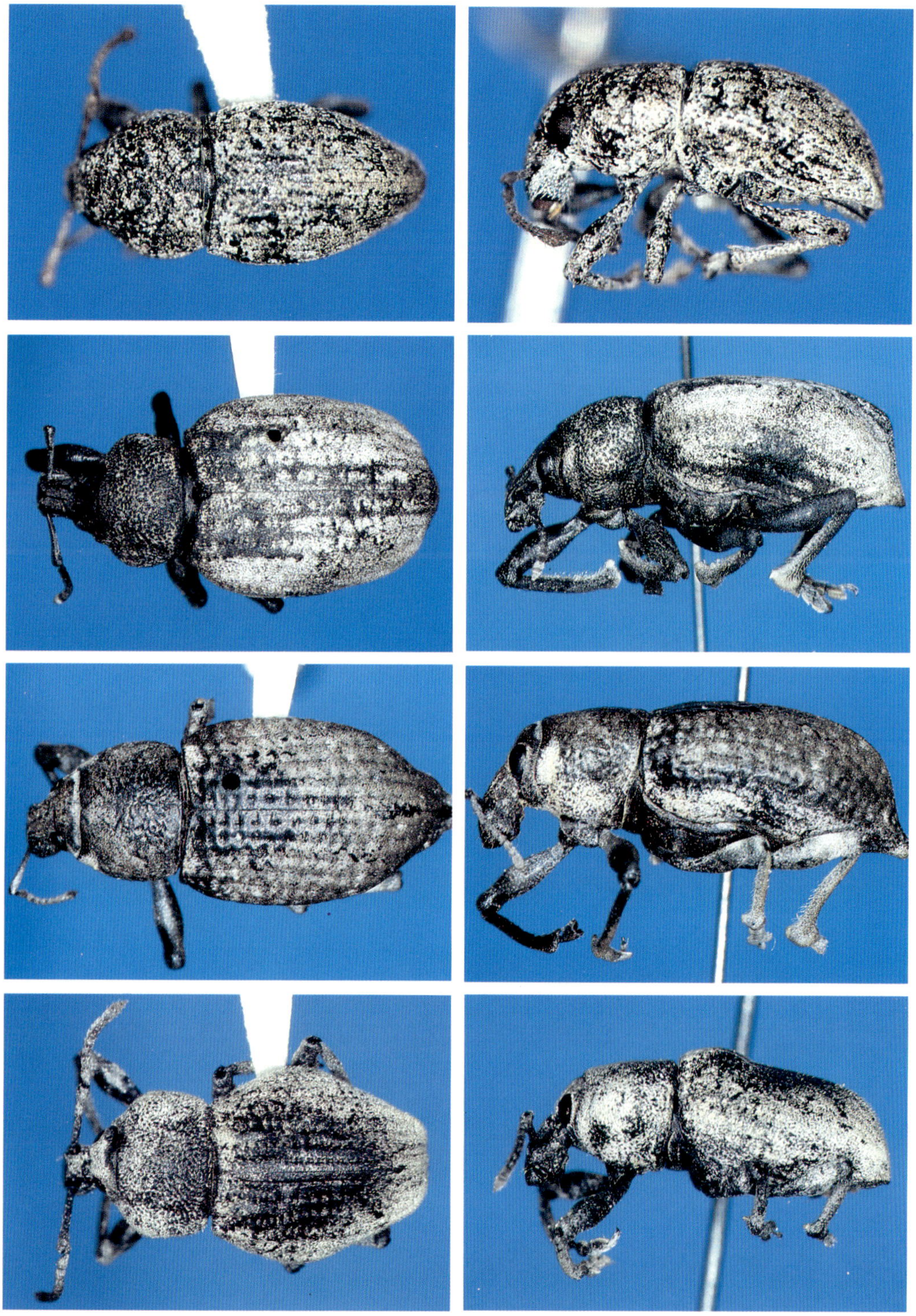

Plate 275 Adelognatha

1, 2 *Polyphrades granicollis* Lea
Mount Lofty, South Australia
Holotype female; 6.1 mm

3, 4 *Polyphrades halmaturinus* Lea
Kangaroo Island, South Australia
Paratype female; 4 mm

5, 6 *Polyphrades hartmeyeri* Lea
Dirk Hartog Island, Western Australia
Paratype female; 5.9 mm

7, 8 *Polyphrades inconspicuus* Blackburn
Victoria
Paratype male; 3.1 mm

1	2
3	4
5	6
7	8

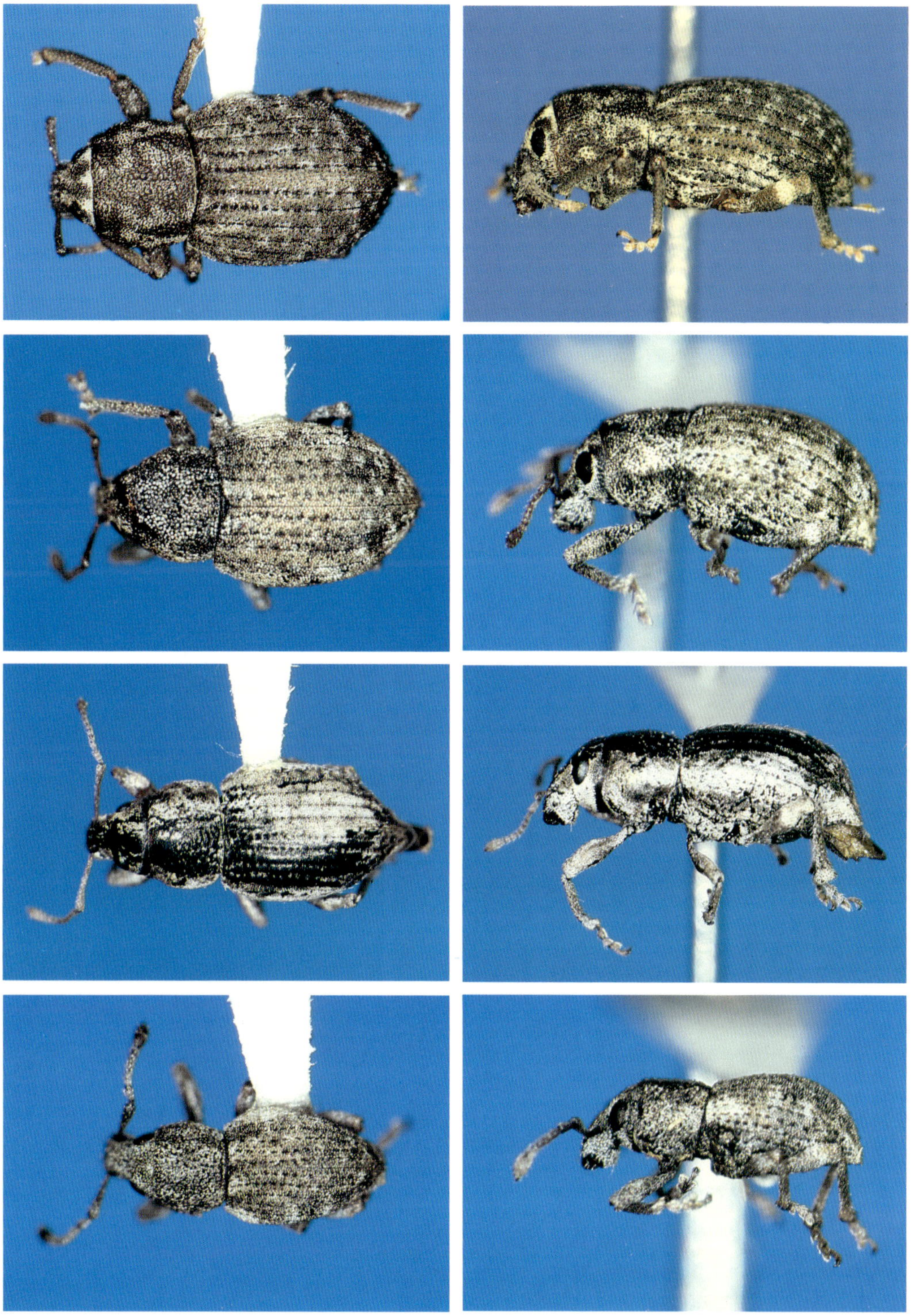

Plate 276 Adelognatha

1, 2 *Polyphrades insignipennis* Lea
Parry Island, Western Australia
Paratype male; 8.9 mm

3, 4 Paratype female of the same species; same locality; 7 mm

5, 6 *Polyphrades laetus* Blackburn
Streaky Bay, South Australia
Male; 3.4 mm

7, 8 *Polyphrades laminatus* Blackburn
Moonta or Kilerran, South Australia
Male; 6.1 mm

1	2
3	4
5	6
7	8

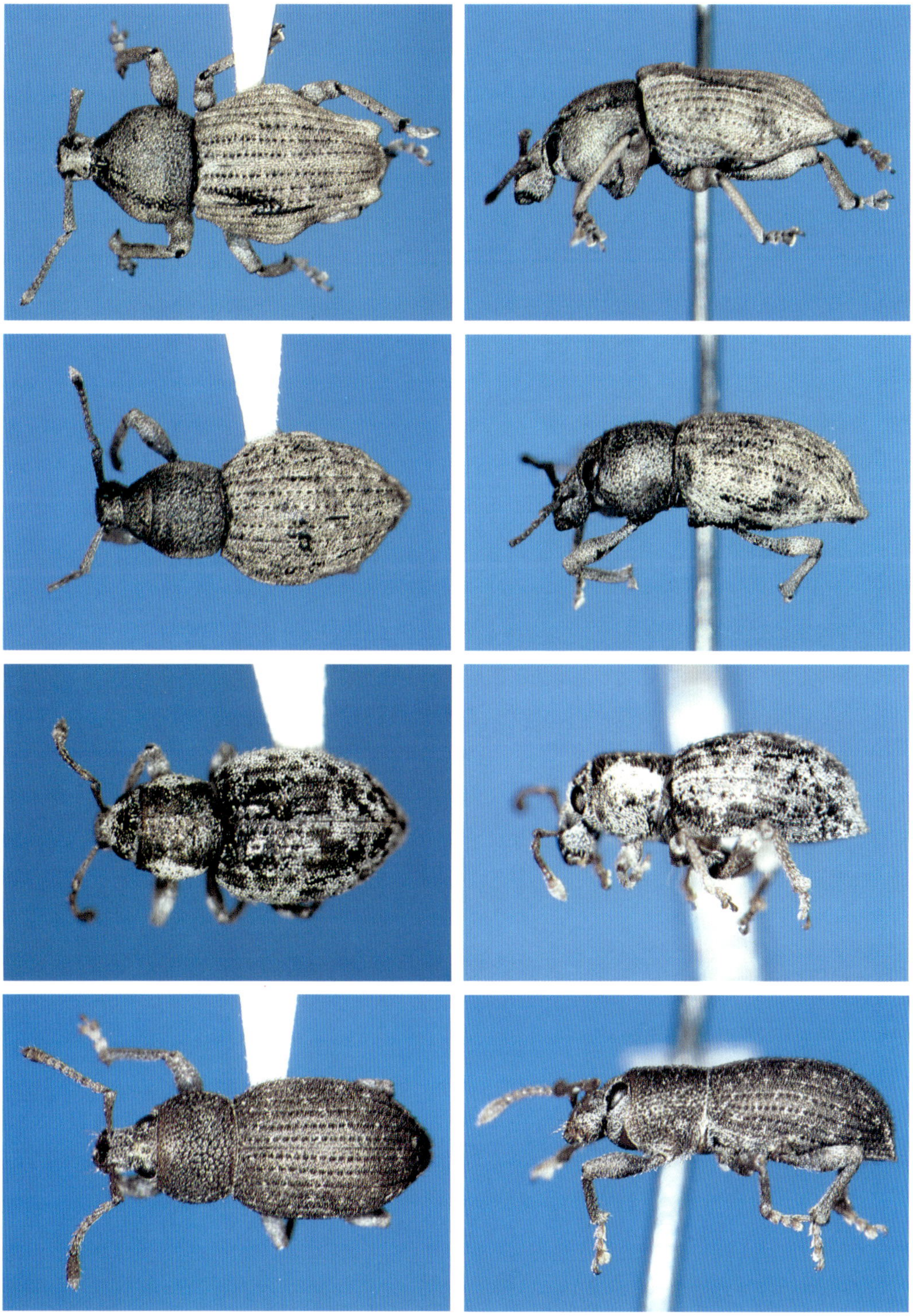

Plate 277 Adelognatha

1, 2 *Polyphrades laticollis* Fahraeus
Swan River, Western Australia
Male; 8.2 mm

3, 4 *Polyphrades latipennis* Pascoe
Champion Bay, Western Australia
Holotype male; 7.5 mm

5, 6 *Polyphrades latus* Lea
Kuranda, Queensland
Paratype male; 7.5 mm

7, 8 *Polyphrades longipennis* Pascoe
Ceduna, South Australia
Male; 10.2 mm

1	2
3	4
5	6
7	8

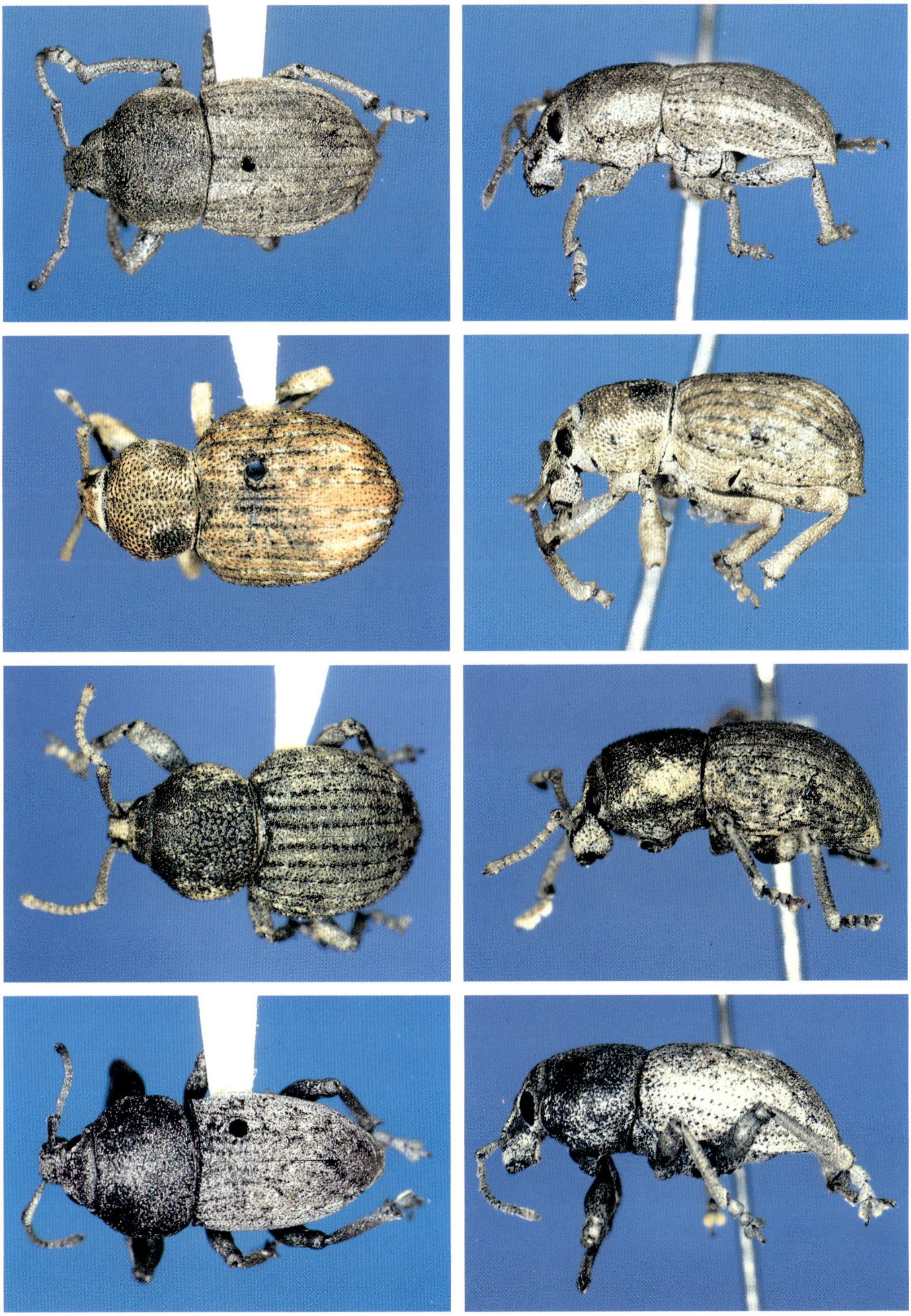

Plate 278 Adelognatha

1, 2 *Polyphrades macrops* Lea
Mount Gambier, South Australia
Holotype male; 2.1 mm

3, 4 *Polyphrades maestus* Lea
Lucindale, South Australia
Holotype male; 7 mm

5, 6 *Polyphrades marmoratus* Lea
Murray Bridge, South Australia
Holotype female; 4 mm

7, 8 *Polyphrades modestus* Blackburn
Port Augusta, South Australia
Paratype male, abraded; 7 mm

1	2
3	4
5	6
7	8

Plate 279 Adelognatha

1, 2 *Polyphrades nanus* (Gyllenhal)
Alps, Victoria
Male, from a paratype of the synonym *fulvus* Blackburn; 7.3 mm

3, 4 *Polyphrades paganus* Boheman
Sydney, New South Wales
Male; 11 mm

5, 6 Female of the same species; same locality; 11 mm

7, 8 *Polyphrades pardalotus* Pascoe
King George Sound, Western Australia
Female; compared with holotype; 5.8 mm

1	2
3	4
5	6
7	8

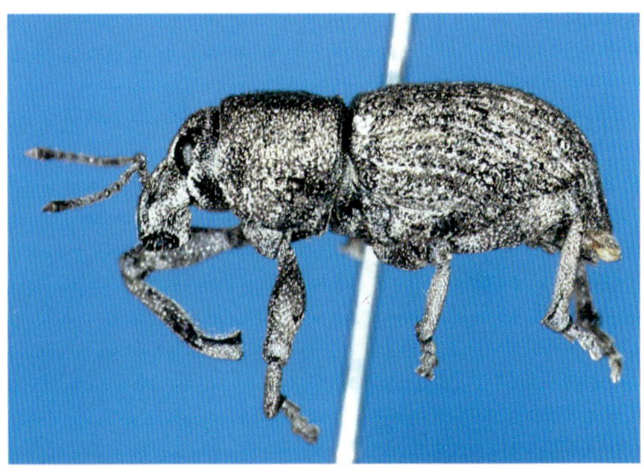

Plate 280 Adelognatha

1, 2 *Polyphrades parvus* Lea
Brisbane, Queensland
Paratype female; 4.3 mm

3, 4 *Polyphrades pictus* Blackburn
Port Lincoln, South Australia
Paratype male; 5.7 mm

5, 6 *Polyphrades planipennis* Lea
Tennant Creek, Northern Territory
Holotype male; 7.1 mm

7, 8 *Polyphrades pusillus* Pascoe
Coolup, Western Australia
Male; 4.3 mm
From a paratype of the synonym *vitis* Lea. Compare the female on Plate 281:1,2.

1	2
3	4
5	6
7	8

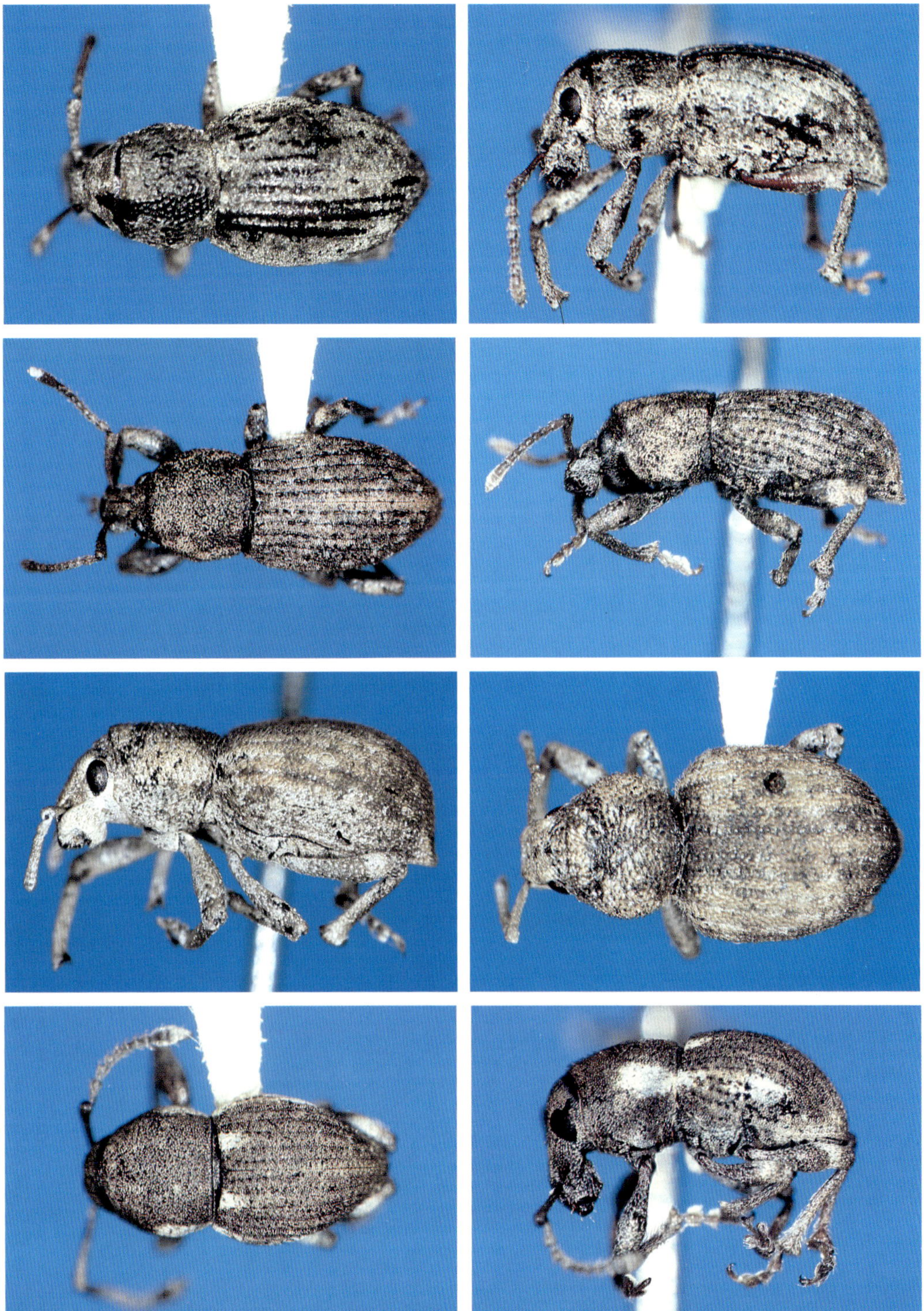

Plate 281 Adelognatha

1, 2 *Polyphrades pusillus* Pascoe
Geraldton, Western Australia
Female; 4.6 mm
From the holotype of the synonym *despicatus* Lea.

3, 4 *Polyphrades raui* Lea
Carraweena, South Australia
Holotype female; 6.5 mm

5, 6 *Polyphrades rugulosus* Blackburn
Victorian Desert, Western Australia
Paratype female; 13 mm

7, 8 *Polyphrades satelles* Blackburn
Eucla, Western Australia
Female; 14 mm

1	2
3	4
5	6
7	8

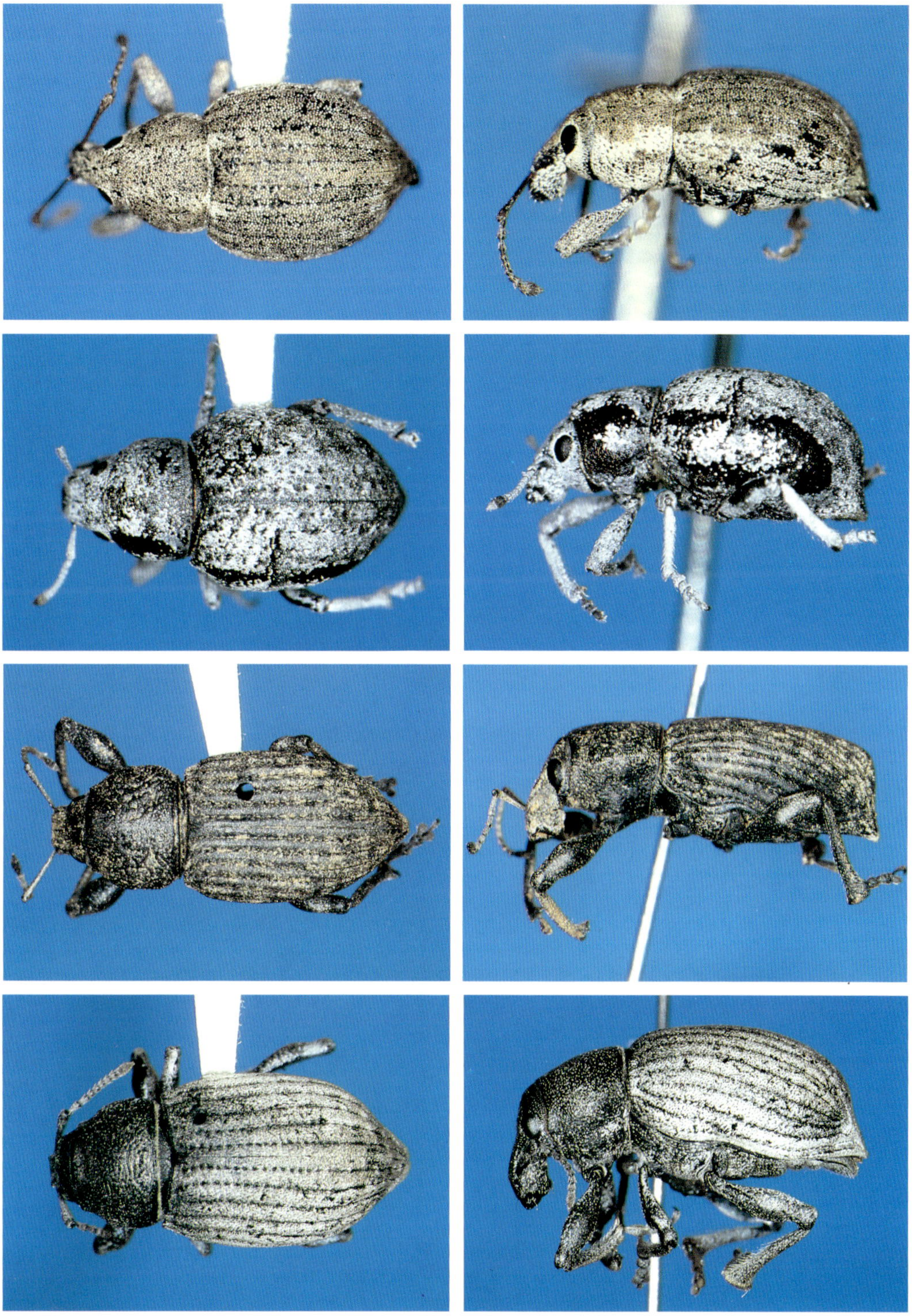

Plate 282 Adelognatha

1, 2 *Polyphrades tumidulus* Blackburn
Everard Ranges, South Australia to Warburton Range, Western Australia
Female; 9.7 mm

3, 4 *Polyphrades uniformis* Lea
Boorabbin, Western Australia
Holotype male; 6.3 mm

5, 6 *Prypnus angustus* Lea
Mount Victoria, New South Wales
Holotype female; 15.5 mm

7, 8 *Prypnus canaliculatus* Gyllenhal
Monaro, New South Wales
Female; 13 mm

1	2
3	4
5	6
7	8

Plate 283 Adelognatha

1, 2 *Prypnus glaber* Lea
South Australia
Holotype male; 14.5 mm

3, 4 *Prypnus quinquenodosus* Gyllenhal
Sydney, New South Wales
Male; 16.5 mm

5, 6 *Prypnus squalidus* (Gyllenhal)
Sydney, New South Wales
Male; 14 mm

7, 8 *Prypnus tenebricosus* (Gyllenhal)
Sydney, New South Wales
Female; 17.5 mm

1	2
3	4
5	6
7	8

Plate 284 Adelognatha

1, 2 *Prypnus tibialis* Lea
South Australia
Paratype male; 15.5 mm

3, 4 *Prypnus tuberculatus* Germar
Adelaide, South Australia
Male, compared with holotype; 15.5 mm

5, 6 *Psapharus areolus* (Pascoe)
Deep Dene, Western Australia
Female; 12.5 mm

7, 8 *Psapharus inconspicuus* (Blackburn)
Bridgetown, Western Australia
Female; 13.5 mm

1	2
3	4
5	6
7	8

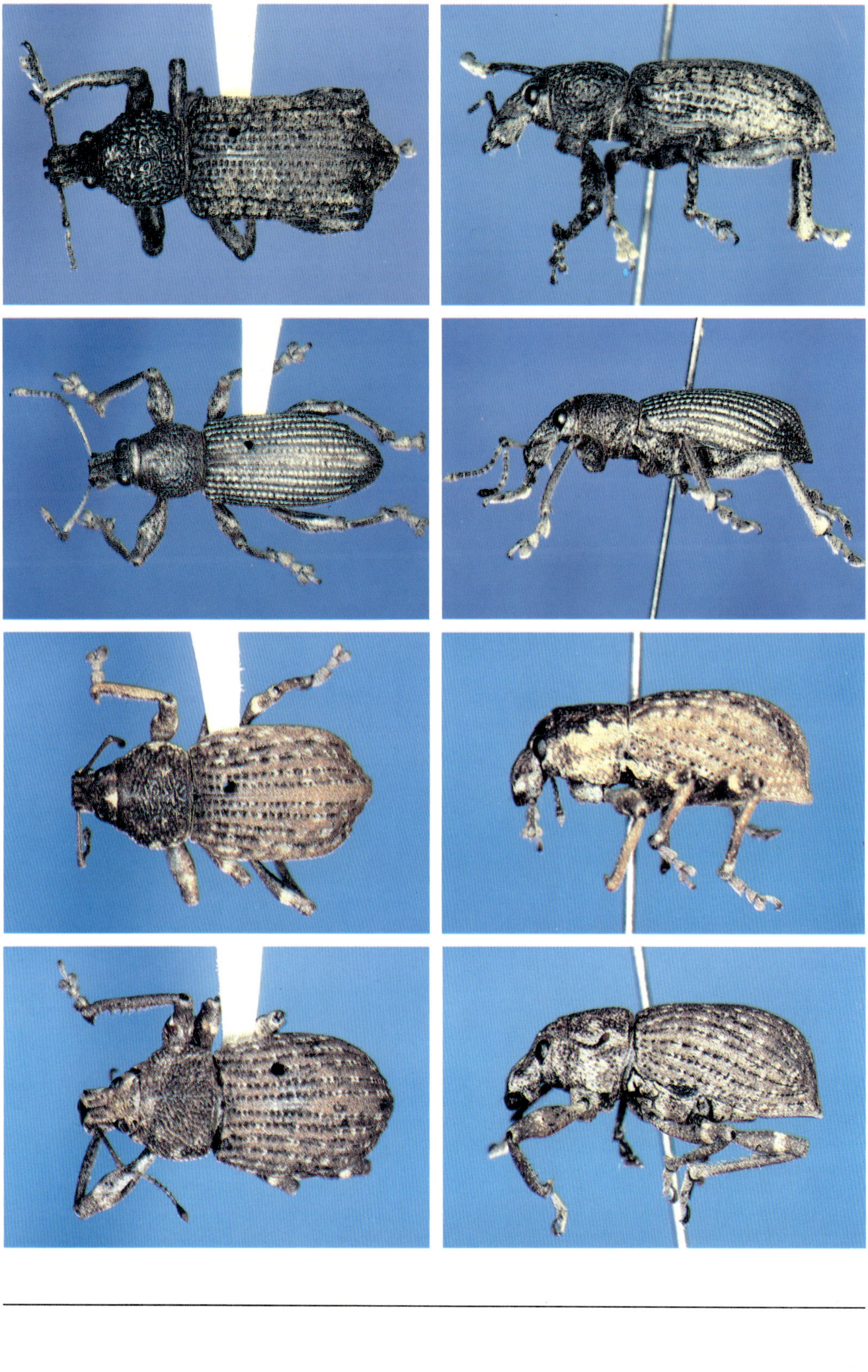

Plate 285 Adelognatha

1, 2 *Psapharus ruficornis* (Blackburn)
Mount Lofty, South Australia
Female; 18 mm

3, 4 *Psapharus strigiceps* (Lea)
South Australia
Holotype male; 11 mm

5, 6 *Pseudosynomus inconspicuus* (Lea)
Northern Territory (Elder Expedition)
Holotype male; 3.3 mm

7, 8 *Pseudotimareta subterranea* (Lea)
King Island, Bass Strait
Holotype male; 4.6 mm

1	2
3	4
5	6
7	8

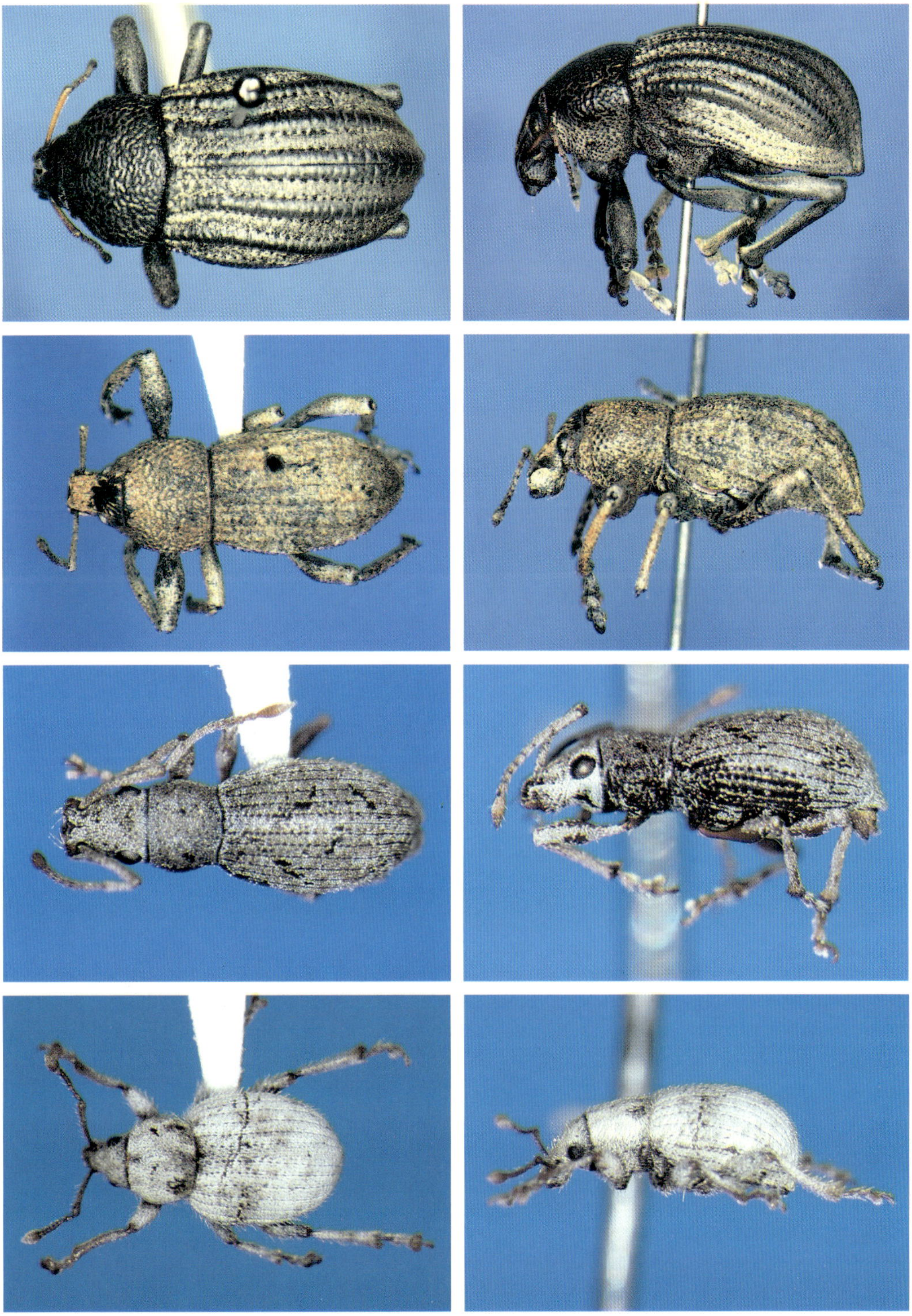

Plate 286 Adelognatha

1, 2 *Rhinoscapha doriae* Pascoe
Townsville, Queensland
Male; 22 mm

3, 4 *Rhyparophilus alternans* Boheman
Sydney, New South Wales
Male; 5.1 mm
From the holotype of the synonym *Mandalotus crassicornis* Lea.

5, 6 *Rhyparophilus ammophilus* (Lea)
New South Wales
Paratype male; 4.5 mm

7, 8 *Rhyparophilus goudiei* (Lea)
Black Rock, Victoria
Paratype female; 2.8 mm

1	2
3	4
5	6
7	8

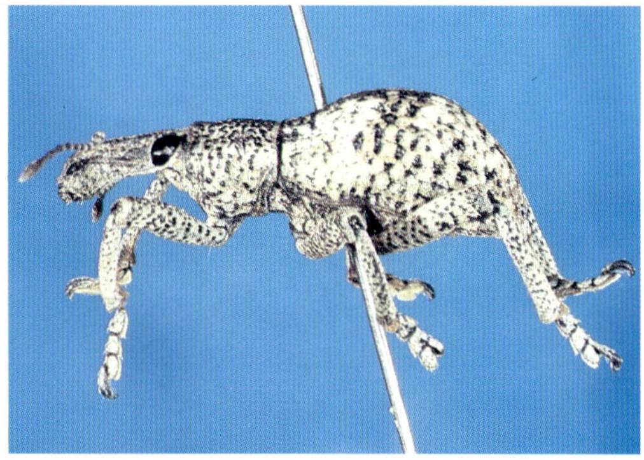

Plate 287 Adelognatha

1, 2 *Rhyparophilus herbivorus* (Lea)
Sydney, New South Wales
Male; 5.1 mm

3, 4 *Rhyparophilus howensis* (Lea)
Lord Howe Island
Paratype female; 5.9 mm

5, 6 *Rhyparophilus pondericornis* (Lea)
Lakes Entrance, Victoria
Female; 3.9 mm

7, 8 *Scotasmus carinirostris* Boheman
Hobart, Tasmania
Male; 6 mm
From the holotype of the synonym *Perperus malevolens* Lea.

1	2
3	4
5	6
7	8

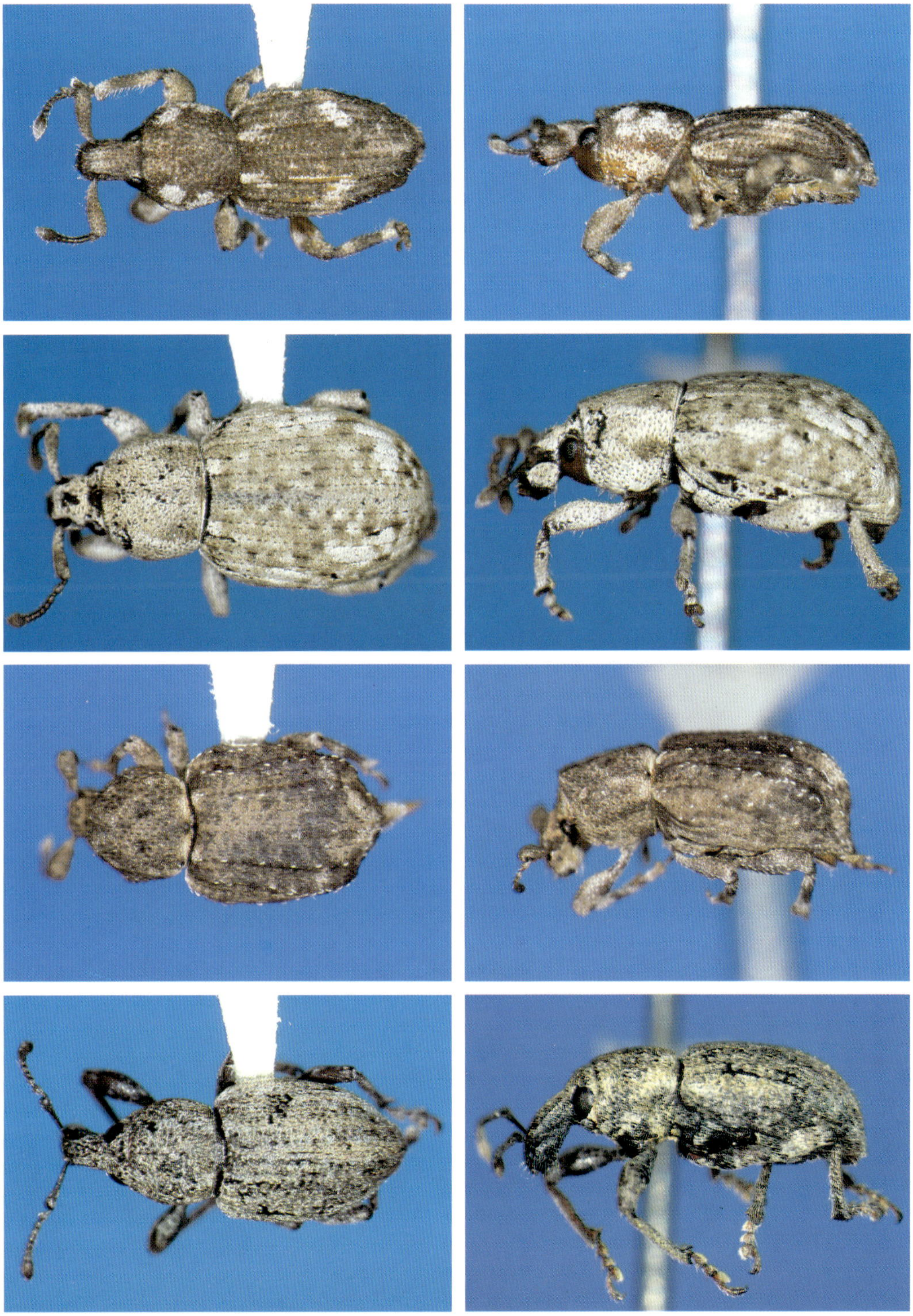

Plate 288 Adelognatha

1, 2 *Scotasmus costirostris* (Lea)
King Island, Bass Strait
Paratype male; 9 mm

3, 4 *Scotasmus litoralis* (Lea)
Tasmania
Male; 6 mm

5, 6 *Scotasmus parvicornis* (Lea)
Hobart, Tasmania
Paratype male; 9.4 mm

7, 8 *Stenocorynus albolineatus* (Macleay)
Cairns, Queensland
Female; 11 mm

1	2
3	4
5	6
7	8

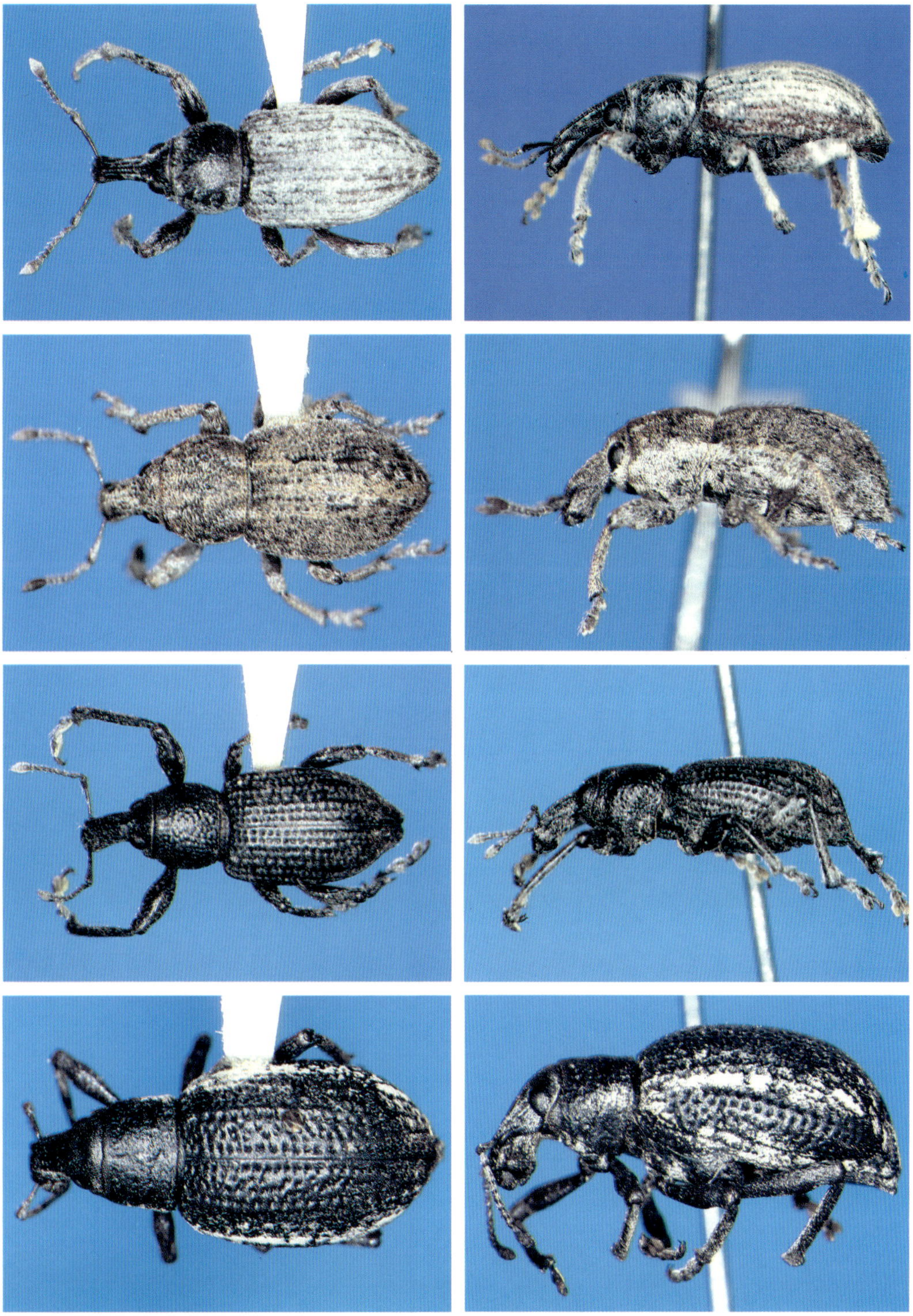

Plate 289 Adelognatha

1, 2 *Stenocorynus alleni* Lea
Cairns District, Queensland
Male; 11 mm

3, 4 *Stenocorynus apicipennis* Lea
Queensland
Holotype female; 12.5 mm

5, 6 *Stenocorynus aridus* Pascoe
North Endeavour River, Queensland
Male; 8 mm

7, 8 *Stenocorynus crenulatus* (Fabricius)
Endeavour River, Queensland
Male; 13.5 mm

1	2
3	4
5	6
7	8

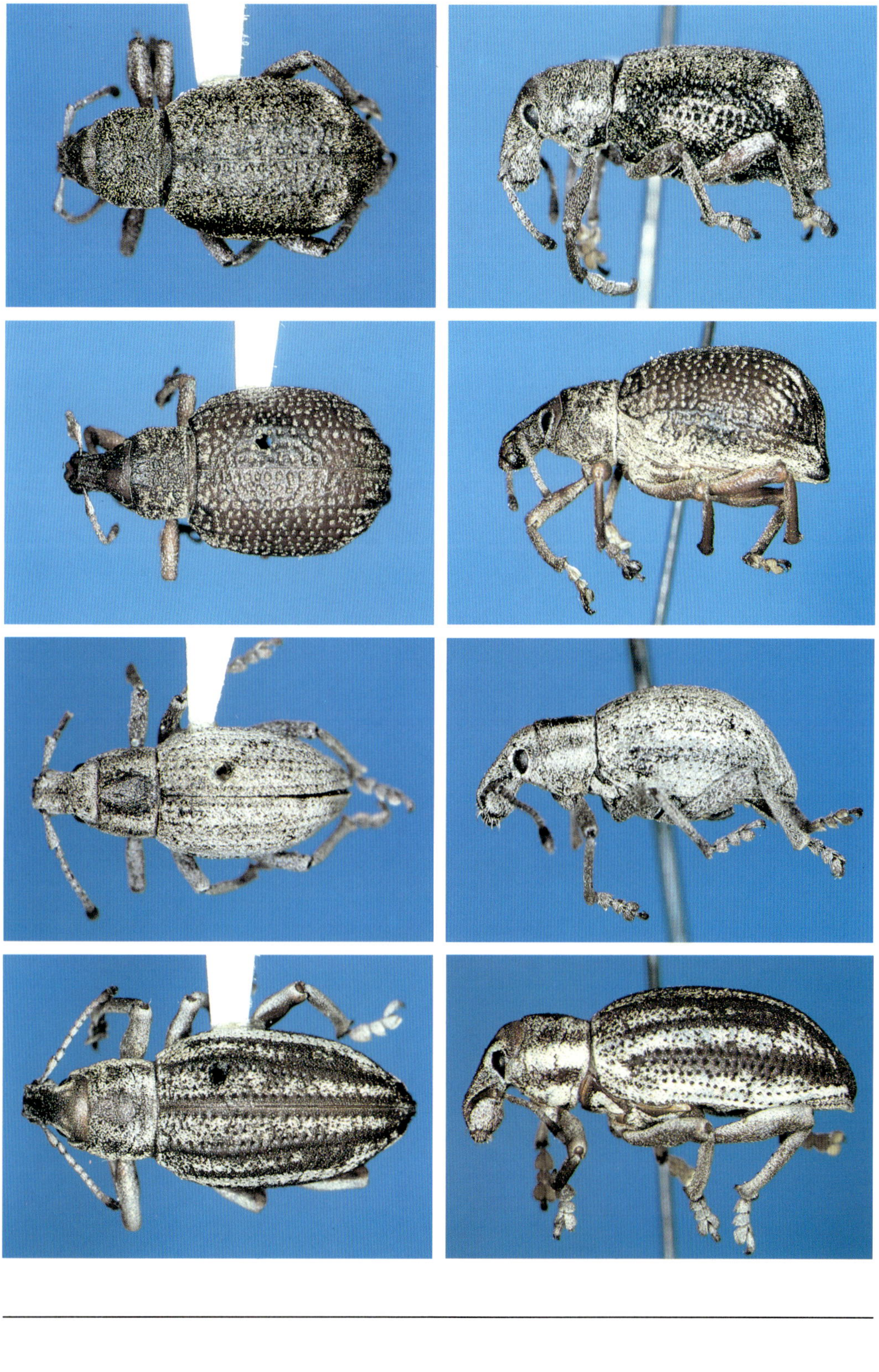

Plate 290 Adelognatha

1, 2 *Stenocorynus cretatus* Lea
Mareeba, Queensland
Female; 11 mm

3, 4 *Stenocorynus lepidurus* Lea
Coen River, Queensland
Holotype female; 15 mm

5, 6 *Stenocorynus minor* Lea
Cairns, queensland
Holotype male; 5.7 mm

7, 8 *Stenocorynus subfasciatus* Pascoe
Queensland
Male, compared with holotype; 11.5 mm

1	2
3	4
5	6
7	8

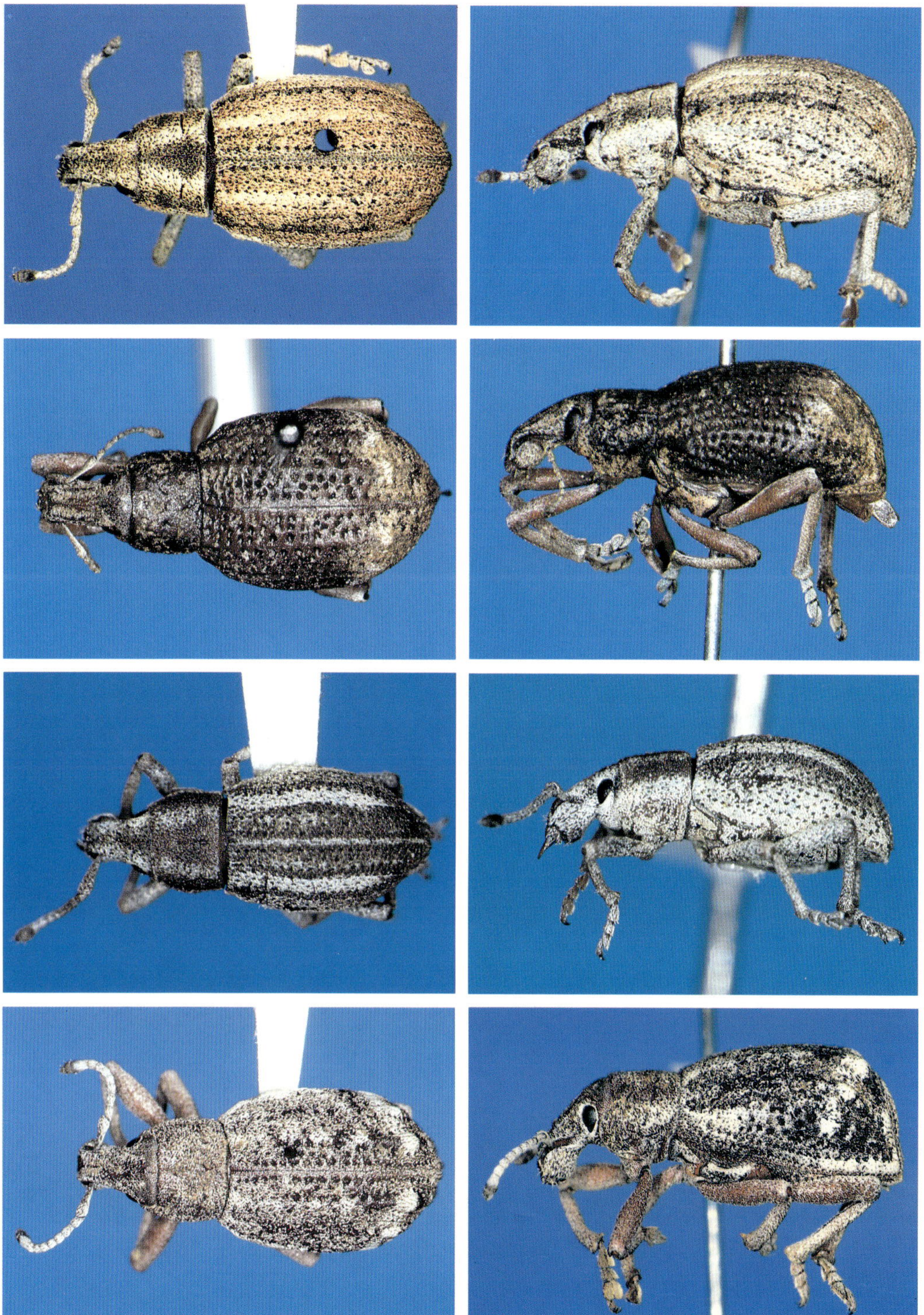

Plate 291 Adelognatha

1, 2 *Stenocorynus variabilis* (Blackburn)
Northern Territory
Holotype female; 11.5 mm

3, 4 *Stenocorynus vittatus* Pascoe
Saddlebag Creek, 16 km north of Mount Molloy, Queensland
Female; 14 mm

5, 6 *Synaptonyx ovatus* Pascoe
King Sound, Western Australia
Male; 6.1 mm

7, 8 *Synomus aeruginosus* Lea
Chillagoe, Queensland
Holotype male; 4.8 mm

1	2
3	4
5	6
7	8

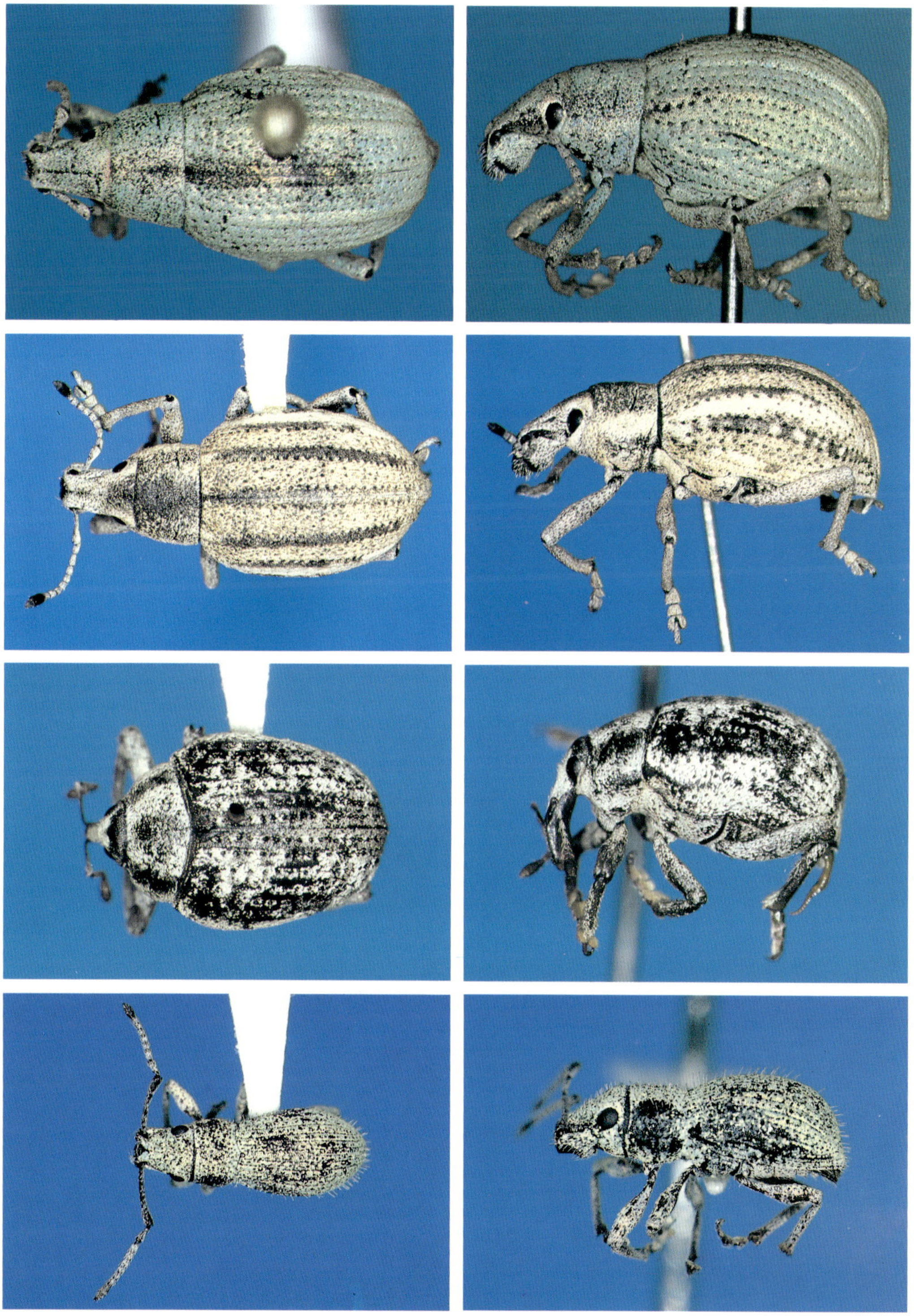

Plate 292 Adelognatha

1, 2 *Synomus ornatus* (Faust)
"Australien"
Male; 6.4 mm
From a paratype of the synonym *setipennis* Lea.

3, 4 *Synomus ovipennis* Lea
Cairns, Queensland
Holotype female (teneral); 3.6 mm

5, 6 *Synomus subapterus* (Lea)
Northwest Western Australia
Holotype female; 6.4 mm

7, 8 *Synomus* new species
Charters Towers, Queensland
Male?; 4.25 mm

1	2
3	4
5	6
7	8

Plate 293 Adelognatha

1, 2 *Syrphax blackburni* (Lea)
North Queensland
Holotype female; 6.4 mm

3, 4 *Syrphax bovilli* (Blackburn)
Mary River, Northern Territory
Female; 6 mm

5, 6 *Syrphax fieldi* (Lea)
Tennant Creek, Northern Territory
Paratype female; 5.9 mm

7, 8 *Syrphax incurvus* (Lea)
Northwest Western Australia
Female; 6 mm

1	2
3	4
5	6
7	8

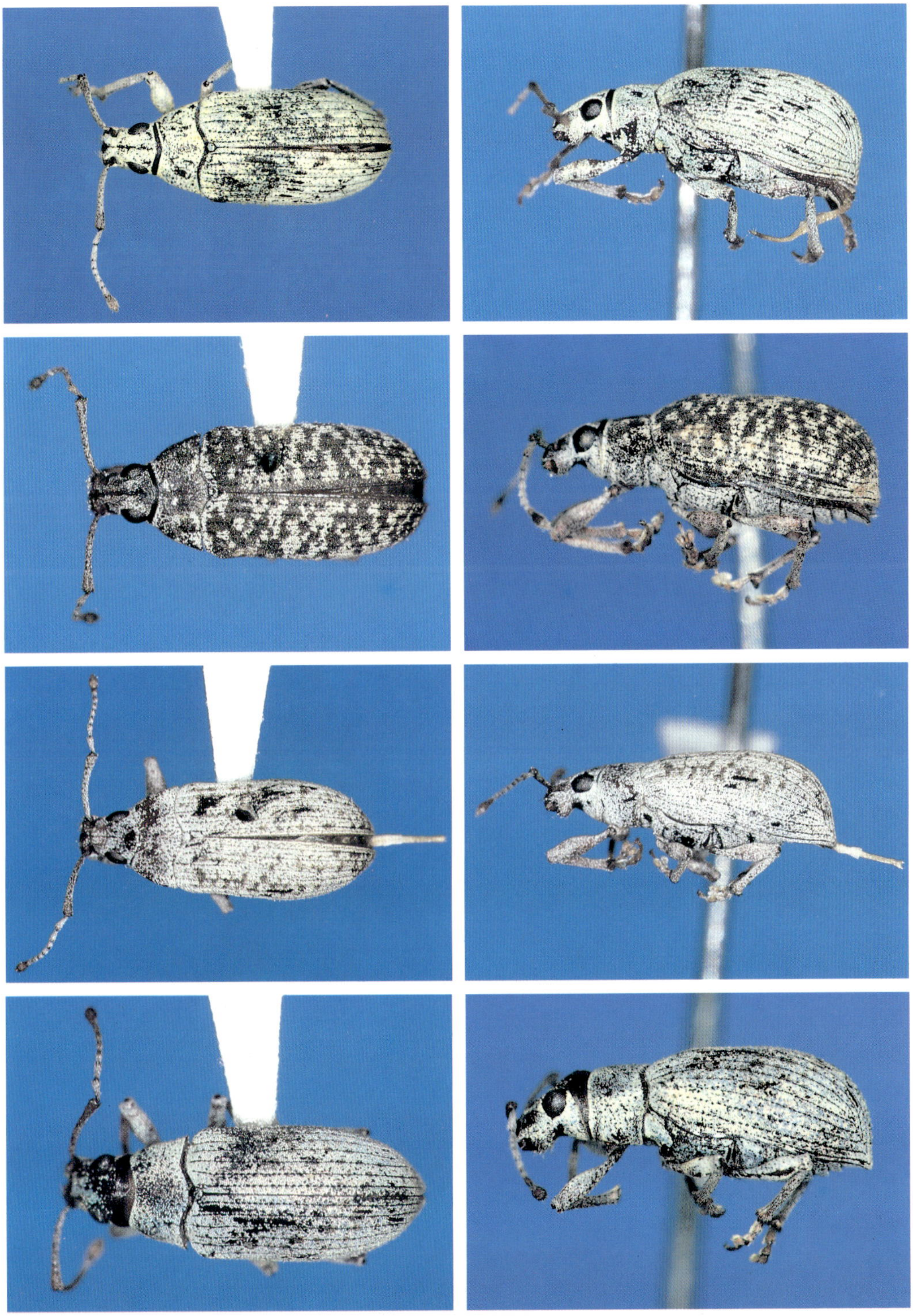

Plate 294 Adelognatha

1, 2 *Tanymecus infestus* Zimmerman
Mary River, Northern Territory
Paratype male; 7.3 mm

3, 4 *Timareta crinita* Pascoe
Geraldton, Western Australia
Male; 6.5 mm

5, 6 *Timareta figurata* Pascoe
Bunbury, Western Australia
Male; 4 mm

7, 8 *Timareta hamata* Lea
Flinders Island, South Australia
Paratype male; 6.5 mm

1	2
3	4
5	6
7	8

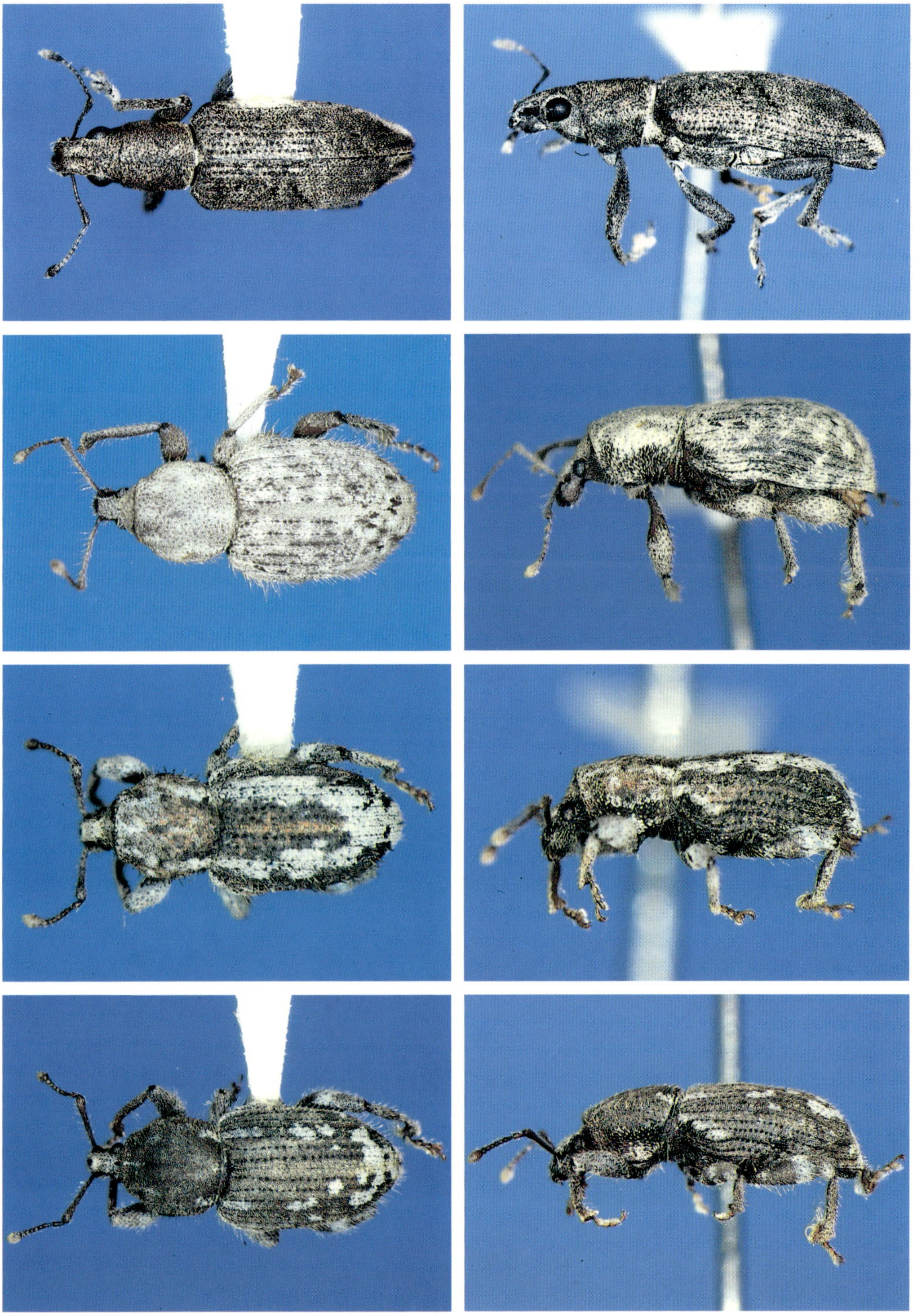

Plate 295 Adelognatha

1, 2 *Timareta incisipes* Lea
St. Francis Island, South Australia
Paratype male; 4.8 mm

3, 4 *Timareta infortunata* Lea
Swan River, Western Australia
Holotype male; 3.2 mm

5, 6 *Timareta pilipes* (Pascoe)
Albany, Western Australia
Male; 7.2 mm
From a syntype of the synonym *Dysostines pustulosus* Pascoe.

7, 8 *Timareta pilosa* (Blackburn)
Hallart Cove, South Australia
Male; 7.3 mm

1	2
3	4
5	6
7	8

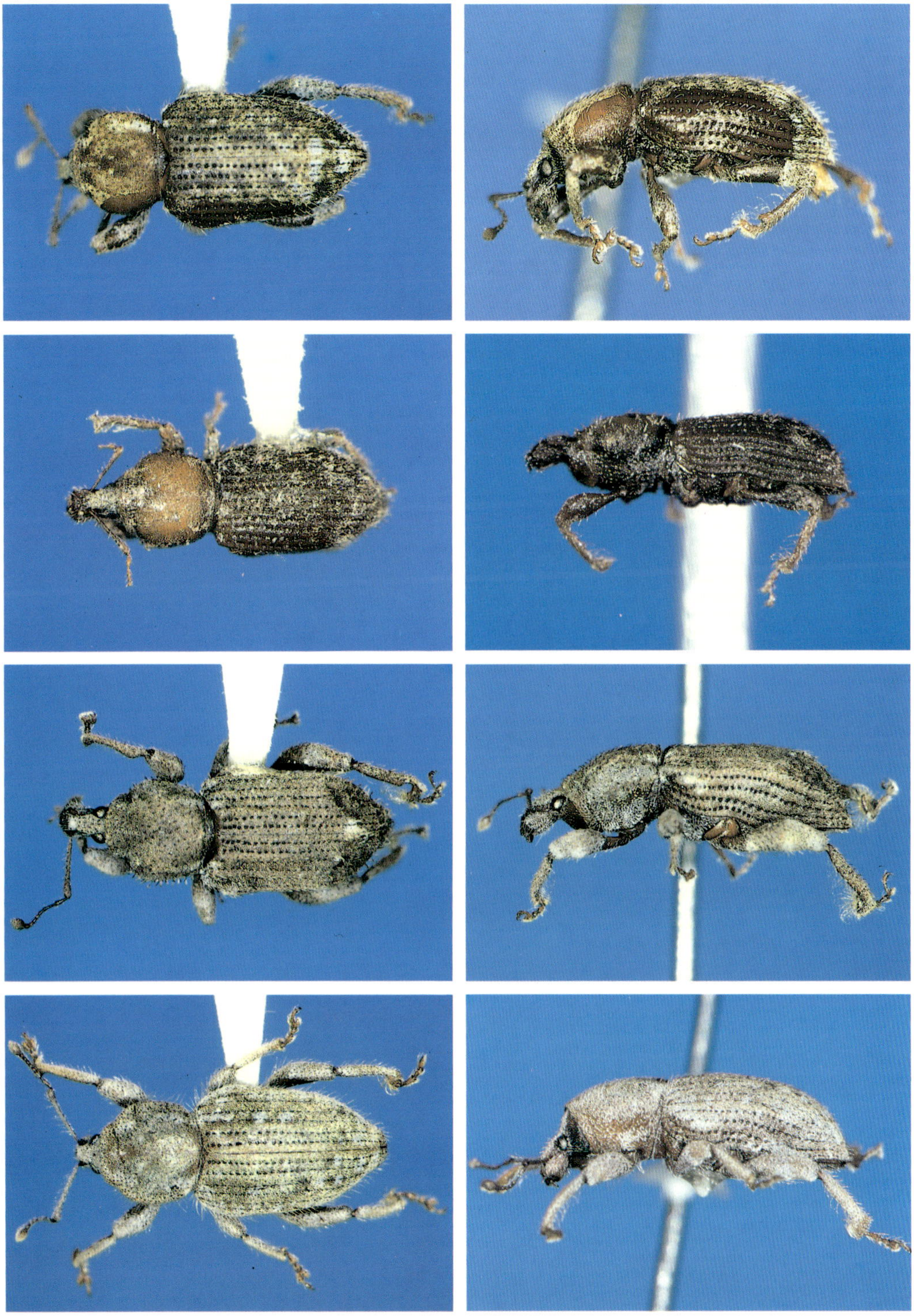

Plate 296 Adelognatha

1, 2 *Titinia brevicollis* (Blackburn)
Western Australia
Female; 4.2 mm

3, 4 *Titinia laeta* Blackburn
Australian Alps
Holotype female; 4.2 mm

5, 6 *Titinia grata* (Pascoe)
Cape York, Queensland
Female; 3.6 mm

7, 8 *Titinia tenuis* (Germar)
South Australia
Female, compared with holotype; 3. 6 mm

1	2
3	4
5	6
7	8

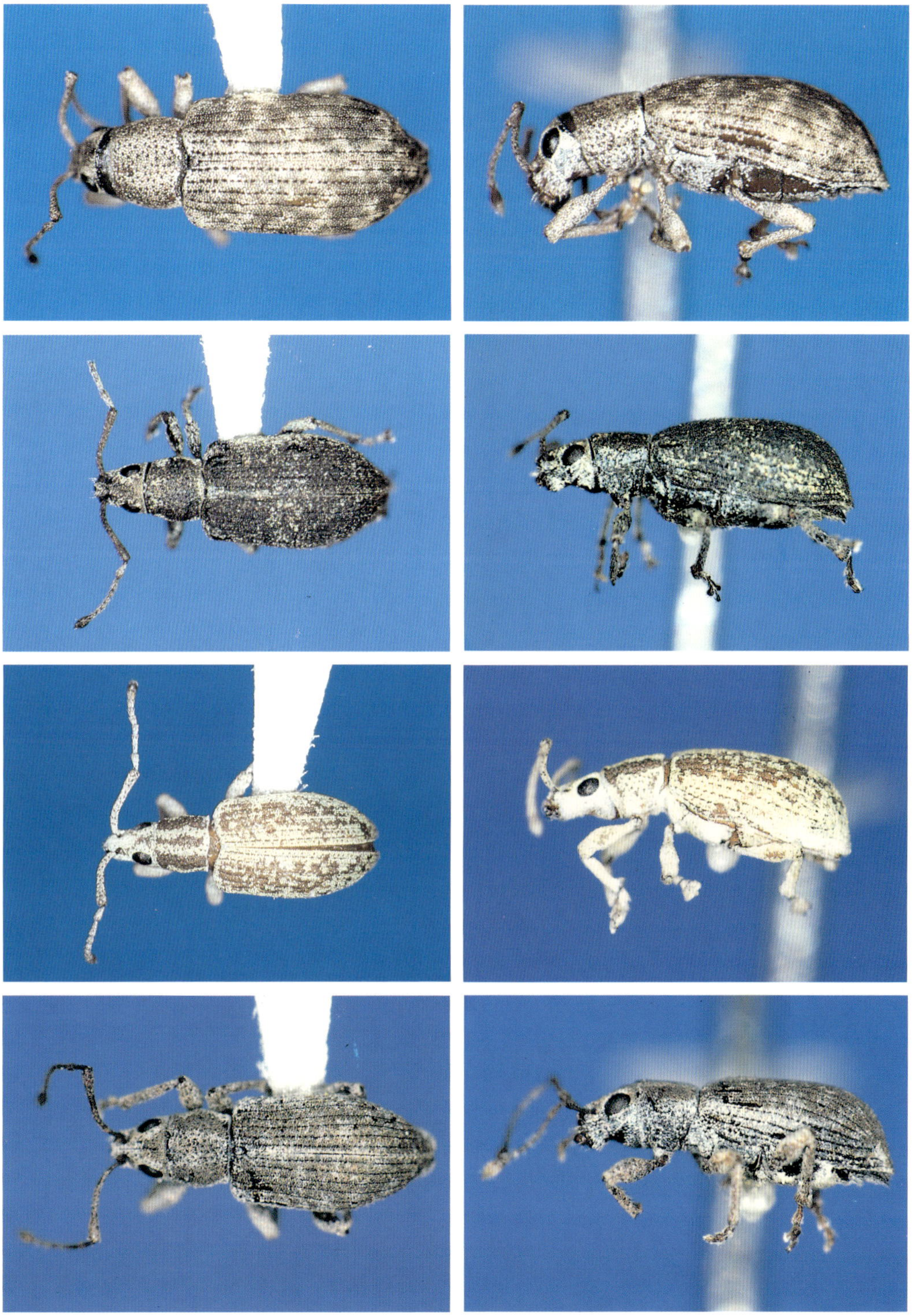

Plate 297 Adelognatha

1, 2 *Paratitinia ceratorhinus* (Lea)
Cue, Western Australia
Paratype female; 6.3 mm

3, 4 *Paratitinia hoplosternus* (Lea)
Stewart River, Queensland
Paratype female; 4.7 mm

5, 6 *Paratitinia obliquicollis* (Lea)
Stewart River, Queensland
Paratype female; 5.2 mm

7, 8 *Paratitinia* new species
13 km east by north of Newman, Western Australia
Male; 3.8 mm

1	2
3	4
5	6
7	8

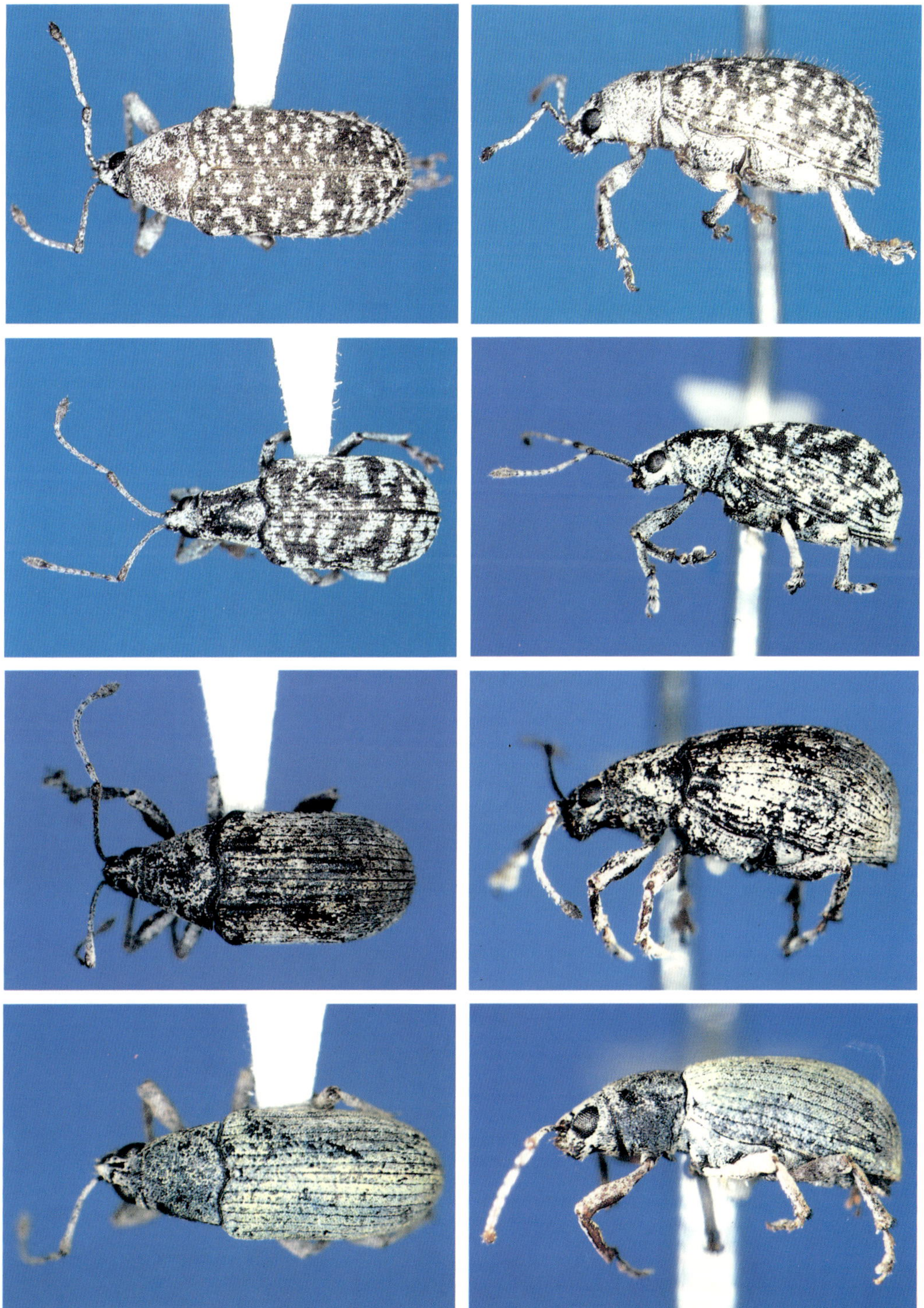

Plate 298 Adelognatha

1, 2 *Titinia eremita* Blackburn
South Australia
Female; 4.5 mm

3, 4 *Titinia microps* Lea
Gayndah, Queensland
Male; 5 mm

5, 6 *Titinia terrea* Pascoe
Ooldea, South Australia
Female, compared with holotype; 4 mm

7, 8 *"Prosayleus" latipennis* Lea
Swan River, Western Australia
Paratype male; 4.5 mm

1	2
3	4
5	6
7	8

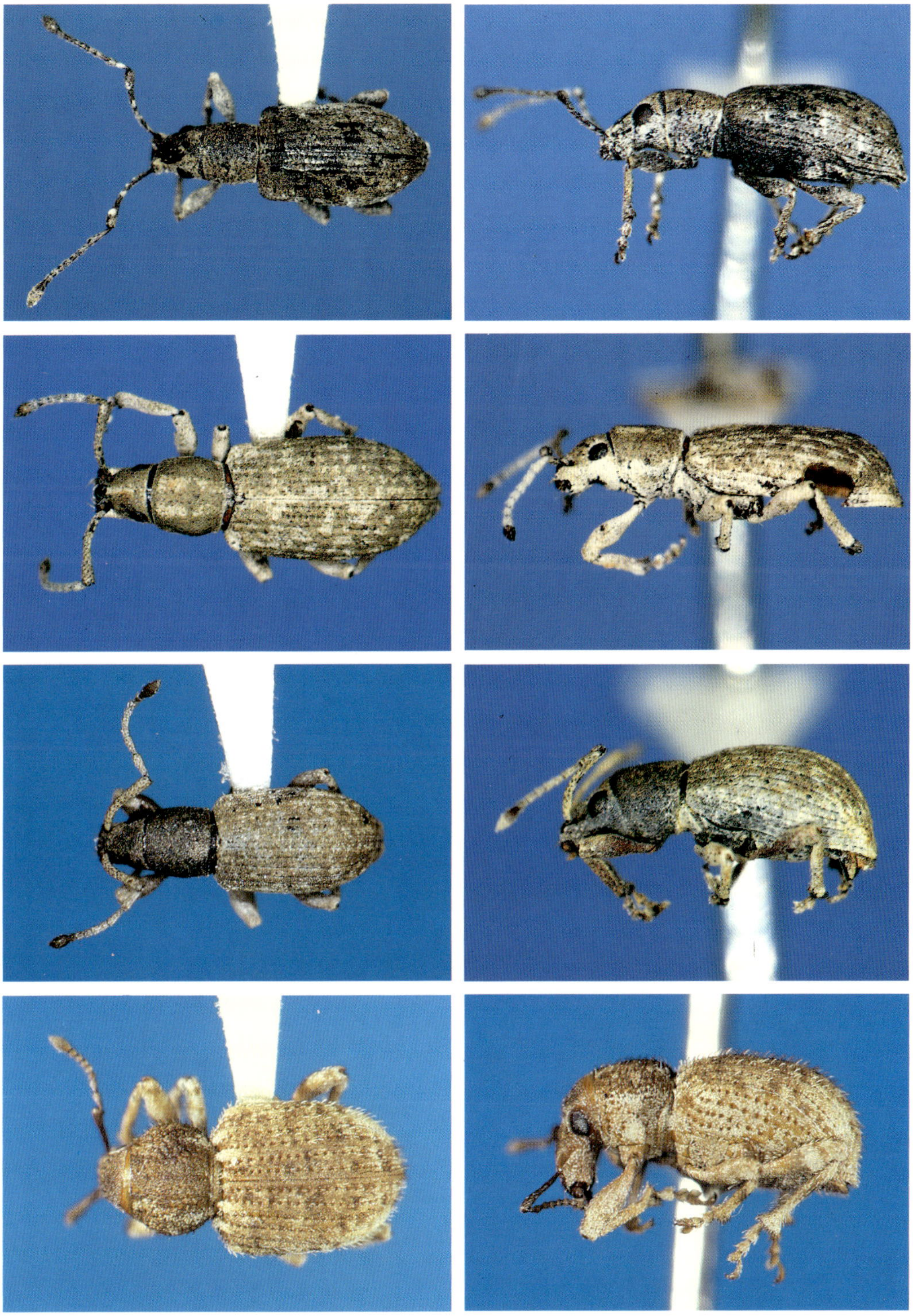

Plate 299 Adelognatha

1, 2 *"Polyphrades" brevirostris* Lea
Mullewa, Western Australia
Male; 4.5 mm

3, 4 *"Polyphrades" crassicornis* Lea
Batchelor, Northern Territory
Holotype male; 5.3 mm

5, 6 A male of the same species from the same locality to show variation in shape.
Paratype female; 2 mm

7, 8 *"Polyphrades" subterraneus* Lea
Hobart, Tasmania
Paratype female; 2 mm

See the Volume IV on the **Adelognatha** for placement of these species.

1	2
3	4
5	6
7	8

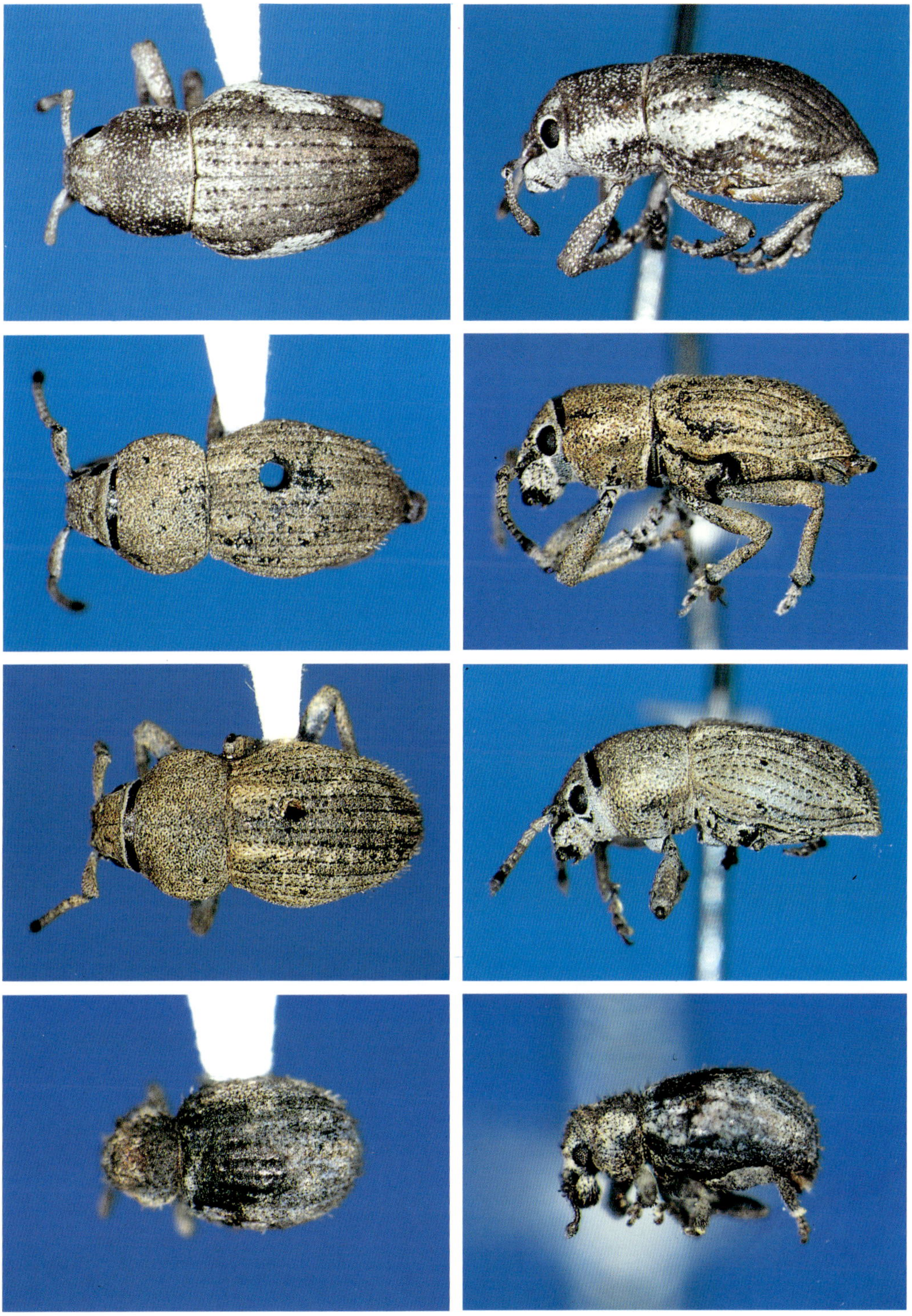

Plate 300 Adelognatha

1, 2 *"Pascoellus" rattulus* Pascoe
Farina, Central Australia
Male, compared with syntype; 5.5 mm

3, 4 *"Pascoellus" umbrata* (Blackburn)
Western Australia
Paratype male; 5.2 mm

5, 6 *"Mandalotus" coxalis* Lea
Mulgrave River, Western Australia
Holotype male; 4.2 mm

7, 8 *"Myllocerus" trivitticollis* Lea
Barton, South Australia
Male; 5 mm

See Volume IV on the **Adelognatha** for new assignments.

1	2
3	4
5	6
7	8

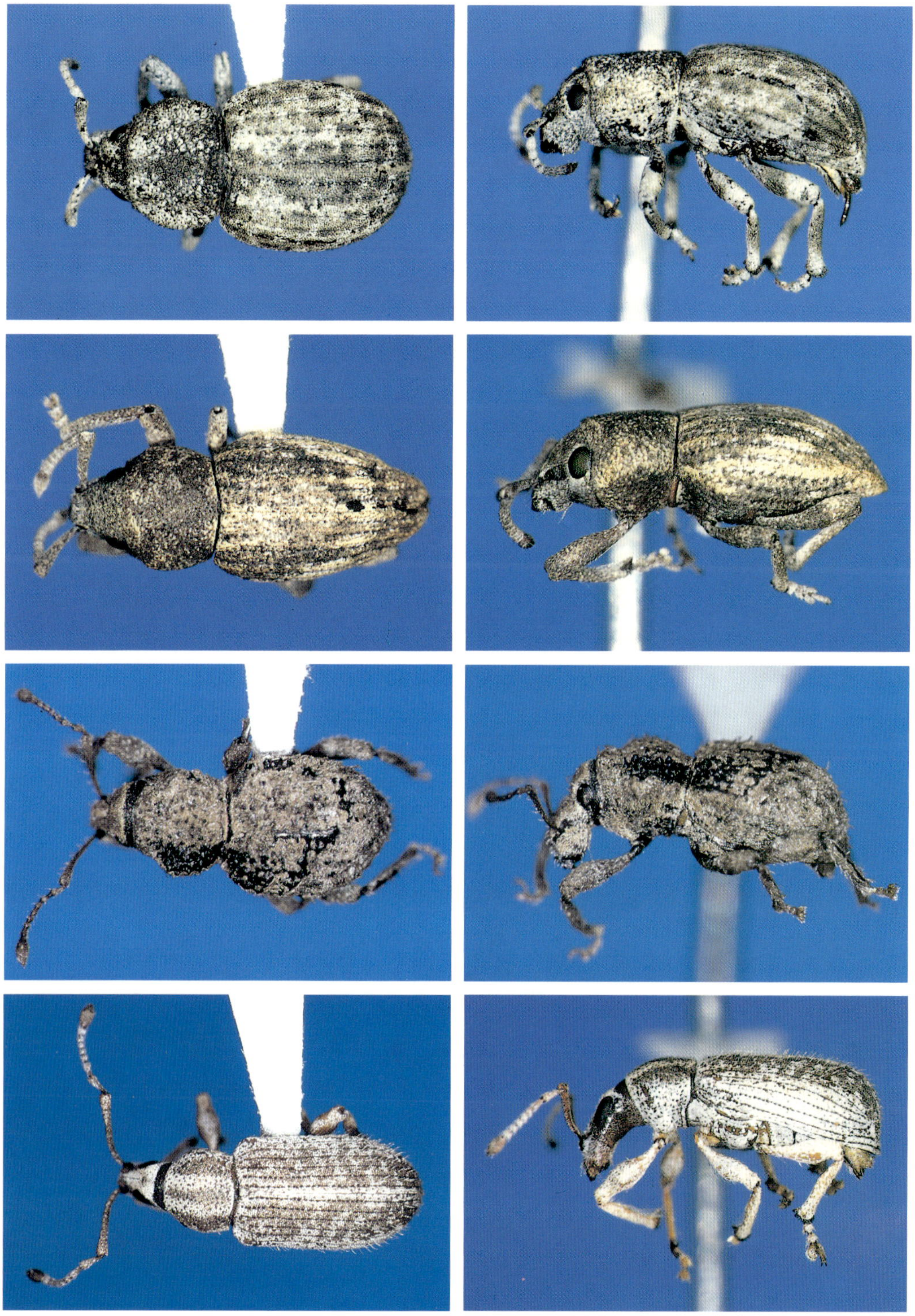

Plate 301 Adelognatha, Apionidae, Amycterinae

1, 2 *Howeocis microps* (Lea)
Lord Howe Island
Paratype female; 2.9 mm

3, 4 *Howeocis nodipennis* (Lea)
Lord Howe Island
Holotype male; 2.4 mm

5 *Myrmacicelus exertus* Pascoe
Geraldton, Western Australia
Female; 4.8 mm
Note the lateral elytral striae.

6 *Myrmacicelus puerulus* Lea
Beverley, Western Australia
Syntype male; 1.8 mm
Note absence of elytral striae.

7, 8 *Talaurinus foveogranulatus* Ferguson
Western Australia
Holotype male; 18 mm

1	2
3	4
5	6
7	8

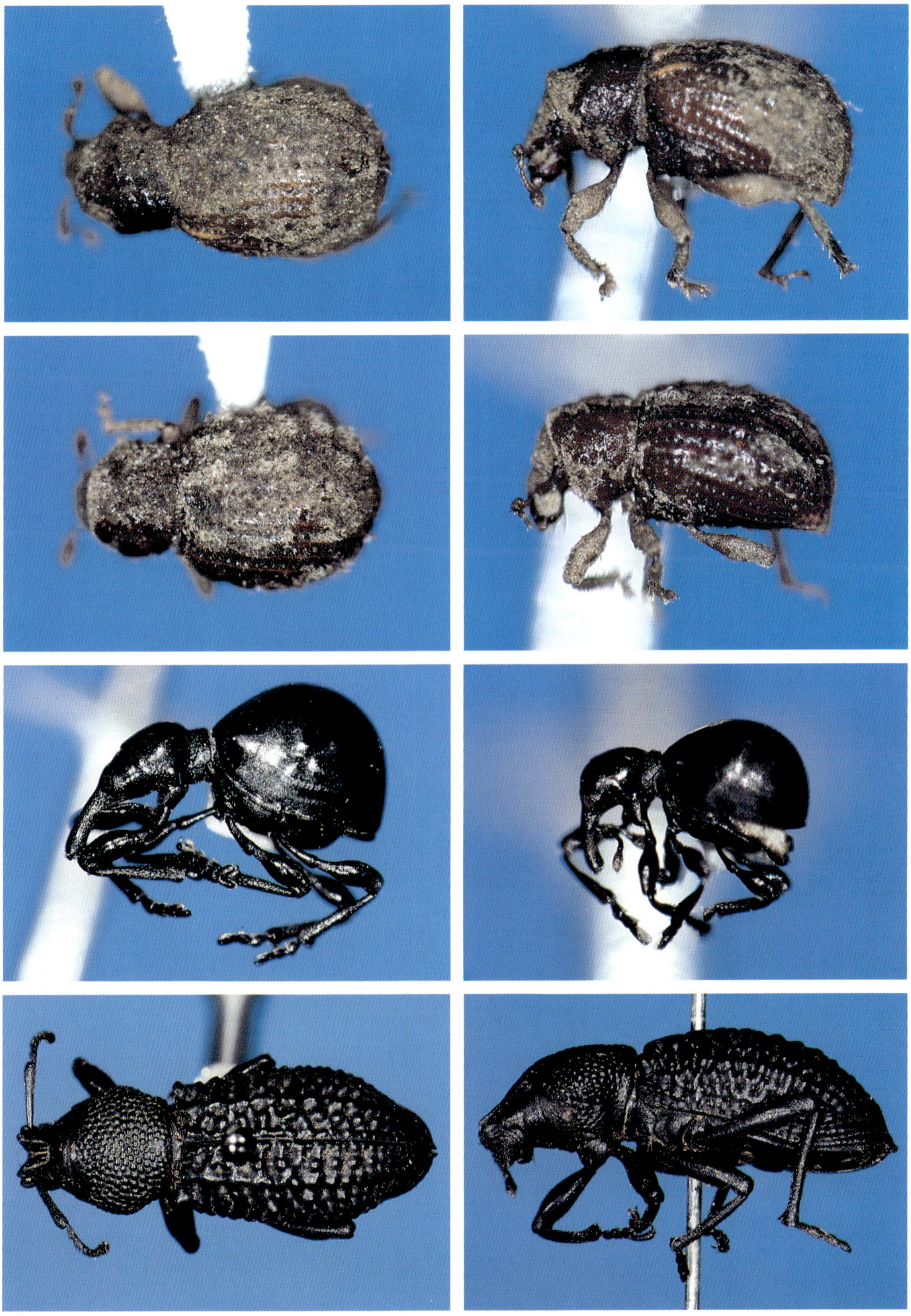

Plate 302 Adelognatha

1, 2 *"Cherrus" mastersii* Pascoe
Donnybrook, Western australia
Male; 21 mm

3, 4 *"Cherrus" vestitus* Pascoe
King George Sound, Western Australia
Male, compared with holotype; 16.5 mm

5, 6 *"Perperus" melancholicus* Boisduval
Vicinity of Jenolan Caves, New South Wales
Male; 9 mm

7, 8 *"Perperus" ziczac* Lea
Mount Spec, north Queensland
Female; 16 mm

See Volume IV on **Adelognatha** for further information on these species.

1	2
3	4
5	6
7	8

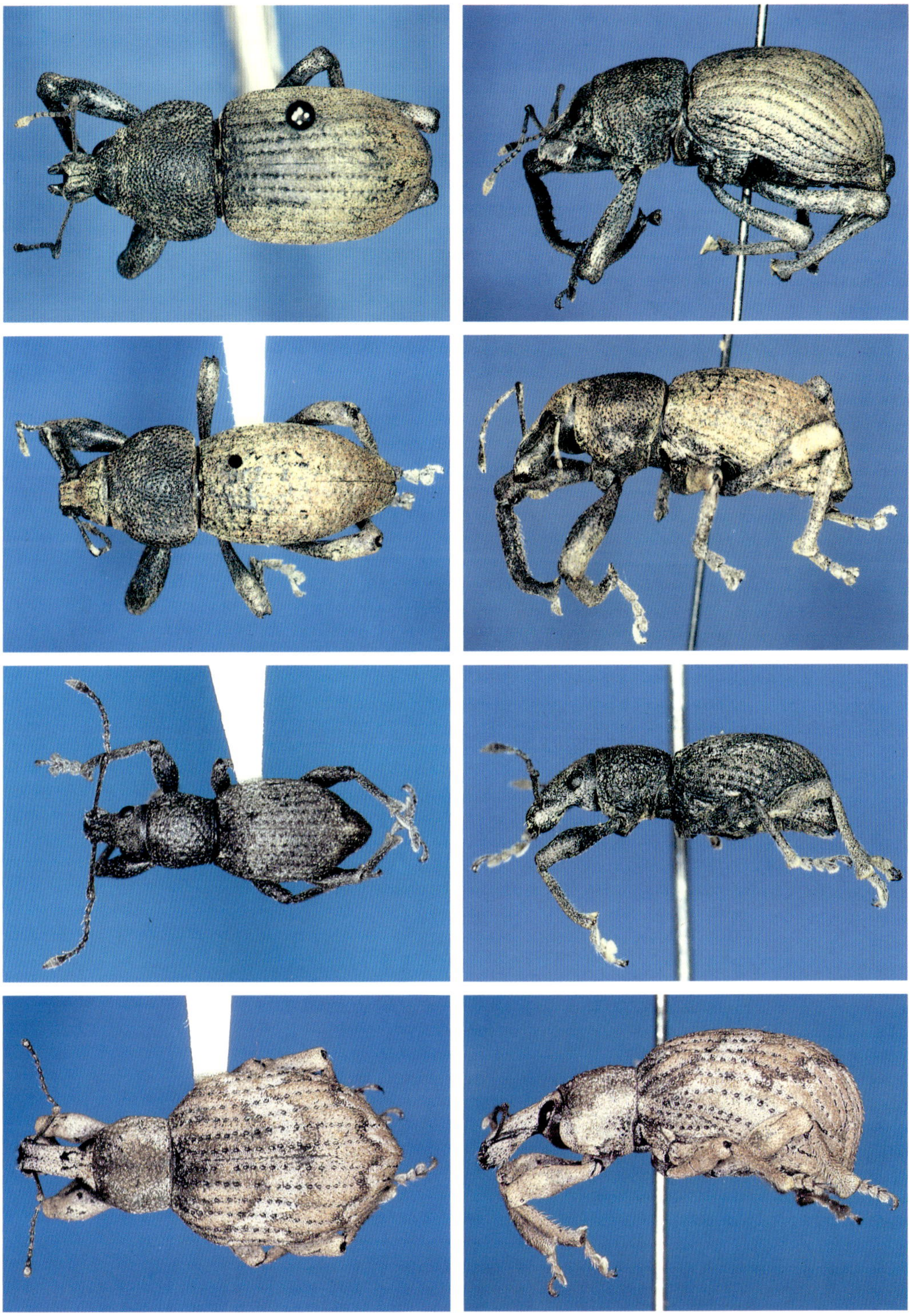

Plate 303 Adelognatha

1, 2 New genus near *Matesia*, new species 1
About 1.6 km north of Karratha, Western Australia
Male; 4.3 mm

3, 4 New genus near *Matesia*, new species 2
Calwynyardah Station, 130 km east of Derby, Western Australia
Female; 5.9 mm

5, 6 New genus
Dandarangan, Western Australia
Male; 3.5 mm

7, 8 New genus near *Enchymus*
4.4 km southeast of Weldborough, Tasmania
Male; 10.5 mm

See Volume IV on **Adelognatha** for further details
See Volume VI for contiuaton of coloured illustrations.

1	2
3	4
5	6
7	8

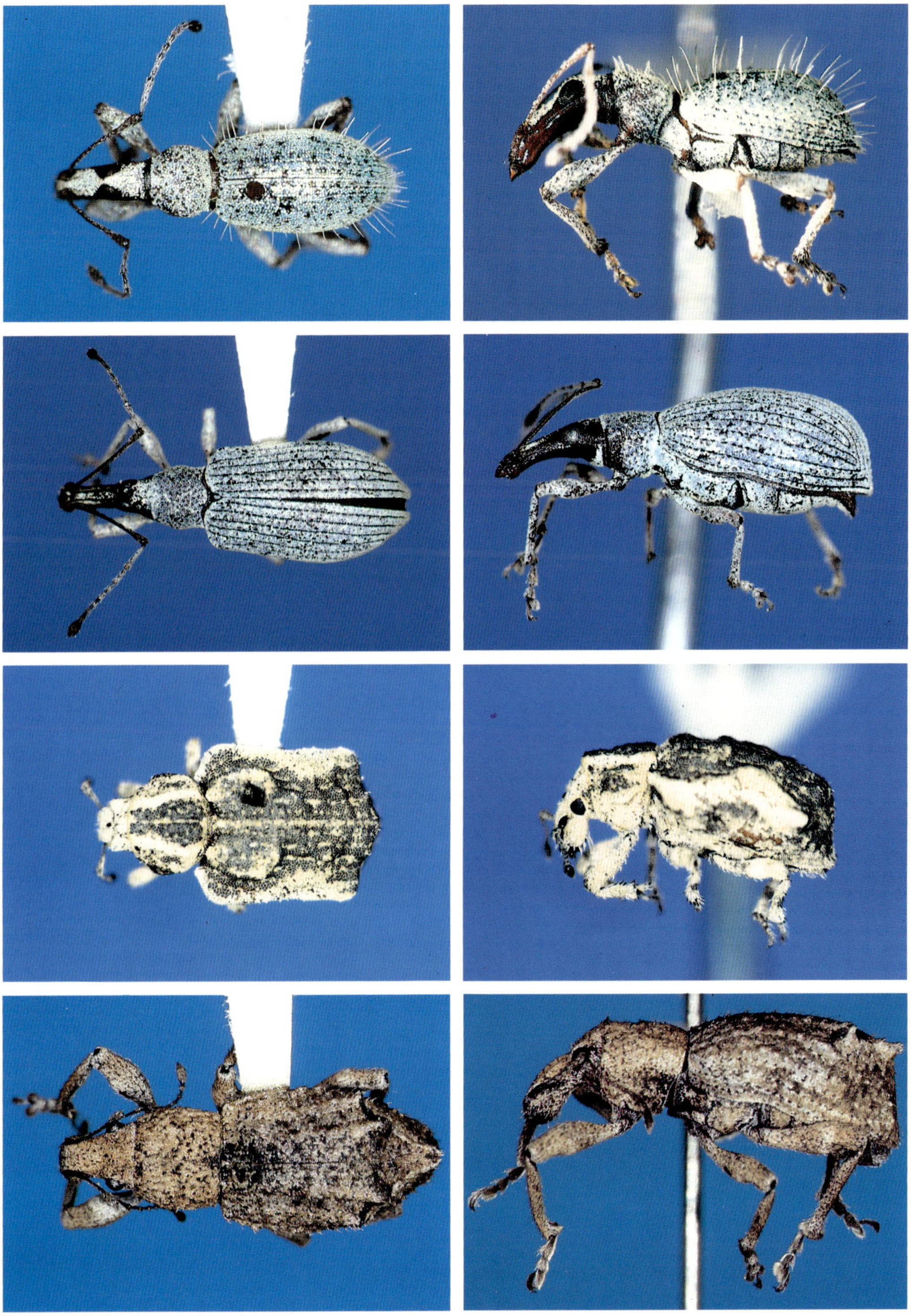

Index A – Genus to Species

Numbers refer to plates

Index B – Species to Genus

Numbers refer to plates

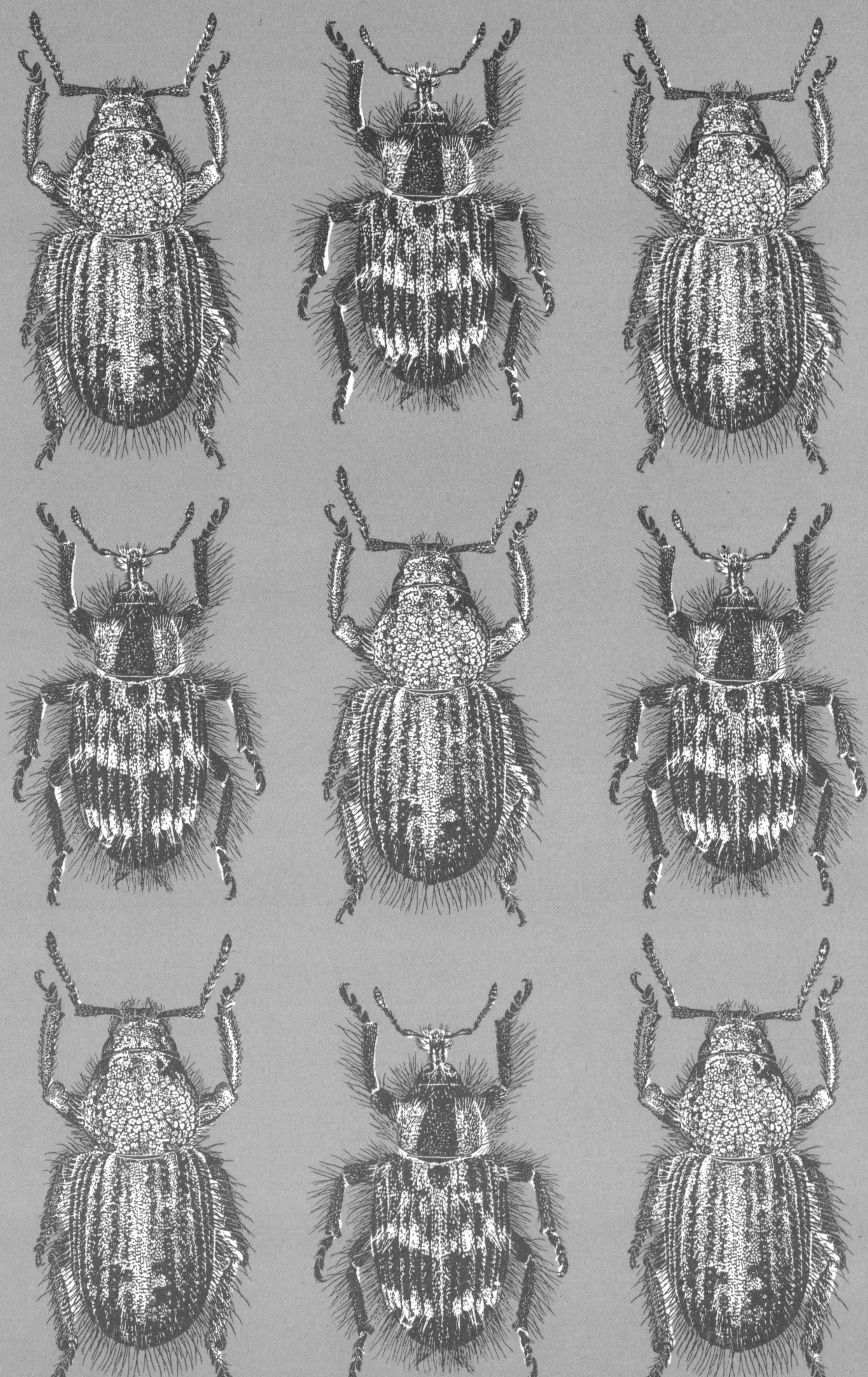